Production Technology

Proce   rials and plan   ·

W. L

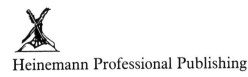

Heinemann Professional Publishing

Heinemann Professional Publishing Ltd
22 Bedford Square, London WC1B 3HH

LONDON   MELBOURNE   AUCKLAND

First published 1988
© W. Bolton 1988

British Library Cataloguing in Publication Data
Bolton, W.
  Production technology : processes, materials
  and planning.
  1. Production management
  I. Title
  685.5   TS155

ISBN 0 434 90173 3

Photoset by Deltatype Ltd, Ellesmere Port
Printed by Redwood Burn Ltd
Trowbridge, Wiltshire

Contents

Preface

This book has been written to present an overview and appraisal:

- of the manufacturing processes used with metals and polymers, so that an informed *choice of process* can be made, taking into account the various alternatives.
- of the materials used in engineering—metals both ferrous and non-ferrous, polymers, ceramics and composites—so that an informed *choice of material* can be made, taking into account the properties that are required.
- of production planning and cost accounting, so that choice of process and material can be *related to quantity, quality and cost requirements*, with an appreciation of the planning needed in the design of work tasks.

The level of discussion of these subjects is about that of HNC/HND for mechanical and production engineering technicians. The book caters for a range of BTEC HNC/HND units—manufacturing technology, engineering materials and associated materials units, production planning and control, and various aspects of other units. The aim has been to produce a text covering the core of most mechanical and production engineering courses at this level.

Parts of this book have been taken from other texts, also published by Heinemann, that I have written for specific BTEC units. Acknowledgements are due to the large number of companies that supplied me with information, also to other publishers for permission to reproduce from their publications. I am indebted to the Controller of HMSO for permission to quote from the Health and Safety of Work Act 1974. Extracts from PD 6470: 1981 are reproduced by permission of BSI; complete copies can be obtained from them at Linford Wood, Milton Keynes MK14 6LE. Every effort has been made to acknowledge sources of material used—if at any time I have not made full acknowledgement I hope that my apologies will be accepted.

PART ONE

Manufacturing Processes

This part of the book consists of four chapters concerning the range of manufacturing processes, their characteristics and suitability for particular types of components.

Chapter 1   Forming processes – metals

Casting processes, hot and cold manipulative processes, powder techniques, cutting and grinding, metal removal by electrochemical, electrical discharge and chemical means, and surface finishing are described in outline. The main emphasis is placed on the characteristics of the processes and their suitability for particular types of operation.

Chapter 2   Forming processes – polymers

Following a preliminary discussion of the characteristics of polymer materials, the forming processes of casting, moulding, extrusion, calendering, forming and machining are described. Consideration is then given to the choice of such processes and the design constraints involved when using polymers.

Chapter 3   Assembly operations

Assembly processes are described and their characteristics considered: for metals – adhesives, soldering, brazing, welding and the use of fastening systems; for plastics – welding, adhesives, riveting, press and snap fits and thread systems. Finally, there is a discussion of the combination of such processes for an entire component, with consideration of limits and fits.

Chapter 4   Automation

This chapter presents an introductory discussion of numerical control of machining, and of robotics, in the automation of manufacturing methods.

1

Forming processes – metals

1.1 Introduction

With metals, the range of forming processes possible for component production can be divided into four main categories:

Casting: shaping of a material by pouring the liquid material into a mould.
Manipulative processes: shaping of materials by plastic deformation methods.
Powder techniques: production of a shape by compacting a powder.
Cutting: production of a shape by metal removal.

In this chapter each of the above types of process will be considered in more detail so that comparisons can be made as to the suitability of a particular process for the manufacture of a component.

One factor that will be referred to is surface roughness. Roughness is defined as the irregularities in the surface texture which are inherent in the production process but excluding waviness and errors of form. Waviness may arise from such factors as machine or work deflections, vibrations, heat treatment or warping strains.

One measure of roughness is the arithmetical mean deviation, denoted by the symbol R_a. This is the arithmetical average value of the variation of the profile above and below a reference line throughout the prescribed sampling length. The reference line may be the centre line, this being a line chosen so

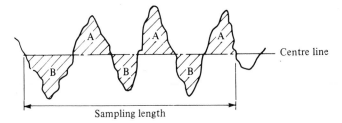

Figure 1.1 The centre line. The sum of the areas marked A equals the sum of those marked B

Table 1.1 *Typical R_a values*

Surface texture	R_a (μm)
Very rough	50
Rough	25
Semi-rough	12.5
Medium	6.3
Semi fine	3.2
Fine	1.6
Coarse-ground	0.8
Medium-ground	0.4
Fine-ground	0.2
Super-fine	0.1

that the sums of the areas contained between it and those parts of the surface profile which lie on either side of it are equal (Figure 1.1). Table 1.1 indicates the significance of R_a values.

Another factor in making comparisons between processes is the tolerance possible. Tolerance is the difference between the maximum limit of size and the minimum limit of size. It is the amount by which deviation may occur from a desired dimension.

1.2 Casting

Essentially, casting consists of pouring a liquid metal into a suitable mould and then permitting it to solidify, thereby producing a solid of the required shape. The products fall into two main categories, those for which the solid shape is just a convenient form for further processing, and those for which the component produced requires only some machining or finishing to give the final product.

Where the products are produced for further processing simple, regular geometrical shapes are generally used, the products being known as ingots, billets, pigs, slabs, or by other descriptive terms. The shape adopted depends on the processes that are to follow.

Where the casting is used to produce the product in almost finished state, a mould is used which has the appropriate internal shape and form. The mould has, however, to be designed in such a way that the liquid metal can easily and quickly flow to all parts. This has implications for the finished casting in that sharp corners and re-entrant sections have to be avoided and gradually tapered changes in section used. Account has also to be taken of the fact that the dimensions of the finished casting will be less than those of the mould, because shrinkage occurs when the metal cools from the liquid state to room temperature.

A number of casting methods are available, the choice of method depending on:

size of casting required;
number of castings required;
complexity of the casting;
mechanical properties required for the casting;
surface finish required;
dimensional accuracy required;
metal to be used;
cost per casting.

Sand casting
This involves using a mould made of a mixture of sand and clay. The mixture is packed around a pattern of the required casting to give the mould, which is usually made in two main parts so that the pattern can be extracted (Figure 1.2). The mould must be designed so that when the liquid metal is introduced into the mould, all air or gases can escape and the mould can be completely filled. After the casting has solidified the mould is broken open, the moulds only being used for the one casting. Some machining is always necessary with the casting, such as the trimming off of the metal in the feeder and riser.

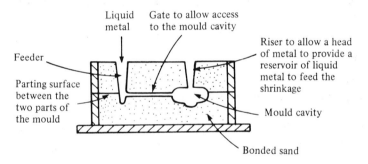

Figure 1.2 Sand casting

Sand casting can be used for a wide range of casting sizes and for simple or complex shapes. Holes and inserts are possible. However, the mechanical properties, surface finish and dimensional accuracy of the casting are limited. Roughness values (R_a) of the order of 12.5 to 25 μm are produced. A wide range of alloys can be used. The cost of the mould is relatively cheap, at least in comparison with metal moulds, but the cost of the mould has to be defrayed against just one casting as it is broken up after being used only once. For small number production, sand casting is the cheapest casting process.

Die casting
This involves the use of a metal mould. With *gravity die casting* the liquid metal is poured into the mould, gravity being responsible for causing the metal to flow into all parts of the mould. With *pressure die casting* the liquid metal is

injected into the mould under pressure. This enables the metal to be forced into all parts of the mould and enables very complex shapes with high dimensional accuracy to be produced.

There are limitations to the size of casting that can be produced with pressure die casting, the mass of the casting generally being less than about 5 kg. Gravity die casting can, however, be used for larger castings. Because the metal mould is expensive, compared with, for example, a sand mould, this process is generally uneconomic for one-off castings or small runs. The mould can be used for many castings and thus the cost defrayed over the larger number of castings. The castings produced by this method have very good mechanical properties, dimensional accuracy and finish, good enough to reduce or even eliminate machining or other finishing processes. Roughness values (R_a) of the order of 0.8 to 1.6 μm are produced. The metals used with this casting method are limited to aluminium, copper, magnesium and zinc alloys.

The main cost factor is the cost of the metal mould. If the cost per casting is to be reasonable, there needs to be a large number production from a particular mould. The metal mould does, however, mean that, with little further machining or finishing necessary, the cost element for such operations is reduced.

Cetrifugal casting
This method involves rotating either consumable or metal moulds, using the forces set up during rotation of the mould to force the liquid metal to cling to the inside surface (Figure 1.3). This enables hollow objects to be produced without the use of an inner core in the mould. This method is used for simple geometrical shapes, e.g. large diameter pipes.

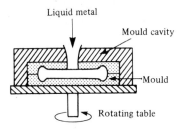

Figure 1.3 Centrifugal casting

Investment casting
This method, sometimes known as *lost wax casting*, can be used with metals that have to withstand very high temperatures and so have high melting points, and for which high dimensional accuracy is required—aero engine blades are a typical product.

With this method the required shape is made from wax or a similar material. The wax pattern is then coated with a ceramic paste. When this coated wax pattern is heated the ceramic hardens, the wax melts and runs out to leave a mould cavity. Liquid metal is injected into the hot mould. After cooling, the ceramic is broken away to leave the casting.

The size of casting that can be produced by this method is not as great as that possible with sand casting. The method does, however, give high dimensional accuracy and a good surface finish. Roughness values (R_a) of the order of 1.6 to 3.2 μm are produced. For large number production it is a more expensive method than die casting, but it is cheaper for small number production.

Choosing a casting process

The following factors largely determine the type of casting process used:

Large, heavy casting. Sand casting can be used for very large castings.
Complex design. Sand casting is the most flexible method and can be used for very complex castings.
Thin walls. Investment casting or pressure die casting can cope with walls as thin as 1 mm. Sand casting cannot cope with such thin walls.
Small castings. Investment casting or die casting. Sand casting is not suitable for very small castings.
Good reproduction of detail. Pressure die casting or investment casting, sand casting being the worst.
Good surface finish. Pressure die casting or investment casting, sand casting being the worst.
High melting point alloys. Sand casting or investment casting.
Tooling cost. This is highest with pressure die casting. Sand casting is cheapest. However, with large number production the tooling costs for the metal mould can be defrayed over a large number of castings, whereas the cost of the mould for sand casting is the same no matter how many castings are made because a new mould is required for each casting.

Design considerations when using casting

Rounded corners, no abrupt changes in section, gradually sloping surfaces are all necessary with casting if there is to be a proper flow of metal and a complete filling up of the mould. During the casting gas bubbles escape from the liquid metal, so corners in which the gas could collect have to be eliminated. Thus for example in Figure 1.4a the design in (ii) is to be preferred to that in (i).

Shrinkage occurs during the cooling and solidification of a casting. The amount of shrinkage depends on the metal being used and so the pattern used for the casting must be designed to take this into account, the pattern being larger than the required casting by an amount depending on the metal concerned. During the solidification the outer surfaces of the metal cool more

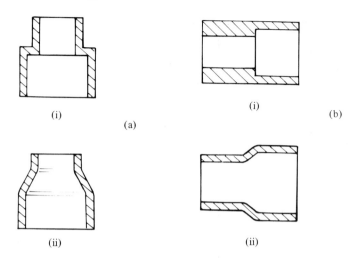

Figure 1.4 (a) The need for no sharp corners in casting means (ii) is preferred to (i) (b) The need for uniform thickness sections in casting means (ii) is preferred to (i)

rapidly than the inside of the metal. This can result in the outer layers solidifying while the inside is still liquid. When this liquid then contracts and solidifies, cavities can be caused if the solidifying inner liquid metal cannot 'pull in' the outer solidified metal. The production of cavities due to this effect is markedly increased where the section thickness shows abrupt changes. Ideally, a section of uniform thickness should be used; however, if this is not feasible a gradual change in thickness should occur rather than an abrupt change (Figure 1.4b).

When to use casting

Casting is likely to be the optimum method for product production under the following circumstances.

1 *The part has a large internal cavity.* There would be a considerable amount of metal to be removed if machining were used. Casting removes this need.
2 *The part has a complex internal cavity.* Machining might be impossible; by casting, however, very complex internal cavities can be produced.
3 *The part is made of a material which is difficult to machine.*
4 *The metal used is expensive and so there is to be little waste.*
5 *The directional properties of a material are to be minimised.* Metals subject to a manipulative process often have properties which differ in different directions.
6 *The component has a complex shape.* Casting may be more economical than assembling a number of individual parts.

Casting is not likely to be the optimum method for parts that are simple enough to be extruded or deep drawn.

1.3 Manipulative processes

Manipulative processes involve the shaping of a material by means of plastic deformation processes. Where the deformation takes place at a temperature in excess of the recrystallisation temperature of the metal the process is said to be *hot working*. Where the deformation is at a temperature below the recrystallisation temperature the process is said to be *cold working*.

When compared with hot working, cold working has the following advantages:

Better surface finish.
Improved strength.
Better dimensional control.
Better reproducibility.
Directional properties can be imparted to the material.
Contamination is minimised.
No heating is required.

Cold working has, however, the following disadvantages when compared with hot working:

Higher forces are needed for plastic deformation.
More energy is needed for plastic deformation.
Work hardening occurs.
The resulting material has less ductility and toughness.
The directional properties given to the material may be detrimental.
The metal used must be clean and scale-free.

Hot working

The main hot-working processes are rolling, forging and extruding. The following notes involve a closer look at these and related processes.

1 Rolling
This is the shaping of the metal by passing it, hot, between rollers. With nominally parallel cylinder rolls, flat sheet or strip can be produced in long lengths. If contoured rollers are used, channel sections, rails, etc can be produced.

Hot rolling is usually done in a number of stages. This could be either a series of passes through one set of rollers or a continuous process with the material passing through a sequence of sets of rollers. Figure 1.5a shows the type of rolling sequence that is adopted in rolling a structural beam section.

Darlington and Simpson Rolling Mills Ltd in their catalogue state that over six hundred different rolled shapes are available ranging from 1 kg to 10 kg per metre length. Figure 1.5b shows some of their rolled sections. Roughness values (R_a) of the order of 12.5 to 25 μm are produced.

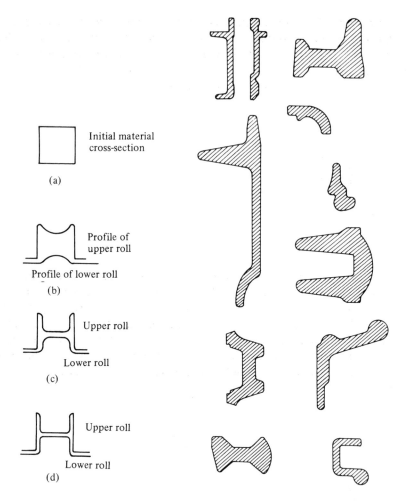

(a) Initial material cross-section

Profile of upper roll

Profile of lower roll

(b)

Upper roll

Lower roll

(c)

Upper roll

Lower roll

(d)

Figure 1.5 (a) The type of rolling sequence used for a structural beam section (b), (c), (d) Examples of rolled sections

2 Forging

Forging is a squeezing process, the hot metal being squeezed by pressing or hammering between a pair of dies. *Heavy smith's forging or open die forging* involves the metal being hammered by a vertically moving tool against a stationary tool, or anvil. Such forgings are fairly crude in form and generally are only the first stage in the forming operation. *Closed die forging* involves the hot metal being squeezed between two shaped dies which effectively form a complete mould (Figure 1.6). The metal in the cavity is squeezed under pressure in the cavity, flowing under the pressure and filling the cavity. In order to fill the cavity completely, a small excess of metal is allowed and this is squeezed outwards to form a flash which is later trimmed away. *Drop forging* is one form of closed die forging and uses the impact of a hammer to cause the metal billet to flow and fill the die cavity. The term *machine forging* (or *press forging*) is used for the closed die forging process where a press is used to squeeze the metal billet slowly between the dies.

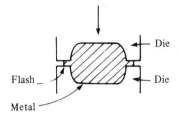

Flash

Metal

Die

Die

Figure 1.6 Closed die forging

The cost of forging is high for small number production because of the cost of the dies. Large number production is necessary to reduce the die cost per unit product produced. For example, with an aluminium alloy being forged, increasing the production run from 100 units to 1000 units can result in the cost per unit decreasing by 50% to 75%. An indication of the factors, and their relative costings, involved in producing a forging is given by the following costs for a drop forging (Iron and Steel Institute, ISI publication No. 138, 1971):

	Percentage of total production cost
Material	52
Direct overheads	15
Direct labour	10
Dies	8
Maintenance	4
Stock heating	3
Other	8

Forging results in a product superior in toughness and impact strength to that given by casting. The process also results in the welding up of shrinkage cavities in the cast ingot used for the forging. The process can be used to give a fibre structure within the material, a directionality of properties, which minimises the chances of crack formation and enhances the properties in service of the product. Roughness values (R_a) of the order of 3.2 to 12.5 μm are produced.

There are few limits to the shape of component that can be forged with a closed die. However, webs and thin sections should not be made too thin because they are likely to cool too quickly and thus necessitate a great forging pressure. Also, sharp corners should be avoided; a larger radius allows a better metal flow during the forging operation.

The flash, see Figure 1.6, that is produced with forging has to be machined off, so to that extent some machining is always necessary after forging.

3 *Extrusions*

With hot extrusion, the hot metal is forced under pressure to flow through a die, i.e. a shaped orifice. Two basic methods of hot extrusion are used, *direct extrusion* and *indirect extrusion*. Figure 1.7 shows the fundamentals of these two methods. Rods are formed by extruding the billet material through a die and give a solid product of uniform cross-section. Tubes and other hollow shapes can be extruded by using an internal mandrel (Figure 1.8a) or a spider mandrel (Figure 1.8b). As the material flows past the spider, a further reduction between the die and the mandrel forces the material to close up and weld together.

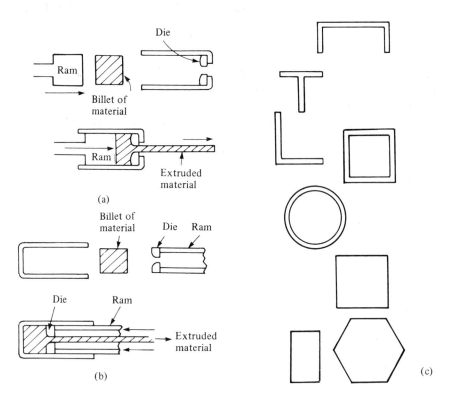

Figure 1.7 (a) Direct extrusion (b) Indirect extrusion (c) Some of the standard sections available with extrusion. Considerably more complex sections are possible

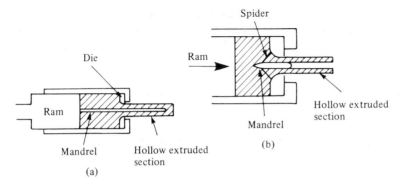

Figure 1.8 Extruding hollow sections

The process is used extensively with aluminium, copper and magnesium alloys to produce very complex shaped sections, and with steel for less complex sections. High production rates are possible and the products have a good surface finish, better tensile strength and more favourable grain structure and directionality than the starting material used for the extrusion. Roughness values (R_a) of the order of 0.8 to 3.2 μm are produced.

To illustrate the types of product sizes available by this method, aluminium tubes up to 150 mm in diameter and up to 2500 mm long have been produced; tubes in mild steel up to 130 mm in diameter and 750 mm long. Minimum tube diameters are about 5 mm. Aluminium cans up to 350 mm in diameter and 1500 mm long are possible.

Extrusion dies are expensive and thus large quantity production is necessary so that the cost can be spread.

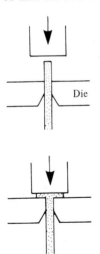

Figure 1.9 Upsetting

4 *Upsetting*

There are many products that require a marked change of section along their length. One possibility is to machine the product from material having the largest diameter required. As the volume of material at the largest diameter may be small in comparison with the rest of the product this can be extremely wasteful of material and involve high machining costs. An alternative is to use upsetting. In this process a billet or extrusion is inserted into a stationary die with part of it protruding from the die (Figure 1.9). A punch moving along the die axis impacts on the protruding part and 'squashes' it out to the required diameter. A second blow may be used to complete the head shape.

The process is used for high volume fastener production, e.g. rivets and bolts. Hot upsetting is used for those materials that have low ductility in the cold. Otherwise cold upsetting is used.

Cold working

1 *Cold rolling*

Cold rolling is the shaping of a material by passing it, at normal temperatures, between rollers. This gives a cleaner, smoother finish to the metal surface than rolling hot, also a harder material is produced. Roughness values of the order of 0.4 to 3.2 μm are produced.

2 *Wire drawing*

This involves the pulling of material through a die (Figure 1.10). Wire manufacture can involve a number of drawing stages in order that the initial material can be brought down to the required size. As this cold working hardens the material there may be annealing operations between the various drawing stages to soften the material and so permit further drawing. The essential difference between drawing and extrusion is that in drawing the material is pulled through the die, in extrusion the material is pushed through the die.

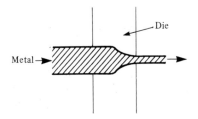

Figure 1.10 Drawing

3 *Shearing*

Shearing, bending and deep drawing are the three methods of working sheet metal in presses. Shearing is the deformation to shear failure of a metal in order to cut various shapes from the metal sheet. It is cutting without the formation of chips or the use of burning or melting. Figure 1.11 illustrates the basic shearing process, the punch descending onto the metal sheet and deforming it plastically into the die. The punch penetrates into the metal, the opposite surface of the metal bulging and flowing into the die. When the punch has penetrated a certain distance into the metal, the applied stress exceeds the shear strength of the metal and the metal suddenly shears through the remainder of the thickness. The depth of penetration that occurs before the metal shears depends on the ductility and shear strength of the material. Thus for annealed mild steel the depth of penetration is of the order of 50% of the sheet thickness, for partially cold worked mild steel this percentage penetration drops to about 40%. On the metal pressed out of the sheet the depth of penetration can be seen as a polishing area on the cut surface (Figure 1.12).

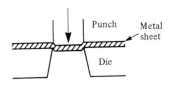

Figure 1.11 The basic shearing process

Figure 1.12 Penetration during shearing

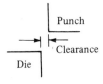

Figure 1.13 Clearance

Figure 1.14 The cut surface with inadequate clearance

The amount of the clearance between the punch and the die (Figure 1.13) also affects the appearance of the cut surface. If an adequate clearance is used, the cut edges may be sufficiently smooth for use without further finishing. An adequate clearance is generally of the order of 5% to 10% of the sheet thickness. Figure 1.12 shows the slightly tapered form of the cut surface when there is adequate clearance and Figure 1.14 shows the type of surface produced when there is inadequate clearance. This occurs because the cracks developed in the sheet during the deformation do not all run together to produce a clean break when the clearance is inadequate, as they do with sufficient clearance.

Piercing and *blanking* are shearing operations. With piercing the piece removed from the sheet or strip is the waste item and the perforated strip or sheet is the required product. With blanking, the piece removed from the sheet or strip is the required product and the perforated strip or sheet is the waste item. In some cases a combination of piercing and blanking may be necessary, this often being accomplished with a single stroke of the punch. An example of a product requiring both piercing and blanking is the simple washer—the piercing being needed for the central hole and the blanking to produce the outer circular cut and so the completed washer (Figure 1.15).

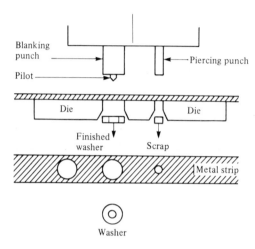

Figure 1.15 Piercing and blanking

A number of other operations are likely to have taken place before piercing or blanking occurs. Thus the material may have been taken from a coil and have to be cut to length and also flattened. This generally takes place by the rolling sheet being fed between rollers, to straighten it, and then sheared to give strip of the right length.

4 *Bending*

When a bar, tube, strip or wire is bent, on one side of the neutral plane there are tensile stresses and on the other side compressive stresses (Figure 1.16). When these stresses are made high enough plastic deformation can occur, in both tension and compression. A problem, however, in bending a piece of material to a particular curvature is that when the bending force has been removed the material has the tendency to *spring back* and so adopt a different curvature. One method for compensating for this spring-back is to overbend slightly.

When a piece of metal is stretched and becomes longer, it suffers a reduction in its width. Similarly, when a piece of material is compressed and becomes shorter, it suffers an increase in width. When bending occurs, there is a reduction in the width on the tensile side of the neutral plane and an increase in width on the compression side. Thus an initially rectangular cross-section is no longer rectangular (Figure 1.16).

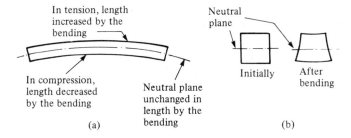

In tension, length increased by the bending

In compression, length decreased by the bending

Neutral plane

Neutral plane unchanged in length by the bending

Initially   After bending

(a)   (b)

Figure 1.16 Bending. (a) Effect on section. (b) Effect on cross-section

One effect of the above change in cross-section is that the neutral axis becomes displaced towards the compressed side of a bend. This affects the calculations that a designer needs to make in determining the length of a blank necessary to produce a bent part of given dimensions. An allowance has to be made for the neutral axis shift, the allowance depending on the thickness of the

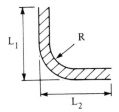

L_1 R

L_2

Figure 1.17 Length of blank $= L_1 + L_2 - D$, where D is the allowance

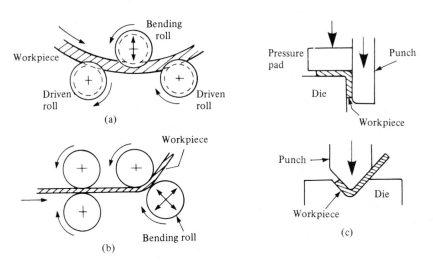

Bending roll

Workpiece

Driven roll   Driven roll

(a)

Workpiece

Bending roll

(b)

Pressure pad   Punch

Die

Workpiece

Punch   Die

Workpiece

(c)

Figure 1.18 (a) Three-roll bending (b) Four-roll bending (c) Bending by means of press-tools

blank and the bend radius. The length of blank required is then calculated, as shown in Figure 1.17, as $L_1 + L_2 - D$, where D is the allowance. Tables are available for the allowances. As an example, for a blank of thickness t and a bend radius R,

where $R = t$, $D = 1.7\,t$

where $R = 2t$, $D = 2.0\,t$

Figure 1.18 illustrates several ways of bending materials. Many products can be made by the relatively simple process of bending, e.g. such common-place items as paper clips, links, hooks, clips, springs and many items used in electrical fitments.

5 *Deep drawing*

With deep drawing, sheet metal is pushed into an aperture by a punch and cup-shaped articles produced. This process had its earliest uses in the production of artillery shells and cartridge cases. Ductile materials, e.g. aluminium, brass, mild steel, must be used.

Figure 1.19 illustrates the sequence of events for a deep drawing operation. An important point to note for this process is that the edges of the blank are not clamped. Thus, as the blank is pushed into the aperture, the edges of the sheet

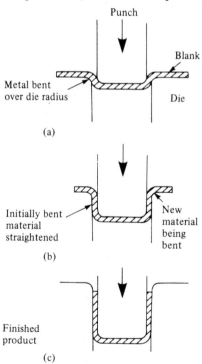

Figure 1.19 Deep drawing

move; when the punch first impacts on the blank, the blank is bent over the edges of the die. As the die continues to move downwards, the previously bent material is straightened against the wall of the die to create the cylindrical side walls of the cup-shaped object. New material is bent over the edge of the die, this material resulting from the movement towards the punch of the edges of the blank. This process of straightening the previously bent material and bending new material continues until the finished product is produced, often when all the blank material has been pulled into the die.

Figure 1.20 shows the types of stresses experienced by the material during deep drawing. The flange material is subject to tensile stress in the axial direction but compressive stress in the circumferential direction. This is because the circumference of the blank is being reduced as it is pushed into the die. The result of this is a thinning of the flange material and a tendency for the material to wrinkle in order to accommodate the change in circumference. For this reason a pressure pad is generally pressed against the flange during the drawing. As the material in the flange passes over the edge of the die it is bent. The wall is subject to vertical and circumferential tensile stresses. The effect of

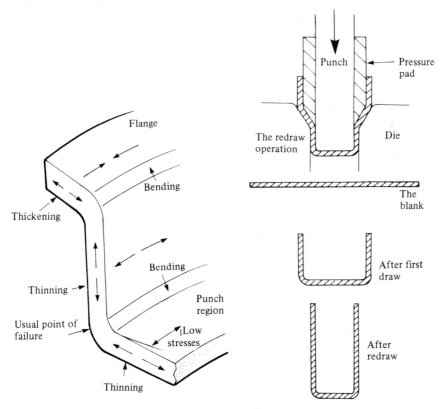

Figure 1.20 Stresses during deep drawing   **Figure 1.21** The redraw process

these stresses is to cause the material to become thinner. Under the base of the punch the material is subject to tensile stresses.

In deep drawing, the region of thinnest material is close to the initial bending region at the punch edge. It is here that failure is most likely to occur. Thus the thickness at this region may be 90% of the original blank thickness, whereas in the flange close to the die edge it may be 115%.

In some circumstances more than one drawing operation may be used in order to achieve the final required product shape. Sometimes the material may be annealed between drawing operations to maintain ductility. Figure 1.21 shows the method adopted for a *redrawing process*.

Deep drawing is not only used to produce cup-like objects but it is also a very widely used process for the shaping of sheet metal. Examples are the production of steel kitchen sink tops and car body pieces. Whether, in all cases, the term 'deep' is appropriate for the drawing operation is irrelevant because the basic principles are the same. The term *pressing* is sometimes used for the operation, because the drawing is carried out using a press.

6 *Ironing*

The length of a draw cup can be extended by thinning the cup walls with an ironing process. The cup wall is squeezed between a punch and the die, the initial wall thickness being greater than the clearance between the punch and the die (Figure 1.22).

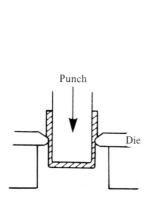

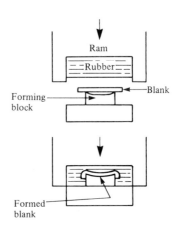

Figure 1.22 Ironing   **Figure 1.23** Forming with rubber

7 *Forming with rubber or fluid pressure*

With the deep drawing process described above, rigid metal dies were used with rigid metal punches. Such press tools can be used to produce many thousands of components, but the cost of such tools can be high. Where small production runs are required, the tooling cost may make the production uneconomic. With smaller production runs, i.e. a thousand or less, a flexible die may be a cheaper and more economic alternative.

Rubber, in a confined space, when acted on by a pressure behaves rather like a fluid and transmits the pressure uniformly in all directions. Figure 1.23 shows how this principle is utilised in a forming process. The metal blank is placed over a forming block. Then a ram presses a rubber pad over the blank and block. The pressure transmitted through the rubber presses the blank over the forming block. The forming blocks can be made of wood and thus are cheaper to produce than conventional metal dies. The process is used for multiple-axis bending and some shallow drawing.

Another flexible die method involves the use of a fluid with a diaphragm. Figure 1.24 shows one form of such an arrangement.

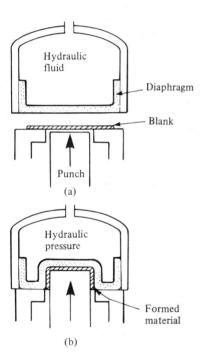

Figure 1.24 Flexible die forming. (a) Initial position. (b) Material being formed

8 *Stretch forming*

This process was developed to enable sheet metal parts, particularly large sheets, to be formed economically in small quantities. The metal sheet, while in tension, is wrapped over a former (Figure 1.25a). Bending without tension would result in a tensile stress on one surface of the sheet and a compressive stress on the other surface (Figure 1.25b). Bending the sheet under tension allows all the stresses in the material to be tensile, as there is an overall

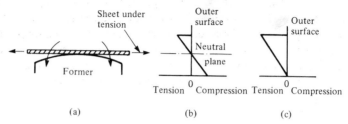

(a)   (b)   (c)

Figure 1.25 Stretch forming. (a) The stretch forming process. (b), (c) The effect on the stresses in the bent material

extension of both surfaces in comparison with the original length. Figure 1.25c shows the stress distribution through the thickness of a sheet bent under tension.

When a sheet is bent without being stretched, the surface that is in compression is likely to buckle or distort. If, however, the material is stretched when bent, this distortion can be avoided.

9 *Spinning*

Spinning can be used for the production of circular shaped objects. A circular blank is rotated and then pressure is applied to it by means of a forming tool in order to bend the blank into the required shape (Figure 1.26). This is an economic method for small numbers of components or for large components where other forming methods would involve high costs. Aluminium, brass and mild steel are examples of materials formed by this process. A typical type of product is a metal reflector.

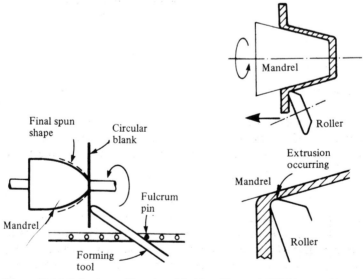

Figure 1.26 Spinning with a hand held tool

Figure 1.27 Flow turning

A variation of spinning is called *flow turning*. Instead of a forming tool being just pressed against the blank to cause it to become bent to the required shape, a roller is used (Figure 1.27). This not only forms the blank to the shape of the mandrel but also progressively extrudes it. The wall thickness of the formed product is thus less than the initial thickness of the blank. Both the shape of the product and its thickness are thus under control, unlike spinning where only the shape is controlled.

10 Impact extrusion

This process involves a punch hitting a slug of the material which is held in a die (Figure 1.28). The impact causes the slug material to flow up the space between the punch and the walls of the die.

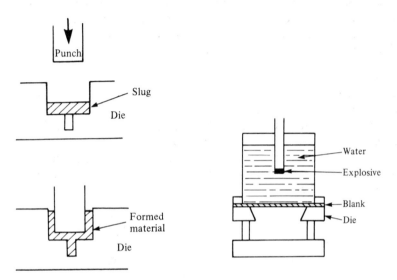

Figure 1.28 Impact extrusion   **Figure 1.29** Explosive forming

Impact extrusion is widely used with the softer materials, i.e. zinc, lead and aluminium, and is used for the production of collapsible tubes or cans, e.g. zinc dry battery cases. A small deep can in these materials can be more easily produced by this method than by deep drawing.

11 Explosive forming

Explosive forming is an example of a relatively new group of forming processes, known as *high velocity metal forming*. Figure 1.29 illustrates the basic principle. An explosive charge is detonated under water and the resulting pressure wave is used to press a metal blank against the die. This method can be used with a wide variety of materials, including some of the high strength materials which may be difficult to form by other methods. Complex parts can be produced at one 'blow'. Large reflectors, cyclinders, domes, etc. can be produced.

12 *Cold forging*

Impact extrusion is often combined with other operations such as upsetting, drawing, etc to convert the billet into the finished product. Because these operations are carried out at room temperature they are collectively known as cold forging.

High production rates can be achieved with the finished products having greater tensile strengths and more favourable grain flow than the starting material. Compared with machining processes there is little waste material. Generally, there is little if any finish machining required as a good surface finish is produced. The process is used for the production of solid objects with differently shaped heads and stepped shafts, hollow objects open at both ends with stepped internal and external diameters, hollow articles with internal and external splines or teeth, etc.

Press capacity

Shearing, bending and drawing are all processes which can use presses. In selecting a press, consideration has to be given to the required press capacity for the process concerned. The press capacity, or size, depends on:

1 The force required.
2 The length of the press stroke required.
3 If a drawing operation, the depth of draw required.
4 The number of press strokes per minute required.
5 The size of press bed needed.
6 The methods and direction of the blank feed.

In shearing the maximum punch force required depends on the edge area to be sheared and the shear strength τ of the material.

Maximum punch force $= \tau \times$ area

$\qquad = \tau \times$ material thickness $\times$ work profile perimeter

Hence the maximum force required to blank a circle of radius 30 mm from a steel sheet of thickness 2.0 mm and having a shear strength of 430 MPa (or MN m^{-2}) is

$$\begin{aligned} \text{Maximum force} &= (430 \times 10^6) \times (2.0 \times 10^{-3}) \\ &\quad \times (2\pi \times 30 \times 10^{-3}) \\ &= 1.62 \times 10^5 \text{ N (or 0.162 MN)} \end{aligned}$$

Figure 1.30 shows how, during shearing, the punch force varies with the penetration into the sheet when normal clearances are used. To a reasonable approximation the work done is the product of the maximum punch force F_{max} and the distance the punch penetrates into the sheet before rupture occurs. But

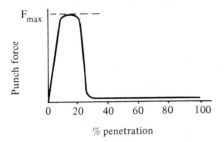

Figure 1.30 Forces during shearing. The area under the graph is the work done

this distance is equal to the percentage penetration multiplied by the sheet thickness.

Work done $= F_{max} \times$ (% penetration) $\times$ (sheet thickness)

Thus in the example given above, if the penetration was 30% then the work done would be

$$\text{work done} = (1.62 \times 10^5) \times (30/100) \times (2.0 \times 10^{-3})$$
$$= 97.2\text{J}$$

This work may be spread over a larger length of the press stroke by putting *shear* on the punch or the die. This is particularly useful if the blank is of substantial thickness. Figure 1.31a illustrates the form of single and double

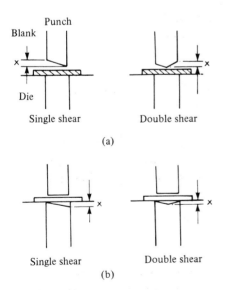

Figure 1.31 (a) Shear on the punch. (b) Shear on the die. The amount of shear has been exaggerated, x is less than the blank thickness

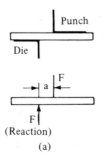

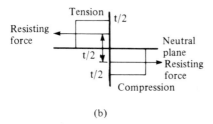

Figure 1.32 (a) Bending moment on sheet. (b) Stress distribution across section

shear on the punch. This type of shear tends to be used when piercing or cropping so that only the punched-out slug is deformed. Figure 1.31b shows the shear applied to the die. This is used when blanking, to give a flat blank.

In using a press for bending, as in Figure 1.18c, the bending moment is given by (Figure 1.32).

bending moment $= Fa$

where F is the punch load and a the clearance, i.e. the distance between the edge of the punch and the side of the die. This clearance is, however, virtually the same as the sheet thickness t. Hence

bending moment $= Ft$

Because the entire section of the sheet is to be plastically deformed it must all reach the yield stress and so the stress distribution across the bent section is likely to be something like that shown in Figure 1.32. Hence, to a reasonable approximation, the moment of resistance to the bending is given by

$$\text{moment of resistance} = (\text{resisting force}) \times \tfrac{1}{2}t$$
$$= (\sigma_y \times b \times \tfrac{1}{2}t) \times \tfrac{1}{2}t$$
$$= \tfrac{1}{4}bt^2\sigma_y$$

where σ_y is the yield stress and b the breadth of the sheet. Hence

$$Ft = \tfrac{1}{4}bt^2\sigma_y$$
$$F = \tfrac{1}{4}bt\sigma_y$$

Thus for a bend in mild steel, with a yield stress of 200 MPa, sheet of thickness 3.0 mm and width 20 mm, the force needed is about

$$\text{force} = \tfrac{1}{4} \times (20 \times 10^{-3}) \times (3.0 \times 10^{-3}) \times 200 \times 10^{-6}$$
$$= 3000 \text{ N } (3.0 \text{ kN})$$

With deep drawing, as in Figure 1.19, one way of considering the punch force required is to consider a vertical strip of the drawn cup wall (Figure 1.33). The effect of the drawing has been to take a piece of the original blank material, length L_0 and cross-sectional area A_0, and stretch it so that it has an extended length of L_1 and a cross-sectional area of A_1. The punch in producing this extension has applied a force of F through the distance L_1. Hence the work done by the punch is FL_1.

Work done by punch $= FL_1$

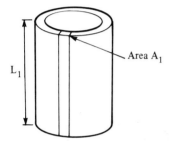

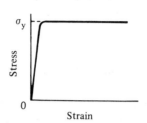

Figure 1.33 A strip on a drawn cup wall

Figure 1.34 The idealised stress–strain graph

If the stress–strain graph for the material, a ductile material, can be considered to be given by Figure 1.34 then the energy needed per unit volume to extend the material is the area under the stress-strain graph and to a reasonable approximation this is

energy per unit volume = yield stress × strain

But the strain $= (L_1 - L_0)/L_0$, hence

$$\text{energy per unit volume} = \sigma_y \times \frac{(L_1 - L_0)}{L_0}$$

The volume of the material is unchanged during the drawing and thus

$$A_0 L_0 = A_1 L_1$$

Hence $\text{energy} = A_1 L_1 \times \sigma_y \times \dfrac{(L_1 - L_0)}{L_0}$

and so $FL_1 = A_1 L_1 \times \sigma_y \times \dfrac{(L_1 - L_0)}{L_0}$

$$F = A_1 \sigma_y \left(\frac{L_1}{L_0} - 1 \right)$$

and as $A_0/A_1 = L_1/L_0$,

$$F = A_1 \sigma_y \left(\frac{A_0}{A_1} - 1 \right)$$

If a circular blank of diameter D and thickness t is used to produce a cup of diameter d and with the material thickness unchanged at t, then

initial cross-sectional area = circumference $\times$ t

Hence $A_0 = \pi D t$

The new cross-sectional area is

$$A_1 = \pi d t$$

Hence $F = \pi d t \sigma_y \left(\frac{D}{d} - 1 \right)$

The ratio D/d is known as the drawing ratio. The maximum value of this ratio likely to be achieved in practice is about 2.0. Hence the maximum drawing force is when this ratio occurs, i.e. when

$F = \pi d t \sigma_y$

Thus if a cup of diameter 50 mm is to be drawn from a blank of diameter 100 mm and sheet of thickness 2.0 mm and having a yield stress of 200 MPa, then the force required is

$$F = \pi \times (50 \times 10^{-3}) \times (2.0 \times 10^{-3}) \times (200 \times \left(\frac{100}{50} \right) - 1 \ 10^{-6}) \times$$

$$= 6.28 \times 10^4 \, \text{N} \ (62.8 \text{ kN})$$

All the calculations in this section have made approximations. They have also ignored frictional effects. Thus in the case of the deep drawing an allowance for friction between the cup and the die walls might increase the required force by 30%.

Types of press

There are several types of press, the main two types being fly presses and power presses.

Fly presses are hand-powered machines. The operator rotates a horizontal lever, weighted at its ends, to transfer energy to the rotating lever and thence via a screw mechanism to the ram. The tool, or punch, fits into the lower end of the ram. Limited amounts of energy are available by this method. The force is limited and, for this reason, the use of fly presses is restricted to blanking, piercing, bending and forming of small thickness material where only small forces are required.

Power presses are electrically driven. An electric motor is used to store energy in a flywheel; the energy being subsequently taken from the flywheel to activate the ram. Power presses can generate large amounts of energy and produce large forces. Hence the limitation to the work that can be handled is determined only by the size of the machine.

Compared with fly presses, power presses are expensive, not only in initial capital cost but also in tooling cost. The speed of a power press is, however, greater than that of a fly press and continuous production is possible.

Combination and progression press tools

In a single-stage operation the press tool is used to carry out one operation—blanking. However it is often possible to incorporate two or more operations into one tool so that the tool is used to carry out these operations in combination. Thus the *combination tool* may blank and draw in the one stroke of the ram. Figure 1.35 illustrates this.

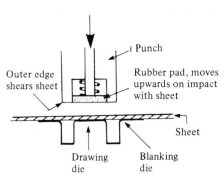
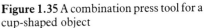

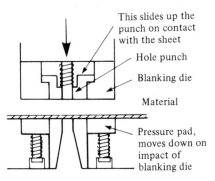

Figure 1.35 A combination press tool for a cup-shaped object

Figure 1.36 A combination press tool for a washer

Another type of combination tool is used for the forming of washers. The sheet is both blanked and pierced. Figure 1.36 illustrates this. With combination tools the increased cost of the tooling is generally more than offset by the increased production rate, provided that large production runs are required.

With a combination tool the various punches and dies are incorporated on the same axis; a *progression tool*, however, has more than one press operation but they are placed at different positions along the tool. Figure 1.15 shows the production of a washer by progressive piercing and blanking. As the strip material progresses through the tool, the piercing punch cuts the hole for the centre of the washer. This part of the strip then progresses to the blanking punch which cuts the outer edge of the washer and so the complete washer is produced. The sequence is a continuous operation with both punches being operated by the same stroke of the ram. The strip of material is thus progressively worked on as it is fed through the tool.

Progression tools have the advantage over combination tools that damage to any one punch or die does not mean the replacement of the entire set. However, combination tools are preferable if accurate alignment is required for the various operations. Progression tools are usually cheaper than combination tools. Progression tools may incorporate a combination tool at one station.

When combination or progression tools are used, the capacity of the press has to be greater than for a single operation. The following calculation illustrates this.

The maximum force required to blank a circle of radius 30 mm from a steel sheet of thickness 2.0 mm and having a shear strength of 430 MPa (or MN m^{-2}) is given by (as earlier in this chapter):

$$\text{Maximum punch force} = \tau \times \text{material thickness} \times \text{work profile perimeter}$$
$$= (430 \times 10^6) \times (2.0 \times 10^{-3})$$
$$\times (2\pi \times 30 \times 10^{-3})$$
$$= 1.62 \times 10^5 \text{ N}$$

The maximum force required to pierce a hole of radius 20 mm from the above sheet is given by:

$$\text{Maximum punch force} = (430 \times 10^6) \times (2.0 \times 10^{-3})$$
$$\times (2\pi \times 20 \times 10^{-3})$$
$$= 1.08 \times 10^5 \text{ N}$$

If single-stage operations are used for the production of a washer in the above material with an external radius of 30 mm and an internal radius of 20 mm, the maximum punch force required is:

$$1.62 \times 10^5 \text{ N}$$

The maximum punch force required if a combination or a progressive tool is used is

$$1.62 \times 10^5 + 1.08 \times 10^5 = 2.70 \times 10^5 \text{ N}$$

When to use manipulative processes

Manipulative processes are likely to be the optimum method for product production when:

1. The part is to be formed from sheet metal. Depending on the form required, shearing, bending or drawing may be appropriate if the components are not too large.

2. Long lengths of constant cross-section are required. Extrusion or rolling would be the optimum methods in that long lengths of quite complex cross-section can be produced without any need for machining.

3. The part has no internal cavities. Forging might be used, particularly if better toughness and impact strength is required than is obtainable with casting. Also directional properties can be imparted to the material to improve its performance in service.

4. Directional properties have to be imparted to the part.

5. Seamless cup-shaped objects or cans have to be produced. Deep drawing or impact extrusion would be the optimum methods.

6. The component is to be made from material in wire or bar form. Bending or upsetting can be used.

1.4 Powder technique

This process is essentially the production of the required shape by pressing fine powders together and then heating them so that they bond together. This last part of the process is known as **sintering**. The various stages in the production can be summarised as follows:

1 *Blending*

This involves mixing the appropriate powder, or powders, with a lubricant. The purpose of the lubricant is to reduce friction between the grains of the powder when they are compacted, as well as to reduce die wear.

2 *Compacting*

The powder is placed in the die and subjected to pressure to compact it, pressures of the order of 100 to 400 MPa commonly being used (Figure 1.37). A variety of methods is used to apply this pressure, the method depending on the shape of the component being produced.

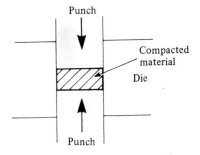

Figure 1.37 Compacting

3 Sintering

The particles in the shaped object are then bonded together by heating at a high temperature for a period of time; usually from 30 minutes to several hours. During this time the volatile lubricant materials are driven off. The temperature used is below the melting point of the metal powder, generally being about 70–80% of the melting temperature.

The process involves no waste materials and usually no machining is needed, since the product has the required shape and surface finish. The process can use materials which cannot be shaped by other methods, because they have high melting points or are extremely brittle. Cobalt-bonded tungsten cutting tools are made this way, and the material is very difficult to form by any other process. Also products of complex shape that would require considerable machining can, with advantage, be manufactured in this way.

Components can be produced with a calculated amount of porosity. Thus, a filter might be made with fine pores about 0.002 mm dia., filling about 80% of the material. Porous bearings can be manufactured containing oil to give self-lubrication.

High rates of production are possible, and the operation does not involve highly skilled labour. The shaped dies and punches are expensive due to the use of expensive materials for the dies and punches. These materials are necessary because of the high pressures and the severe abrasion involved. A large number of components need to be made in order that the die and punch costs can be widely spread and not be too significant an element in the total product cost. This often means production runs in excess of 10 000 components.

In most cases the strength properties of the product are inferior to those that would be obtained using forging or casting techniques. Often, however, the materials that are used would present problems with other methods.

The process poses design restrictions regarding the shapes of the components that can be produced; as uniform density is required in the product. Without uniform density there cannot be uniform strength. Holes having axes perpendicular to the direction of the press force are not possible. Multiple stepped diameters, grooves and undercuts also present problems. The main types of components made by this method have an almost uniform cross-section throughout the length of the component.

1.5 Cutting and grinding

Some of the most common ways of cutting metals involve machine tools, the operations being known as **machining**. All such operations involve a tool being in contact with the workpiece, the machine moving one relative to the other. The operation results in the removal of unwanted metal from the workpiece, the pieces of unwanted material being known as chips (Figure 1.38).

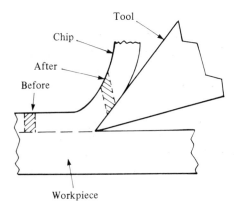

Figure 1.38 Machining

The energy supplied to the tool during the cutting operation is used in four ways:

1 The energy used to break the bonding between the atoms of the metal and create new exposed surfaces;
2 The kinetic energy given to the chip as it moves away from the workpiece and the tool;
3 The energy required to slide the chip along the face of the tool, frictional forces having to be overcome;
4 The shear energy required in deforming the material, the shaded areas in Figure 1.38 showing the shape of the material before and after cutting and the shearing action of the forces involved in cutting.

Most of the energy used in the cutting operation is involved in the shear energy required to deform the material, this often representing some 75% of the total energy. The other main element is the energy involved in sliding the chip along the face of the tool. The other energy terms are generally insignificant.

The outcome of this energy is, in the main, a rise in temperature of the workpiece. Very high temperatures can be produced, even where liquid coolants are used. During the machining it is the material of the workpiece that should be deformed while the tool remains rigid. This means that the tool material must be stronger than the workpiece material, especially at the high temperatures that can be reached at the tool–workpiece interface. Basically three types of tool materials are used—tool steels, metal carbides and ceramics.

Although the material of the workpiece must be deformable and less strong than the tool material, problems can arise if it is very ductile. Highly ductile materials are said to have poor machinability. This is because such materials produce a continuous chip and so there is a large area of material in contact with the face of the tool. This results in larger forces, and more energy, being needed; also greater tool wear. Annealed pure metals are examples of such materials.

Less ductile materials are easier to machine, with the more brittle materials being the best. This is because the material produces only short chips, due to the fact that cracks are easily initiated and propagated in such materials. Grey cast iron has good machinability, this being a result of the graphite flakes in the iron acting as crack initiators. Free cutting steels have had sulphur added to them to produce manganese sulphide inclusions. During machining the manganese sulphide forms a lubricant film along the tool face which lowers the frictional force and hence the cutting force and energy required.

There are essentially five basic metal cutting processes—turning, planing, milling, drilling and grinding. In the following notes a brief look is taken at each of these processes and their general characteristics.

Turning

Figure 1.39 illustrates the type of cutting operation involved in turning. A single-point tool is used with a cylindrical workpiece, the tool being held stationary while the workpiece rotates. The product of a turning operation is always cylindrical. The surface roughness that can be expected is of the order of 0.4 to 6 μm (the R_a value), while minimum tolerances of the order of 0.0125 mm can be achieved.

Tool

Workpiece

Figure 1.39 Turning

The machine tool used for turning is the lathe. Other operations are also possible with the lathe. These include

boring, which involves the enlargement of an existing hole (Figure 1.40a);
facing, which is the production of a flat surface as the result of the tool being
     fed across the end of the rotating workpiece (Figure 1.40b)
drilling with the drill fed against the end of the rotating workpiece and
     screwcutting.

There are several types of lathe and the choice of lathe depends on a number of factors. One factor is the number of parts to be produced. The centre lathe is a versatile machine tool but is not ideal for rapid mass production because of the time required for changing and setting tools and making measurements on the workpiece. Also a skilled operator is required.

Turret and capstan lathes are more appropriate to mass production in that a number of tools can be set up on the machine and then brought into use as

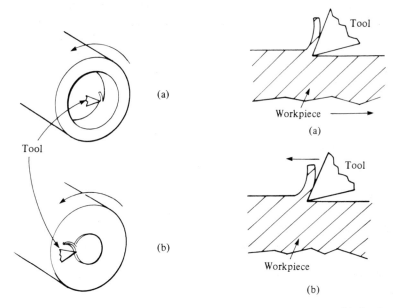

Figure 1.40 (a) Boring, (b) facing   **Figure 1.41** (a) Planing, (b) shaping

required without the need for further adjustments, tool changing or measurements. While a skilled operator is needed for the setting up, a semi-skilled worker can be used for the operation. NC lathes permit even more automatic operation.

Planing and shaping
Planing is used to produce flat surfaces or slots. It involves a single-edge cutting tool (as does turning with a lathe) operating in the way illustrated in Figure 1.41, the tool being stationary while the workpiece moves. Shaping differs from planing in that the tool moves while the workpiece is stationary, otherwise the principle is the same.

Planers tend to be used for larger scale jobs than shapers. However, neither of these tools is very widely used because other machine tools, such as millers or grinders, can do the same job more economically if a large number of parts are to be machined. Shapers tend to be used to some extent in low volume production.

Both planing and shaping tend to give surface roughness values of about 0.8 to 15 μm and minimum tolerances of the order of 0.025 mm.

Milling
A milling machine has multiple cutting edges with each edge taking its share of the cutting as the workpiece is fed past them. A wide variety of different forms of tools exist, milling being a very versatile process. Figure 1.42 illustrates

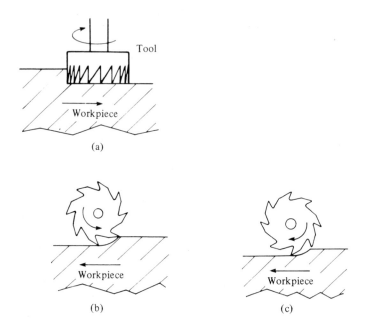

(a)

(b)   (c)

Figure 1.42 (a) Face milling. (b) Up milling. (c) Down milling

some basic forms of tool and the cutting action. The axis of rotation of the tool can be either horizontal or vertical.

Milling machines can be used, with the appropriate tools, to machine:

Plane surfaces, parallel or at right angles to the base face
Plane surfaces at an angle relative to the base surface
Keyways or slots
Helical flutes and grooves
Irregular shaped forms
Circular forms
Holes
Thread forms
Gear teeth

Milling produces roughness values of the order of 0.8 to 6.5 μm and minimum tolerances of the order of 0.25 mm.

Drilling
Drilling, like milling, involves multiple cutting edges. Because the cutting action takes place within the workpiece, the chips have to come out of the hole past the drill which itself largely fills the hole. This can make lubrication and cooling of the tool difficult. Add to this the friction between the body of the drill and the sides of the hole and there is a heat problem which can

significantly affect the accuracy with which a hole can be drilled. Drilling gives surface roughnesses of the order of 1.6 to 8.0 μm and minimum tolerances of the order of 0.05 mm.

Drilling machines can also be used for reaming. This is the process whereby a hole is brought to a more exact size by using a multi-edged tool rotating within the hole.

Grinding

Grinding is a multi-edge cutting operation employing what could be considered a self-sharpening tool. The grinding wheel has abrasive particles, such as carborundum, bonded in a matrix. As the wheel rotates, these small, hard, brittle particles cut small chips from the workpiece. As the edges of these particles become blunted, the forces acting on them due to friction increase and eventually the force may become large enough to fracture, or tear, the particle free from the matrix. The result, either way, is to expose a new cutting surface. Hence the concept of the grinding wheel as being a self sharpening tool.

Because each cutting edge is very small, and the edges are numerous, close dimensional control is possible and very fine surfaces can be produced. Also, because the abrasive particles consist of very hard materials, the grinding wheel can be used to machine very hard material workpieces. Grinding is used:
to remove surplus material from a workpiece;
to improve the dimensional accuracy of the workpiece;
to obtain the required surface finish;
to machine very hard materials which are not so readily turned or milled.
Grinding can give surface roughnesses of the order of 0.1 to 1.6 μm and minimum tolerances of the order of 0.002 mm. Grinding removes only small amounts of metal at each cut, i.e. very small chips are produced. While this allows manufacture to close tolerances it does, however, mean that the time taken to produce a unit area of a finished surface is relatively long. The cost is thus high per unit volume of metal removed. For this reason grinding is often used as a finishing process following machining by another method, e.g. milling.

Figure 1.43 illustrates some of the basic grinding methods used. Centre-type external grinding is used for producing external cylindrical surfaces where the workpiece can be held between centres and rotated against the grinding wheel. Internal cylindrical grinding, as in Figure 1.43b, involves the workpiece being held in a chuck and rotated. With centreless grinding there is no necessity to hold the workpiece between centres or in a chuck. This can reduce total operation time by eliminating the need for centre holes in the workpiece and the time taken to mount the workpiece.

With external centreless grinding two abrasive wheels are used. The larger of these wheels does the actual grinding and rotates at the regular grinding speed. The smaller of the wheels is the control wheel; it rotates at a slower

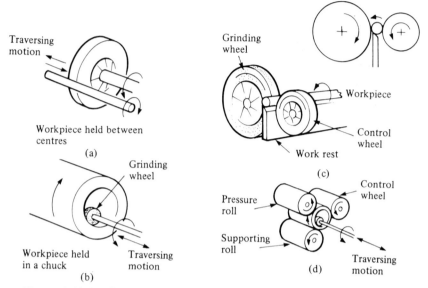

Figure 1.43 (a) Centre-type external cylindrical grinding. (b) Internal cylindrical grinding. (c) Centreless external cylindrical grinding. (d) Centreless internal cylindrical grinding

speed and controls the rotation of the workpiece. A work rest is used to keep the workpiece pressing against the face of the control wheel.

With internal centreless grinding, three rolls are used to support the workpiece and impart rotation to it (Figure 1.43d).

There are three types of feed operation used with centreless grinding. In through-feed grinding the workpiece has a constant diameter and is fed through the wheels which have been set at a constant separation (Figure 1.44a). In order to impart an axial feed to the workpiece the control wheel is tilted slightly, generally less than 7°.

In-feed grinding is used for multidiameter work or any form of work that cannot be passed completely through the wheels. The work rest and the control wheel are retracted and the workpiece placed in position. Then the rotating control wheel advances the workpiece up to the grinding wheel. The control wheel is usually tilted slightly to hold the workpiece against an end stop.

End-feed grinding is used with work which is multidiameter but where the length to be ground is too long for the in-feed grinding operation. The workpiece is loaded as with in-feed grinding but then a feed motion is imparted to the workpiece and it advances through until it meets the end stop. It is thus a mixture of in-feed and through-feed grinding.

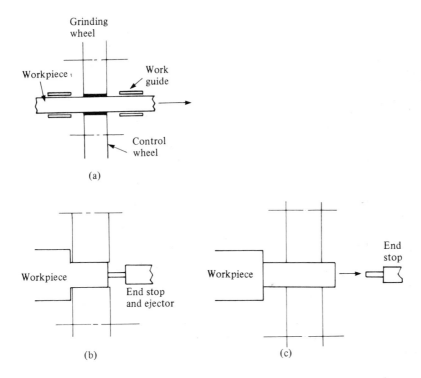

Figure 1.44 (a) Through-feed grinding. (b) In-feed grinding. (c) End-feed grinding

Finishing processes

After machining, some components undergo a further process in order to produce an improvement in accuracy and an improvement in surface finish. The following methods are used.

Grinding

The type of grinding machine used will depend on whether the surface is flat or curved and whether it is an internal or external surface. Thus, for example, a surface grinder may be used for a flat surface and possibly a centreless-type grinder for a cylindrical surface. Specially shaped grinders may be used for threads and gears, and other shapes.

The type of finish obtained will depend on the grinding wheel chosen and the operating conditions used. Typically, roughness values of the order of 0.1 to 1.6 μm can be obtained, though with care 0.025 μm is possible.

Lapping

Lapping consists of rubbing the surface concerned against a softer material surface, there being a fine abrasive in oil between the two surfaces. The operation can be carried out by a machine. The process is very slow as only small pieces of metal are removed in the process. Typically, roughness values of the order of 0.05 to 0.4 μm can be obtained, though with care 0.0125 μm is possible.

Honing

Honing is mainly used for internal cylindrical surfaces. A number of abrasive sticks are mounted on an expanding mandrel (Figure 1.45) which is inserted into the hole and adjusted to bear against the walls. As well as rotating, the tool also has a reciprocating motion. A lubricant, such as paraffin, is used. Typically, roughness values of the order of 0.1 to 0.8 μm are produced, though with care 0.025 μm can be produced.

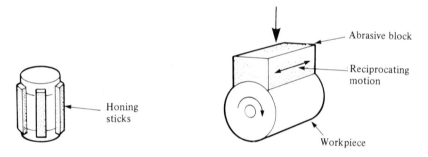

Figure 1.45 Basic format of a honing head   **Figure 1.46** Superfinishing

Superfinishing

An abrasive block reciprocates across the face of the workpiece, Figure 1.46 illustrating the type of process involved. A lubricant is used between the surfaces. A controlled amount of pressure is applied to the abrasive block so that when the appropriate amount of surface cutting has taken place, the lubricant film separates the two surfaces and prevents further cutting action. Typically, roughness values of the order of 0.05 to 0.2 μm can be achieved, though with care values as low as 0.0125 μm are possible.

Burnishing

Burnishing involves the rubbing of a smooth hard object under considerable pressure over the surface concerned. The high pressure causes the surface protrusions to suffer plastic flow. The term 'roll burnishing' is used when internal or external cylindrical surfaces are burnished using hard rollers.

Because the process involves cold working, the resulting surface has better wearing properties. Typically, roughness values of the order of 0.2 to 0.4 μm are obtained, though with care values as low as 0.1 μm are possible.

The selection of cutting processes

When selecting a cutting process, the following factors are relevant in determining the optimum process or processes:

1 Operations should be devised so that the minimum amount of material is removed. This reduces materials costs, energy costs involved in the machining and costs due to tool wear.

2 The time spent on the operation should be a minimum to keep labour costs low.

3 The skills required also affect labour costs.

4 The properties of the material being machined should be considered; in particular the hardness.

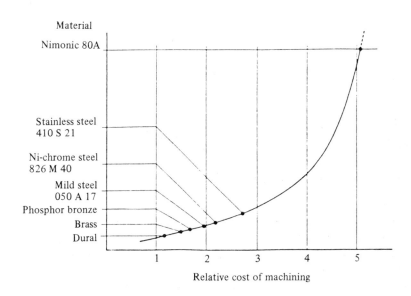

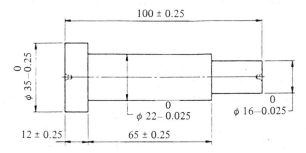

Figure 1.47 Relative cost of machining different materials. (Reproduced with the permission of the British Standards Institute, being taken from *Manual of British Standards in Engineering Drawing and Design* (Hutchinson 1984))

5 The process, or processes, chosen should take into account the quantity of products involved and the required rate of production.
6 The geometric form of the product should be considered in choosing the most appropriate process or processes.
7 The required surface finish and dimensional accuracy also affect the choice of process or processes.

Figure 1.47 illustrates how the relative cost of machining a component is affected by the choice of material for that component. In general, an important factor determining the time taken to cut a material, and hence the cost, is the material hardness; the harder a material, the longer it is likely to take to cut. The hardness also, however, affects the choice of tool material that can be used, and also, in the case of very hard materials, the process. Thus, for instance, grinding is a process that can be used with very hard materials, because the tool material, the abrasive particles, can be very hard.

Where a considerable amount of machining occurs, the use of free-machining grades of materials should always be considered in order to keep costs down by keeping the cutting time to as low a value as possible.

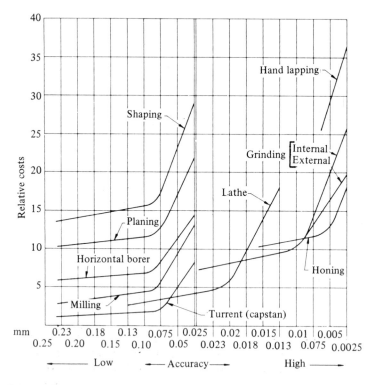

Figure 1.48 Cost of various machine and hand processes for achieving set tolerances. (Reproduced with the permission of the British Standards Institution, being taken from *Manual of British Standards in Engineering Drawing and Design* (Hutchinson 1984))

Machining operations vary quite significantly in cost, particularly if the operation is considered in terms of the cost necessary for a particular machine to achieve particular tolerances. Figure 1.48 shows how the relative costs vary for achieving set tolerances. Thus to achieve a tolerance of 0.10 mm, the rank order of the processes is:

Shaping   most expensive
Planing
Horizontal borer
Milling
Turret (capstan)   least expensive

The cost of all the processes increases as the required tolerance is increased. At high tolerances, grinding is one of the cheapest processes. The different machining operations also produce different surface finishes:

	R_a (μm)	
Planing and shaping	15 –0.8	Least smooth
Drilling	8 –1.6	
Milling	6.3–0.8	
Turning	6.3–0.4	
Grinding	1.6–0.1	Most smooth

The choice of process will depend on the geometric form of the product being produced (Table 1.2).

Table 1.2 *Choice of machining process*

Type of surface	Suitable processes
Plane surface	Shaping
	Planing
	Face milling
	Surface grinding
Externally cylindrical surface	Turning
	Grinding
Internally cylindrical surface	Drilling
	Boring
	Grinding
Flat and contoured surfaces and slots	Milling
	Grinding

Machining, in general, is a relatively expensive process when compared with many other methods of forming materials. The machining process is, however, a very flexible process which allows the generation of a wide variety of forms. A significant part of the total machining cost of a product is due to setting-up times when there is a change from one machining step to another. By reducing the number of machining steps and hence the number of setting-up times, a significant saving becomes possible. Thus the careful sequencing of machining operations and the careful choice of machine to be used is important.

Cutting speed

Cutting speed is defined as the rate at which the cutting edge of a tool passes over the work and is generally expressed in metres per minute. Thus in the case of a drilling operation, one rotation of the drill moves the cutting edge over the workpiece a distance equal to the circumference of the drill, i.e. πD, where D is the drill diameter. Hence in one minute, if the drill is rotating at N rev/min, then the distance covered is $N\pi D$. The cutting speed is therefore $N\pi D$. The following example illustrates the use of this relationship.

At how many revolutions per minute should a 12 mm diameter drill rotate if a cutting speed of 25 m/min is to be obtained?

In order to be consistent in use of units, distance is expressed in metres and time in minutes. Hence

Cutting speed = $25 = N \times \pi \times 12 \times 10^{-3}$
Hence $N = 663$ rev/min

Another continuous rotary machining process is milling. For such a process the same type of relationship holds as for drilling. Thus, if D is the diameter of the cutter and N the number of revolutions of that cutter per unit time, then the cutting speed is $N\pi D$. Thus the rate of revolution of the cutter necessary to give a cutting speed of 20 m/min when the cutter has a diameter of 100 mm is given by

Cutting speed = $20 = N \times \pi \times 100 \times 10^{-3}$
        $N = 63.7$ rev/min

Machine operation times

With a continuous rotary cutting process,

cutting speed $V = N\pi D$

where N is the number of spindle revolutions per unit time and D is the diameter of the cutter. If V is in units of m/min, D in millimetres and N in rev/min, then

$$V = N\pi D \times 10^{-3}$$

or

$$V = \frac{N\pi D}{1000}$$

The movement that is measured in machining per revolution of the spindle, or in non-rotary situations per reversal of the work, is known as the *feed rate*.

Thus, in a drilling operation, the rate at which the drill travels through the workpiece is the feed rate. In milling, the feed rate is the rate at which the work advances past the cutter. If f is the feed rate, then for N revolutions per unit time of a spindle the distance moved by the drill through the workpiece, or the work past the cutter, in unit time is Nf. If N is in rev/min and f in mm/rev, then the distance moved in a minute is Nf mm. If L is the required length of cut then the time taken to achieve this cut is L/Nf.

$$\text{Time for one cut} = \frac{L}{Nf}$$

The following example illustrates the use of the above equation.

Calculate the time, t, taken to drill a hole 12 mm diameter and 30 mm deep with a cutting speed of 20 m/min and a feed rate of 0.18 mm/rev.

$$V = N\pi D$$

$$\text{Hence, } t = \frac{L\pi D}{Vf} = \frac{30 \times \pi \times 12}{20 \times 1000 \times 0.18}$$

Note that in this calculation all distances are in mm and all times in minutes.

$$\text{Hence, } \qquad t = 0.31 \text{ min}$$

In milling the cutting speed is given by $V = N\pi D$, but an allowance called the approach distance A must be added to the length of the machined surface L_m to give the total length of cut L.

$$L = L_m + A$$

$$\text{Hence, for one cut, } t = \frac{L}{Nf} = \frac{L_m + A}{Nf}$$

For peripheral milling

$$A = \sqrt{[d(D - d)]}$$

where D is the diameter of the cutter and d the depth of cut.

For face milling

$$A = \tfrac{1}{2}D - \sqrt{(\tfrac{1}{4}D^2 - \tfrac{1}{4}b^2)}$$

where b is the width of cut and D the diameter of the cutter. Hence, using the above equations, the time take for one cut by milling can be determined.

Similarly the times taken for other cutting operations can be determined. Thus it is possible, for the machining of some component involving perhaps a number of machining operations, to estimate the total machining time by working out the times for each part of the machining.

Cutting times are, however, only that time for which the tool is actually in contact with the workpiece during the metal removal process. No allowance is included for handling the workpiece, tool changing, relaxation time for the operator, etc. To find the standard time for the entire operation these other times must be taken into account. Only then does it become possible to work out how many components can be processed in any given time.

In general, the time for all the handling, machining and manipulating is made up of the following elements:

1 Initial setting up of the machine
2 Loading and unloading the workpieces
3 Manipulation of the machine during its operation, e.g. gear changing
4 Machining time
5 Measuring and gauging
6 Changing and resetting tools
7 Tool regrinding
8 Contingency allowance for such factors as operator fatigue etc

For more information on the calculation of times taken, the reader is advised to consult the book *Principles of Engineering Production* by A. J. Lissaman and S. J. Martin (Hodder and Stoughton, 1982).

Tooling costs

The longer the life of a tool the greater the period of time over which the tool cost can be spread and so the smaller the tooling cost overhead per job. If T is the *tool life* then the cost of tooling for a job is inversely proportional to T, i.e.

$$\text{tooling cost} \propto \frac{1}{T}$$

Figure 1.49 shows the relationship graphically.

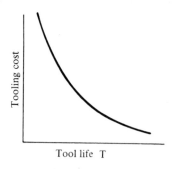

Figure 1.49 Tooling costs and tool life

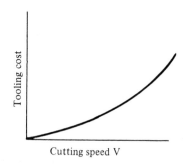

Figure 1.50 Tooling costs and cutting speed

Tool life, however, depends on the cutting speed V. In cutting operations the relationship between cutting speed and the tool life is given by *Taylor's equation,*

$$VT^n = C$$

where n is an index related to the type of material used for the cutting tool and C is a constant. Thus as $T = (C/V)^{1/n}$ the tooling cost is related to the cutting speed by

$$\text{tooling cost} \propto \frac{1}{T} \propto \frac{1}{(C/V)^{1/n}} \propto (V/C)^{1/n}$$

as C is a constant. Figure 1.50 shows this relationship graphically.

The value of n depends on the cutting material used, typically it is about 0.1 to 0.15 for high-speed steel tools, 0.2 to 0.4 for tungsten carbide tools and 0.4 to 0.6 for ceramic tools. To illustrate the effect on tooling cost of cutting speed, consider a tool for which n is 0.2 and the effect of changing the cutting speed from, say, 180 m/min to 90 m/min.

Tooling cost for 90 m/min $\propto 90^{1/0.2}$
Tooling cost for 180 m/min $\propto 180^{1/0.2}$

Hence, $\dfrac{\text{Tooling cost for 90 m/min}}{\text{Tooling cost for 180 m/min}} = \dfrac{90^{1/0.2}}{180^{1/0.2}}$

$$= \left(\frac{90}{180} \right)^{1/0.2} = 0.5^5 = 0.031$$

The tooling cost at 90 m/min is thus considerably less than that at 180 m/min, being only 0.031 of the 180 m/min cost. Lest it be thought that slower cutting speeds inevitably mean lower production cost, account has to be taken of the smaller number of components that can be machined per hour at the lower speed.

The higher the cutting speed, the shorter the time required to remove the metal. As the cost of carrying out the cutting will be proportional to the time taken for the machining, the higher the cutting speed the smaller the machine cost.

Cost of cutting $\propto$ time taken for cutting

But, time taken for cutting $\propto 1/V$

where V is the cutting speed.

If the cutting speed is doubled then the time taken is halved. Thus

Cost of cutting $\propto 1/V$

Figure 1.51 shows how the cost of cutting depends on the cutting speed. Also on the same graph the tooling cost as a function of cutting speed has been plotted (this is Figure 1.50). The total cost is the sum of the costs of cutting and the tooling cost and so can be obtained by adding the two graphs together. The composite graph shows a minimum at a particular cutting speed, this being the most economic cutting speed to use.

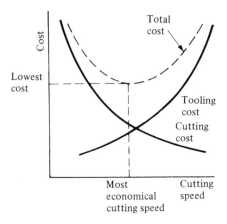

Figure 1.51 Total tooling cost

The following example illustrates the use of Taylor's equation in a costing. For a particular machining operation $n = 0.3$ and $C = 140$. The time taken to change the tool is 5 minutes and the time taken to regrind it, 4 minutes. The machine time per component is 8 minutes. The machine cost is £0.05 per minute. Ignoring workpiece handling times, setting times and tool depreciation, what is the cost of machining 100 components at a cutting speed of 60 m per minute?

$$\text{Tool life } T = \left(\frac{C}{V}\right)^{1/n} = \left(\frac{140}{60}\right)^{1/0.3} = 16.4 \text{ min}$$

Therefore the number of components per regrind is

$$\frac{16.4}{8} = 2.05$$

The tool change cost $= 5 \times 0.05 = £0.25$

The tool change cost per component $= \dfrac{0.25}{2.05} = £0.12$

The tool regrind cost $= 4 \times 0.05 = £0.20$

The tool regrind cost per component $= \dfrac{0.20}{2.05} = £0.10$

The machining cost per component $= 8 \times 0.05 = £0.40$

Hence the total cost per component $= £0.12 + £0.10 + £0.40 = £0.62$

The cost therefore of a batch of 100 components $= £62$

Most economical cutting speed

The following is a derivation of the relationship for the most economic cutting speed.

Time to machine a unit volume of metal $\propto \dfrac{1}{V} = \dfrac{K}{V}$

where V is the cutting speed and K a constant.

If H is the machining cost per minute (the time unit used is generally minutes rather than seconds), then

$$\text{Cost of machining a unit volume of metal} = \frac{HK}{V} \tag{1}$$

with V being the speed in metres per minute.

In the time taken of (K/V) to machine unit volume of metal the number of tool changes will be

$$\text{Number of tool changes} = \frac{K}{TV}$$

where T is the tool life.

If $\mathcal{J}$ is the total cost per tool change then

$$\text{Tooling cost per unit volume of metal cut} = \frac{\mathcal{J}K}{TV}$$

But, according to Taylor's equation, $VT^n = C$ or $T = (C/V)^{1/n}$.

Hence,

$$\text{Tooling cost per unit volume of metal cut} = \frac{\mathcal{J}K}{V(C/V)^{1/n}} = \frac{\mathcal{J}K(V)^{(1-n)/n}}{C^{1/n}} \tag{2}$$

The total cost per unit volume of metal cut is the sum of the machining cost, equation (1), and the tooling cost, equation (2).

Hence,

$$\text{Total cost per unit volume of metal cut} = \frac{HK}{V} + \frac{\mathcal{J}K(V)^{(1-n)/n}}{C^{1/n}}$$

The minimum value of this cost P can be found by differentiating and equating to zero.

$$\frac{dP}{dV} = -\frac{HK}{V^2} + \left(\frac{1-n}{n}\right)\frac{\mathcal{J}K}{C^{1/n}}(V)^{(1-2n)/n}$$

When this is equated to zero, as K is not zero,

$$\frac{H}{V^2} = \left(\frac{1-n}{n}\right)\frac{\mathcal{J}}{C^{1/n}}(V)^{(1-2n)/n}$$

$$H = \left(\frac{1-n}{n}\right) \frac{\mathcal{J}}{C^{1/n}} (V)^{1/n}$$

Hence

$$V = C \left(\frac{n}{(\mathcal{J}/H)(1-n)}\right)^{n}$$

The above is the condition for the most economical cutting speed.

The following example illustrates a use of the expression. The cost of operating a particular machine is £12 per hour and the cost of a tool change is £3. For the machining conditions concerned $C = 150$ and $n = 0.25$. What is the most economical cutting speed?

$$V = C \left(\frac{n}{(\mathcal{J}/H)(1-n)}\right)^{n}$$

$\mathcal{J} = £3$ and $H = £12$ per hour $= £12/60$ per minute. Hence $\mathcal{J}/H = 15$ min. Thus

$$V = 150 \left(\frac{0.25}{15 \times 0.75}\right)^{0.25}$$

$$= 57.9 \text{ m/min}$$

1.6 Metal removal

The machining methods previously considered have all involved removing metal from the workpiece by cutting by means of a tool, the tool being harder than the material of the workpiece. There are however other methods of removing metal which do not involve cutting using a tool. These methods can be summarised as electrochemical machining, electrical discharge machining and chemical machining. All these methods have the great advantage over traditional machining that the hardness of the material of the workpiece is of no consequence, this property not affecting the rate of metal removal.

Electrochemical machining (ECM)

Electrochemical machining involves an electric current being passed between two electrodes dipping into an electrolyte. The workpiece is used as the positive electrode, known as the anode (the electrode connected to the positive side of the d.c. supply) and the tool is the negative electrode, known as the cathode. The tool is not in contact with the workpiece and carries out no cutting action. The tool is merely a piece of suitable conducting metal that is shaped as the reverse image of the form required of the workpiece. Thus if a hollow is required in the workpiece the tool has a corresponding hump. The workpiece and the tool are positioned very close together, with generally less than 1 mm between them. The space between the workpiece and the tool is filled with electrolyte which is constantly being replenished. Figure 1.52 illustrates the arrangement.

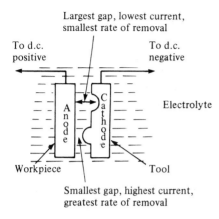

Figure 1.52 Electrochemical machining

When a d.c. supply is connected to the two electrodes a current passes between them. Very high currents are used. The term current density is used to describe the current in relation to the surface area of the electrode concerned. Thus current densities of the order of 50 to 1900 A/cm^2 are used. The result of this is that metal is removed from the workpiece and goes into solution. The metal is removed at the greatest rate from that part of the workpiece which is closest to the tool. Thus where there is a hump on the tool surface, the gap between it and the workpiece is smallest and it is here that there is the greatest rate of metal removal from the workpiece. Where there is a hollow on the tool, the gap between it and the workpiece is greatest and the rate of removal of metal the least. The effect of this is that the workpiece has metal removed at different rates from the different parts of it and a reverse image of the profile of the tool is produced.

During the metal removal the tool is advanced towards the workpiece at a rate which maintains a constant separation between the two; the electrolyte being pumped between the two surfaces at a speed of generally about 10 to 50 m/s. This speed is necessary in order to sweep away the waste products that are produced and allow the current density, and so metal removal, to be maintained. Metal removal rates of the order of 10 to 300 mm^3 per second are possible, this rate being independent of the hardness of the material, being only determined by the rate at which metal atoms in the workpiece go into solution. This is essentially a function of the current.

Electrochemical machining has the following advantages when compared with traditional machining:

(a) The tool does not wear (provided a suitable material is used which is not corroded by the electrolyte);
(b) There are no thermal or mechanical stresses on the workpiece during the metal removal;
(c) By using shaped electrodes on either side of the workpiece, three-dimensional machining is possible.

During electrochemical machining, metal removal occurs not only from that part of the workpiece directly opposite the tool but also from those parts of the workpiece round the side of the tool (Figure 1.53). The result of this is an overcutting, e.g. in the production of a hole in the workpiece the hole will have a greater diameter than the diameter of the tool. Typically the overcut is of the order of 0.02 to 0.1 mm.

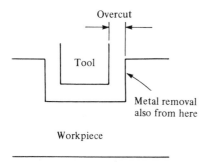

Figure 1.53 Overcutting

The surface roughness produced is of the order of 0.2 to 0.4 μm, a fine surface, and tolerances of the order of 0.05 mm are achieved.
The disadvantages of electrochemical machining are:

(a) The basic cost of the equipment is fairly high.
(b) The tool has to be made of non-corrosive material and has to be rigidly mounted, as also has the workpiece, bearing in mind the high rate of flow of electrolyte between tool and workpiece.

Electrochemical machining can produce small deep holes, odd-shaped holes and cavities, three-dimensional shapes, and can deal with high-strength, high-hardness materials. The process finds application in the machining of high-temperature alloy forgings to the required finish and tolerances, jet engine blade aerofoils, turbine wheels with integral blades.

Electrical discharge machining (ECM)

In electrical discharge machining (ECM) material is removed by the action of electrical discharges—sparks— on the surface of the workpiece. The tool and the workpiece are submerged in a fluid, such as paraffin or light oil. This fluid normally acts as an insulator. A voltage is then applied between the workpiece and the tool and increased until the insulating properties of the fluid break down and a massive pulse of current flows between the tool and the workpiece. This causes part of the workpiece to be vaporised and hence metal to be removed, as the vaporised atoms of metal are swept away by the flow of the fluid.

The fluid is continuously pumped through the assembly, a filter being used in the pumping line to extract the metal that has been removed from the workpiece and the sludge of black particles, mainly carbon, that is produced from the breakdown of the fluid. Figure 1.54 illustrates the basic principle of electrical discharge machining.

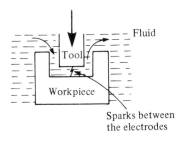

Figure 1.54 Electrical discharge machining

The electrical discharges between the tool and the workpiece occur where the distance between the tool and the workpiece is a minimum. The tool is thus a reverse image of the shape required from the workpiece. During machining by this method the tool is moved so that the distance between the tool and workpiece remains constant, generally about 0.02 mm.

It is not only the workpiece that has metal removed during the discharge, the tool also loses material. The rate of loss of material from the tool can be minimised by suitable choice of the material for the tool and the choice of operating conditions. A high melting point for the tool material is one of the conditions for low tool wear; for this reason, graphite (melting point 3500°C) is widely used.

The metal removal rate with electrical discharge machining is fairly low, i.e. of the order of 1 mm^3 per second. Tolerances of the order of 0.05 mm can be achieved with surface roughness values of the order of 1.6 to 3.2 μm. As with electrochemical machining, metal removal occurs from the workpiece not only directly underneath the tool but also round the sides of the tool. This overcut is typically about 0.005 to 0.2 mm.

The advantages of electrical discharge machining when compared with traditional machining are:

(a) The hardness of the material being machined is not a factor; as long as the material can conduct a current, it can be machined.

(b) Any shape that can be produced for the tool can be reproduced, thus very complicated machining is possible.

(c) There are no mechanical stresses in the workpiece during the metal removal.

The disadvantages of electrical discharge machining are:

(a) Tool wear occurs and in, for example, the production of deep holes the tool is likely to need redressing.

(b) After cutting, the workpiece will have a surface layer different from the parent metal. Typically there are three parts to the layer—an inner annealed layer of hardness less than that of the parent material; a layer which has reached melting point but has not been removed and which remains as a recast layer; an upper layer formed by molten particles being redeposited on the surface. In steel workpieces this layer may be much harder and more brittle than the parent material and can reduce the fatigue endurance limits. Some finishing process may thus be used, after electric discharge machining, in order to remove the layer.

(c) The process is relatively slow.

An important modification of the electrical discharge machining principle is the continuous wire electrode machine (Figure 1.55). Instead of using a shaped tool as one electrode, a wire of diameter about 0.25 mm or less is used. It is continuously fed to the workpiece at a velocity of about 0.1 to 8 m per minute. Enough wire is generally available for the operation to continue for twenty-four hours or more. Complex shapes can be cut in materials up to about 15 cm thick, at rates of the order of 0.5 mm per minute. The process is used in NC and CNC machining.

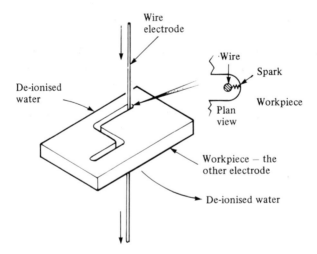

Figure 1.55 The continuous wire electrode electric discharge machine

Chemical machining

In chemical machining the material is removed from the workpiece by exposing it to a chemical reagent. The terms 'chemical milling', 'photo-fabrication' and 'chemical blanking' are sometimes used to describe the process. The metal removal is fairly simple, the component being either immersed in or exposed to a spray of chemical reagent. The type of chemical reagent used depends on the metal concerned. The component remains exposed to the reagent until the required amount of metal has been removed. The more complex part of the operation is, however, the method by which parts of the component are masked so that metal removal only takes place in certain areas.

A common masking method involves the use of photosensitive resists. The workpiece is coated with a light-sensitive emulsion. When this has dried, a photographic negative of the master pattern that is required is placed against the workpiece and light is passed through it and onto the sensitive emulsion on the workpiece. When the workpiece is immersed in a developing agent, those areas that were not exposed to the light are removed, leaving a mask over the exposed part of the workpiece. When the chemical reagent is then used, those areas that are not masked have metal removed but the masked areas are not affected.

This method is used in the production of electronic circuit boards and other components which are often very complex, small and rather thin, and would present problems if tackled in any other way.

Chemical machining can be applied to almost any metal and can be used with both small and large surface areas. The process does not produce any mechanical or thermal stresses in the material, but it may release existing residual stresses already present and so some warping may occur. Surfaces typically have a roughness of 1.6 to 6.3 μm and tolerances of the order of 0.075 mm can be achieved.

The disadvantages of chemical machining are that the rate of metal removal is fairly slow per unit area exposed. If the metal of the workpiece is not homogeneous, different removal rates can occur for different parts of the workpiece. Another problem is that the chemical reagent acts on all exposed surface. Consequently, as the depth of metal removal into the workpiece

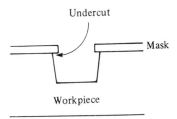

Figure 1.56 Undercutting with chemical machining

increases there will be an increasing tendency to undercut and remove metal from beneath the masked surface (Figure 1.56).

1.7 Surface finishes

A large proportion of products are given a form of decorative or protective surface treatment to complete their manufacturing process. In some cases there may be a need for only cleaning the surface, but this is also likely to be necessary before applying any other form of finish. Cleaning can be accomplished by mechanical, chemical or ultrasonic means.

Mechanical cleaning can involve various methods of abrasive cleaning in which an abrasive agent, such as sand or shot, is directed against the surface to be cleaned. Another mechanical method involves high speed rotary wire brushes being applied to the surface. A further form of mechanical cleaning involves placing the parts to be cleaned in a drum, along with some abrasive material such as sand or special abrasive pellets. The drum is then either rotated, to give a tumbling action, or vibrated back-and-forth.

Chemical cleaning is commonly used at some stage in the production process to remove dirt, scale oil, grease, etc. Various forms of chemical cleaning are used. Alkaline cleaning uses such materials as caustic soda and is widely employed. Solvents are often used when oils and greases are involved and for metals such as aluminium or zinc which would be attacked by alkaline cleaners.

Ultrasonic cleaning is used where fairly small parts have to be made very clean. The parts are placed in a cleaning bath of liquid and very high frequency sound (ultrasonic sound) is passed through the liquid, which causes the liquid to cavitate. This term is used to describe the process in which the sound causes small vacuum pockets to be produced in the liquid. These pockets almost immediately collapse and the resulting action causes a scrubbing action.

Following cleaning, a decorative or protective surface treatment can be applied to the surfaces of a product. Such finishes can be either an additive coating of the surface or a conversion of the surface layers of the product. Common examples of additive coatings are paints and electroplating; an example of conversion layers is anodising.

Painting may be carried out using dipping, hand or automatic spraying and electrocoating. With dipping the parts are dipped into the paint, either manually or as part of an automated process. This method is used where the entire component needs the same coat of paint, often with small components where spraying could result in a considerable loss of paint due to overspraying and for the priming coat of paint which precedes the finishing coats. The disadvantages of this method are that very thin layers cannot be produced and there is a tendency for the paint to run and produce a wavy surface and a drip of paint at the lowest point. In spray painting, paint is atomised and directed against the product. With conventional spraying there can be a very significant wastage of paint and also there is difficulty in obtaining a uniform coverage. Electrostatic spraying gives better results. In this process the atomised paint is given an electrostatic charge. These charged particles are then attracted to the workpiece, which is generally electrically earthed. Under reasonable conditions, more than 90% of the paint can be deposited on the product by this method.

Electroplating involves making the workpiece the negative electrode, i.e. the cathode, in a suitable electrolyte while the metal to be deposited is the positive electrode, i.e. the anode (Figure 1.57). Tin, cadmium, chromium, copper, gold, silver and zinc are examples of metals that are used to electroplate products. In some cases more than one material may be used to coat the workpiece, nickel often being used as an undercoating to gold or silver.

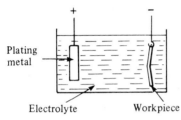

Figure 1.57 Electroplating

Anodising is a conversion process used with aluminium. Unlike electroplating, the workpiece is the anode in an electrolyte and, instead of a layer of material being added to the surface, the result is the conversion of the surface layers to aluminium oxide. Unlike the addition processes, anodising does not cause an increase in size of the product.

This short section has barely touched the subject of finishes, particularly those used to reduce corrosion. Further information can be found in *Metal Corrosion* by T. K. Ross, an Engineering Design Guide, No. 21, published for the Design Council, the British Standards Institution and the Council of Engineering Institutions by Oxford University Press (1977).

Problems

1 Compare sand casting and die casting as methods for the production of a product.
2 Why is die casting not suitable for the production of small quantities of components?
3 Under what conditions would investment casting be more economic than die casting?
4 Suggest a casting process for the following situation. A small one-off

casting is required using aluminium, there is a lot of fine detail which has to be reproduced and a good surface finish is required.

5 State two design features that have to be taken into account when designing a product to be produced by casting.
6 Under what circumstances is casting likely to be the optimum process?
7 What are the benefits and disadvantages of using hot working as opposed to cold working?
8 Compare the surface finishes produced by hot and cold rolling.
9 How are hollow shapes produced by extrusion?
10 It is proposed to use extrusion for the production of seamless aluminium tubing. What is approximately the largest diameter tubing that could be produced? About what size would be the maximum length possible?
11 Compare forging and casting as methods of production for a product.
12 What types of product is the upsetting process used for?
13 Suggest a process for the production of the section used for aluminium window frames.
14 Explain the basic process of shearing that occurs when a punch is used to shear a metal sheet.
15 Explain how piercing and blanking operations can be used for the production of washers.
16 Explain the need for bending allowances in the production of components by bending.
17 How does the cross-section of a bar change when it is bent?
18 Give two examples of products that could be produced using bending.
19 What type of materials have to be used with deep drawing?
20 Describe the process of deep drawing.
21 How is the wrinkling of the material during deep drawing prevented?
22 At what position in a deep drawn product is failure likely to occur during the drawing process?
23 What is ironing?
24 Compare rigid die and flexible die methods of drawing and state the situations under which flexible dies would be more economic.
25 Give an example of products that can be produced using deep drawing.
26 When a sheet of material is bent, the side that is in compression may buckle or distort. How can this be prevented?
27 How does flow turning differ from spinning?
28 What type of materials can be used with impact extrusion?
29 What is high velocity metal forming and what types of product are produced by this process?
30 Suggest processes that might be used for the following products:
(a) a toothpaste tube from a very soft alloy; (b) the reflector concave dish for a satellite TV receiving aerial, using aluminium or mild steel; (c) rivets; (d) an aluminium can for a drink or for food storage; (e) the formed wing of a car; (f) a hollow hexagonal length of brass rod; (g) a spanner; (h) railway lines; (i) a kitchen pan.

31 Calculate the maximum punch force needed to blank a circle of radius 20 mm from a steel sheet of thickness 1.5 mm and having a shear strength of 420 MPa.

32 Calculate the maximum punch force needed for a single-stage piercing and blanking operation in which a washer with an external radius of 20 mm and an internal radius of 12 mm is produced from material of thickness 1.5 mm and having a shear strength of 430 MPa.

33 Estimate the work done in blanking a square of side 80 mm from sheet of thickness 2.0 mm if the penetration is 25% and the material has a shear strength of 430 MPa.

34 Why in a blanking or piercing operation with a press might the punch or the die have shear on it? Explain what shear is.

35 What force will need to be applied by the punch in a bending operation with a press if a right angle bend is to be produced in mild steel sheet of thickness 3.0 mm and width 100 mm if the material has a yield stress of 200 MPa.

36 What is the force necessary to draw a cup 30 mm dia. from a blank of 60 mm dia. and sheet of thickness 2.0 mm if the material has a yield stress of 200 MPa?

37 How do combination press tools differ from progression press tools?

38 Outline the processes used to produce a sintered product.

39 What are the advantages and the limitations of powder techniques for producing products?

40 What makes some materials easy to machine and others difficult?

41 Explain why free cutting steels have good machinability.

42 Rank the cutting operations of turning, planing, milling, drilling and grinding in the order of surface finish, rough to fine.

43 Why are turret and capstan lathes preferred to centre lathes for mass production work?

44 List five types of form that milling machines, with appropriate tools, can be used to produce in a workpiece.

45 State three types of work that can be carried out by grinding.

46 How does centreless grinding differ from centre-type grinding?

47 What are the three types of feed operation used with centreless grinding?

48 Which type of feed operation would be used with centreless grinding if the workpiece was a long shaft of constant cross-section that had to be uniformly brought to the required tolerance.

49 Rank the finishing operations of grinding, lapping, honing, superfinishing and burnishing in order of the surface finish produced, rough to fine.

50 State five factors that have to be considered in determining which cutting process to use for a particular operation.

51 Rank the cutting operations of turning using a capstan lathe, milling, planing and shaping in the order of least expensive to most costly to produce work to a set tolerance (say 0.05 mm).

52 State the advantages and disadvantages of electrochemical machining when compared with traditional machining.

53 State a typical type of product that could be produced by electrochemical machining.

54 State the advantages and disadvantages of electric discharge machining when compared with traditional machining.

55 Explain the operation of the continuous wire electrode machine version of electric discharge machining and state the type of operation for which it is used.

56 State the type of product that is produced by chemical machining.

57 What is the difference between an addition type and a conversion type of finish that will affect the design of a component?

58 State the type of sequence that is likely to be adopted in applying a finish, such as paint to a product.

59 State three different types of cleaning that can be used with metals.

60 State three reasons why a finish may be applied to the surfaces of a product.

2

Forming processes – polymers

2.1 Polymer materials

Polymers can be divided into two groups—thermoplastics and thermosets. Thermoplastic materials can be softened, allowed to harden and then resoftened indefinitely by the application of heat, provided the temperature is not so high as to cause decomposition. Such materials can be formed into different shapes by the application of heat. Examples of such materials are polyethylene (polythene), polyvinyl chloride (PVC), polyamide (nylon), polycarbonate and cellulose acetate.

Thermosetting materials undergo a chemical change when they are subjected to heat; this change cannot be reversed by the application of further heating. Typical thermosetting materials are phenol formaldehyde (Bakelite), urea formaldehyde and melamine formaldehyde.

Polymers are produced by a process called polymerisation, a process in which many small molecules combine to form a more complex large molecule. Such a molecule has a chain-like structure with hundreds or thousands of identical groups of atoms all linked together. With thermoplastics, the process often stops at the polymerisation stage and the resulting material then has a large number of these long molecular chains. With thermosetting materials the process continues, with crosslinks forming between different chains and the establishment of a three-dimensional network of interlocked molecules.

Both thermoplastic and thermosetting polymers have relatively low densities, relatively low strength and stiffness, low electrical conductivity, low thermal conductivity, low specific heat capacity, high coefficient of thermal expansion and are softened or degraded at temperatures not far removed from the boiling point of water.

For most purposes other materials are added to the polymer to make the material for processing into products, the term *plastic* generally being used for this 'mixed' material. The additives may be in the form of solids, liquids or gases. The purpose of the additives can be to improve the properties, reduce the cost, improve the mouldability, or add colour. In many instances the additive may constitute more than half the weight, or volume, of the plastic.

The following are some typical examples of additives. Glass fibres can be added to improve the strength of the material, and to make it less ductile but stiffer. Mica may be added to improve the electrical resistance. Liquids may be added to improve the flow characteristics of the material during processing, the liquid acting as a lubricant between the polymer chains. One form of additive is a gas, the result being foamed or expanded materials.

The flow behaviour of polymers

A material can be considered to be elastic when the deformation produced in the material is wholly recovered after the removal of the force causing the deformation. For example, this is the type of behaviour that we might expect of a metal in the solid state. However if we consider liquids, deformation means flow. The term used to describe the flow properties is *viscosity*. It can be considered to be the resistance of a fluid during flow. The higher the viscosity, the more resistance there is to flow. A fluid is said to be Newtonian when its viscosity is independent of shear stress and time. Water, for example, is a Newtonian fluid, whereas many paints are non-Newtonian in that, if you press the brush harder against the paint, its viscosity decreases and it flows more easily. This enables the paint to be spread over a surface. When, however, the brush pressure is removed, the viscosity increases and the paint thus does not flow so readily and hence does not drip.

Polymers exhibit both elastic and viscous responses to applied forces and are therefore described as being visco-elastic. Thus, when a force is applied they can show:

1 An instantaneous elastic deformation.
2 A delayed and recoverable elastic deformation; in other words, the material continues to become deformed for some time after the force has been applied. They show a *creep* which is *recoverable*.
3 An irrecoverable deformation due to the flow of polymer chains past each other.

The above behaviour of polymers, in the solid state, has to be taken into account in the use made of the material. This behaviour also has implications for the processing methods used to form products. Polymers also are non-Newtonian in their flow behaviour, the viscosity decreasing in much the same way as in paint. The following are some of the ways in which this flow behaviour affects processing methods.

1 In injection moulding a screw or a ram can be used to compress the polymer and force it out into a mould. A screw, however, applies higher shear rates to the polymer than a ram and so the polymer flows more easily.
2 An alternative to applying shear to a polymer to reduce its viscosity is to heat it—the higher the temperature, the lower the viscosity. However, in using a thermoplastic in, say, injection moulding the mould has to be cooled in order

to cool the thermoplastic component and enable it to be removed from the mould. The higher the temperature at which the polymer is introduced into the mould, the longer the time interval before it is cold enough to remove. Shearing thus has an advantage over heating in that the time taken to obtain the product is less.

3 When extruded material emerges from the die, a phenomenon called *die swell* occurs. This, as the name implies, is a swelling of the material so that it is not the same size as it was when in the die. This can be considered to be due to the delayed and recoverable elastic deformation mentioned earlier. Because the swelling is not necessarily the same in all directions, the shape of the extruded component can change.

2.2 Forming processes

The processes used with polymers can be considered to be divided into two groups—primary processes and secondary processes. Primary processes involve the fabrication of the component in a single operation from the polymer. Casting, moulding, extrusion and calendering are examples of such processes. With secondary processes, a product of a primary process, e.g. a sheet of polymer, is transformed into a finished product. Forming and machining are examples of such processes.

Casting

A number of casting methods are employed, but for polymers casting is not like the casting of metals where the hot liquid metal is poured into the mould. Few polymers can be heated to a high enough temperature to flow well enough to be used in such an operation.

One form of casting involves mixing, in a mould, substances of relatively short molecular chains, with any required additives, so that polymerisation (i.e. the production of long-chain molecules) occurs during the chemical reaction which leads to the solid product being produced. For such polymers the term *cold-setting* is used. This process is used for encapsulating small electrical components, producing tubes, rods and sheets. It can be used with both thermoplastics and thermosets.

Powder casting involves the melting of a powdered polymer inside a heated mould which may be either stationary or rotated during the operation. The term *rotational moulding* is often used to describe the process when the mould is rotated. The effect of the rotation is to cause the polymer to coat the inside walls of the mould and so produce a hollow article. Holes are not possible though inserts can be used. The process is used with thermoplastics for producing containers, e.g. large milk storage tanks, canisters and crates. Powder casting can also be used to coat surfaces with films of polymers, e.g. non-stick surfaces of cooking pans.

Slush casting involves a fine polymer powder suspended in a liquid (this being the slush) being coated onto the inside of a heated mould. The liquid evaporates and the polymer particles fuse together to form a thin layer of solid polymer. This method is used to produce hollow articles which have fine details, e.g. some plastic toys and plastic gloves.

Moulding

A widely used process for thermoplastics, and to some extent for thermosetting materials, is injection moulding. In this process the polymer is fed into the cold end of the injection cylinder (Figure 2.1). A rotating screw, or a ram, then compresses the material and passes it through a heated section and then injects the polymer into the mould. In the case of thermoplastic materials the mould has water cooling. When the material has sufficiently cooled, the component is ejected.

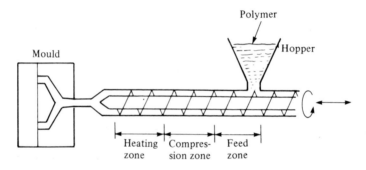

Figure 2.1 Injection moulding

With a thermosetting material the same procedure is adopted but the mould is heated and the component can be ejected without waiting for the component to cool, though the material must be held in the mould for a sufficient time for the curing to be completed. The production rate possible with thermoplastics is faster than that possible with thermosetting materials.

Complex shapes with inserts, holes, threads, etc can be produced but enclosed hollow shapes are not possible. Typical products are beer or milk bottle crates, toys, control knobs for electronic equipment, tool handles and pipe fittings. The size of the products can vary from the control knob size, a mass of perhaps 15 g, up to pallets perhaps 1.0 m square and having a mass of many kilogrammes.

The cost of the moulds used with injection moulding is high and thus it is only with large production runs that the process becomes economic. However there is little waste of material in the process and the parts taken from the mould are finished products.

Foam plastic components can be produced by injection moulding, inert gases being dissolved in the molten polymer. When the polymer cools the gases come out of solution and expand to form a cellular structure. A solid skin is produced where the molten polymer comes into contact initially with the cold mould surface. Typical products are stacking containers, toughened polystyrene tables and other office items.

Compression moulding and transfer moulding are used with thermosetting polymers. In *compression moulding* (Figure 2.2) the powdered polymer is compressed between the two parts of the mould and heated under this pressure. With *transfer moulding* (Figure 2.3) the powdered polymer is heated in a chamber before being tranferred by a plunger into the mould, a process not unlike injection moulding.

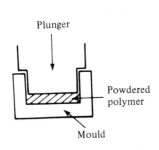

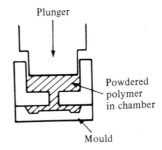

Figure 2.2 Compression moulding   **Figure 2.3** Transfer moulding

Moulds for compression moulding tend to be cheaper than those used for either injection or transfer moulding, because no access is needed for the molten polymer. With compression moulding, inserts may present a problem, being damaged or moved when the pressure is applied. Transfer moulding does, however, allow complex parts with inserts to be readily made.

Compression moulding is used to produce products such as washing machine agitators, electrical plug cases, switch cases, knobs, car instrument panels, etc.

Extrusion

With the extrusion process, molten polymer is forced through a die. This is done by a screw mechanism which takes the polymer through a heated zone before forcing it through the die (Figure 2.4). The operation is continuous, a steady source of molten polymer being forced through the die. Long lengths of constant cross-section are produced. The process is used with thermoplastics for the production of pipes and various profiles such as curtain rails, sealing strips and skirting boards.

If thin film is required, a die can be used which gives an extruded cylinder of

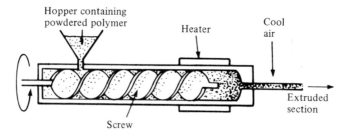

Figure 2.4 Extrusion

the material. This cylinder while still hot is inflated by compressed air to give a sleeve of thin film. An alternative method is to use a slit die and allow the hot material issuing from the die to fall vertically into a cooling system. This may be a water bath or a pair of cooled rollers. The film produced by either of these methods is used for packaging and a variety of decorative and office uses.

Sheet can be extruded by using a horizontal die of the appropriate shape. Such sheet may be used in secondary processing, such as thermoforming, to form products such as dinghy hulls and containers. Corrugated sheet produced this way is used for roofing.

With *extrusion coating*, a thin film of polymer is extruded from a slit die, drawn down to the required thickness and then pressed into contact with the required substrate. High coating speeds are possible, though the initial capital cost is high.

An important extrusion process is *extrusion blow moulding*. This process is used for the production of hollow articles, particularly plastic moulds. The operation involves extruding a hollow tube. A mould is placed round the hot tube, known as a parison, and closed up on it (Figure 2.5). The mould seals the lower end of the tube and the top end is cut off with a knife. The mould then moves on to have air injected into the parison. The air causes the still hot parison to fill the cold mould where it hardens before being ejected when the

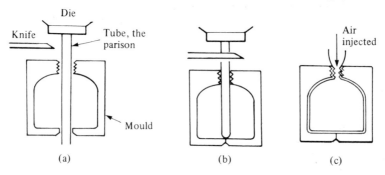

Figure 2.5 Blow moulding. (a) Tube extruded, (b) mould closed up and tube cut off, (c) air injected forces the parison to fill the mould

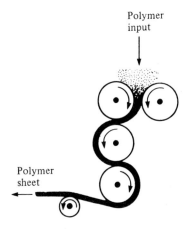

Polymer
input

Polymer
sheet

Figure 2.6 Calendering

mould is opened. The entire process is automated and is capable of high production speeds.

Calendering

This process is used for the production of continuous lengths of sheet thermoplastic materials. The calender consists of essentially three or more heated rollers (Figure 2.6). The heated polymer passes between and round rollers to emerge as an output of sheet. The process is continuous.

Forming

Forming processes are secondary processes in that they are used to form articles from sheet polymer. The heated sheet is pressed into or around a mould. Figure 2.7 illustrates one form of this process, this version being known as *vacuum forming* in that a reduction in air pressure between the sheet and the mould is used to cause the sheet to adopt the shape of the mould. With *pressure moulding* an increase in pressure is used on one side of the sheet in order

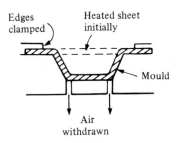

Edges
clamped

Heated sheet
initially

Mould

Air
withdrawn

Figure 2.7 Vacuum forming

to force the sheet to adopt the shape of the mould. With *matched moulds forming*, the heated sheet is formed by pressing between a pair of matched male and female moulds.

Thermoforming can have a reasonable output rate, but dimensional accuracy is not very good and holes and threads are not possible. The method can be used with very large sheets and is used for forming machine housings, pallets, car bodies, dinghy hulls, and small items such as drinking cups for vending machines, egg cartons and margarine tubs. The moulds may be made out of wood, metal or plaster. Mould costs can be relatively low.

Machining

In the production of many products, machining can be avoided by a careful design and choice of the manufacturing process used. Conventional machining processes can be used with polymers but there are some problems associated with the facts that polymers are poor heat conductors and have low melting or degradation temperatures.

Because of the poor heat conduction, little of the heat developed during the machine operation is conducted away through the material. Thus the tool used tends to run very hot. Also the polymer, if a thermoplastic, tends to soften, swell and bind against the tool. This results in yet more friction and heat development, as well as tool wear. Hence, in machining polymers, correct machining conditions are vital. Some polymer materials are brittle and so present problems during machining, shock loads having to be avoided if cracking is not to occur.

2.3 Choosing a process

Injection moulding and extrusion are the most widely used processes. Injection moulding is generally used for the mass production of small items, often with intricate shapes. Extrusion is used for products which are required in continuous lengths or which are fabricated from a continuous length of material of constant cross-section. The following are some of the factors that are involved in choosing a process.

1 *Rate of production.* Injection moulding has the highest rate, followed by blow moulding, then rotational moulding, compression moulding, transfer moulding and thermoforming, with casting being the slowest.

2 *Capital investment required.* Injection moulding requires the highest capital investment, with extrusion and blow moulding requiring less capital. Rotational moulding, compression moulding, transfer moulding, thermoforming and casting require the least capital investment.

3 *Most economic production run.* Injection moulding, extrusion and blow moulding are economic only with large production runs. Thermoforming, rotational moulding, casting and machining are used for the small production runs.

4 *Surface finish*. Injection moulding, blow moulding, rotational moulding, thermoforming, transfer and compression moulding, and casting all give very good surface finishes. Extrusion gives only a fairly good finish.

5 *Metal inserts, during the process*. These are possible with injection moulding, rotational moulding, transfer moulding and casting.

6 *Dimensional accuracy*. Injection moulding and transfer moulding are very good. Compression moulding good with casting; extrusion is fairly poor.

7 *Very small items*. Injection moulding and machining are the best.

8 *Enclosed hollow shapes*. Blow moulding and rotational moulding can be used.

9 *Intricate, complex shapes*. Injection moulding, blow moulding, transfer moulding and extrusion.

10 *Threads*. Injection moulding, blow moulding, casting and machining.

11 *Large formed sheets*. Thermoforming.

Design considerations when using polymers

detail design

Polymers are not direct substitutes for metals and a component design that was a good one with a metal may be a bad one if a polymer is used instead of the metal. Polymers behave differently under load, there is a higher rate of creep, they are more affected by temperature changes, etc. The main advantage of polymers is that fairly accurate components can be produced, with an excellent surface finish and in colour, at a fairly low cost and at high production rates.

In a process involving the casting of metals or the casting or moulding of polymers, essentially the same design considerations prevail. They are the problems associated with introducing a fluid or semifluid into a mould. The right amount of fluid has to be introduced and the mould must be completely filled without any air trapped in the mould. The design must therefore ensure that there is a free flow of the fluid to all parts of the mould and that air cannot become trapped. As with metals, polymers shrink when they cool and solidify. Thus allowances have to be made for such shrinkage when the mould is designed. Also, in designing the mould, care has to be taken to ensure that the component can be removed from the mould; this means that the product needs to have a taper, generally of the order of a few degrees.

Thinner wall sections cool more rapidly than thicker wall sections, also the thicker sections shrink more than the thinner sections. Thus distortion can occur where there are changes in wall thickness. For this reason wall thicknesses should be kept as constant as possible. This generally means keeping walls fairly thin. Extra stiffness can be provided without increasing wall thickness by the use of ribs, flanges or webs (Figure 2.8a).

Polymers have low values for the modulus of elasticity and so large flat areas of sheet tend to bend easily and are difficult to keep rigid. The sheet can be made considerably more rigid, without increasing its thickness, by the use of ribbing or doming (Figure 2.8b).

Threads, either external or internal, can generally be moulded into the

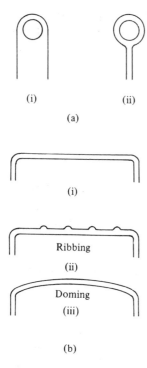

(i)   (ii)

(a)

(i)

Ribbing

(ii)

Doming
(iii)

(b)

Figure 2.8 (a) To obtain even shrinkage and less distortion (ii) is preferred to (i). (b) For more rigidity (ii) and (iii) are preferred to (i)

component. However, such threads are generally not very hard wearing and it may be preferable, if frequent assembly and disassembly is required, to use an insert. Inserts can be moulded-in during the production of the component or introduced during later assembly by just push-fitting the insert, while hot, into the polymer. In either method, care has to be taken that the insert does not pull out of the component when in service—it is held only by a mechanical bond. For this reason, inserts may be knurled or grooved. Screwdriver blades are inserts into plastic handles. So as to give a good grip when the handle is rotated, the metal part in the handle is usually winged or flattened.

Problems

1 How do thermosetting materials differ in properties from thermoplastics?
2 Explain the principles of the rotational moulding process and state types of products that can be produced by this process.
3 How does compression moulding differ from transfer moulding? What is the main effect of the difference on the type of mould that can be used?
4 State two properties of polymers that significantly affect the way they can be machined.

the polymer. In either method, care has to be taken that the insert does not pull out of the component when in service—it is held only by a mechanical bond. For this reason, inserts may be knurled or grooved. Screwdriver blades are inserts into plastic handles. So as to give a good grip when the handle is rotated, the metal part in the handle is usually winged or flattened.

Problems

1 How do thermosetting materials differ in properties from thermoplastics?
2 Explain the principles of the rotational moulding process and state types of products that can be produced by this process.
3 How does compression moulding differ from transfer moulding? What is the main effect of the difference on the type of mould that can be used?
4 State two properties of polymers that significantly affect the way they can be machined.
5 Suggest processes suitable for the following:
   (a) High production rate required for a small plastic toy to be made from a thermoplastic material.
   (b) A 1 litre bottle for soft drink, a thermoplastic material to be used. High production rates are required.
   (c) A switch cover in a thermosetting material, as high a production rate as possible being required.
   (d) Milk churns, about 340 mm diameter and 760 mm high, from polyethylene.
   (e) A thermoplastic strip for use as a draught excluder with windows, long lengths being required.
   (f) Mass production of polythene bags.
   (g) The body for a camera, fairly high production rates being required.
   (h) Mass production of the bodywork of an electric drill, threaded holes are needed.
   (i) A nylon gear wheel, a high production rate being required.

3

Assembly operations

3.1 Assembly processes for metals

The main processes for assembly can be summarised as:

Adhesive bonding. The types of adhesives used include natural adhesives, elastomers, thermoplastics, thermosets and two-polymer types.
Soldering and brazing. The joining agent used is different from the two materials being joined but alloys locally with them.
Welding. Heat or pressure is used to fuse the two materials being joined together.
Fastening systems. Fasteners provide a clamping force between the two pieces of materials being joined, e.g. nuts and bolts, rivets.

Adhesives for metals

The use of adhesives to bond materials together can have advantages over other joining methods, i.e.

(a) Dissimilar materials can be joined, e.g. metals to polymers
(b) Jointing can take place over large areas
(c) A uniform distribution of stress over the entire bonded area is produced—with a minimum of stress concentration
(d) The bond is generally permanent
(e) Joining can be carried out at room temperature or temperatures close to it
(f) A smooth finish is obtained

Disadvantages are that optimum bond strength is usually not produced immediately: a curing time has to be allowed. The bond can be affected by environmental factors such as heat, cold and humidity. These adhesives generally cannot be used at temperatures above about 200°C.

Adhesives can be classified according to the type of chemical involved. The main types are as follows:

Natural adhesives

Vegetable glues made from plant starches are typical examples of natural adhesives. These types are used on postage stamps and envelopes. However, such adhesives give bonds with poor strength which are susceptible to fungal attack and are also weakened by moisture. They set as a result of solvent evaporation.

Elastomers

Elastomeric adhesives are based on synthetic rubbers; they also set as a result of solvent evaporation. Strong joints are not produced as they have low shear strength. The adhesive is inclined to creep. These adhesives are mainly used for unstressed joints and flexible bonds with plastics and rubbers.

Thermoplastics

These include a number of different setting types. An important group are those, such as polyamides, which are applied hot, solidify and bond on cooling. They are widely used with metals, plastics, wood, etc and have a wide application in rapid assembly work such as furniture assembly and the production of plastic film laminates.

Another group are the acrylic acid diesters which set when air is excluded, the reaction being one of a build-up of molecular chain length. Cyanoacrylates, the 'super-glues', set in the presence of moisture, in a similar way, with the reaction taking place in seconds. This makes them very useful for rapid assembly of small components. Other forms of thermoplastic adhesives set by solvent evaporation, e.g. polyvinyl acetate.

In general, thermoplastic adhesives have a low shear strength and under high loads are subject to creep, so they are generally used in assemblies subject to low stresses. They have poor to good resistance to water but good resistance to oil.

Thermosets

These set as a result of a build-up of molecular chains to give a rigid cross-linked. Epoxy resins, such as Araldite, are one of the most widely used thermoset adhesives. These are two-part adhesives, in that setting only starts to occur when the two components of the adhesive are brought together. They will bond almost anything and give strong bonds which are resistant to water, oil and solvents.

Phenolic resins are another example of thermoset adhesives. Heat and pressure are necessary for setting. They have good strength and resistance to water, oil and solvents and are widely used for bonding plywood.

Two-polymer types

Thermosets by themselves give brittle joints, but combined with a thermoplastic or elastomer a more flexible joint can be produced. Phenolic resins with nitrile or Neoprene rubbers have high shear strength, excellent peel strength, good resistance to water, oils and solvents and good creep properties. Phenolic resins with polyvinyl acetate, a thermoplastic, give similar bond strengths but with even better resistance to water, oils and solvents. These adhesives are used for bonding laminates and metals. Joints using them can be subjected to high stresses and can often operate satisfactorily up to temperatures around 200°C.

For the maximum-strength bonds to be realised with an adhesive, the maximum area of bonding should be used. Figure 3.1 shows some typical joints.

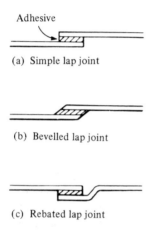

Adhesive

(a) Simple lap joint

(b) Bevelled lap joint

(c) Rebated lap joint

Figure 3.1 Adhesive joints

Soldering and brazing

Soldering involves heating the joining agent (the solder) together with the materials being joined until the solder melts and alloys with their surfaces. On cooling, the alloy solidifies and forms a bond between the two materials. The joining process requires temperatures below about 425° and often below 300°C.

Solders are only weak structural materials when compared with the metals they generally join; thus there is a need to ensure that the strength of the soldering does not rely on solder strength and is designed so that preferably the materials interlock or overlap in some way (Figure 3.2). The hot solder must wet the surfaces being joined otherwise it will not alloy with them. Preparation of the surfaces is thus vital, generally involving abrasion, degreasing and the use of a flux. Fluxes, when heated, promote or accelerate the wetting of the surfaces by the solder. They remove oxide layers from both metal and solder and prevent them reforming during soldering.

Brazing is a process similar to soldering but involves temperatures above

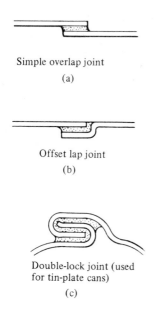

Simple overlap joint

(a)

Offset lap joint

(b)

Double-lock joint (used
for tin-plate cans)

(c)

Figure 3.2 Solder joints

425°C, but below the melting points of the materials being joined. The term
'braze' derives from the use of brass to make the joint. However, nowadays,
other alloys are also used. As in soldering, it is preferable for the materials to
interlock or overlap in some way, so that the joint is designed for shear rather
than direct tension. Figure 3.3 shows some of the joint forms that are
preferable with brazing. Brazing gives stronger joints than soldering.

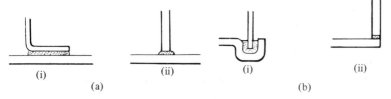

(i)   (ii)   (i)   (ii)

(a)   (b)

Figure 3.3 So that the joints are in shear (i) is preferred in each case to (ii). (a) A T
junction. (b) A corner

Welding metals

With brazing and soldering, the joint is effected by inserting a metal between
the two surfaces being joined, the inserted metal having a lower melting point
than that of the materials being joined. With welding, the joint is effected
directly between the parts being joined by the application of heat or pressure.
In fusion welding, an external heat source is used to melt the interfaces of the
joint materials and so cause the materials to fuse together. With solid state

welding, pressure is used to bring the two interfaces into intimate contact and so fuse the two materials together.

With adhesive bonded, soldered or welded joints there needs to be an overlap of the materials at the joints in order that the joint material is in shear rather than direct tension. This is because the joint material is weaker than the materials that are joined and is necessary to achieve the maximum strength. This is not the case with welding. The two materials being joined are fused together and so the material at the joint is the same as the parent material. Hence there is no need for an overlap; butt joints can be made and are strong (Figure 3.4).

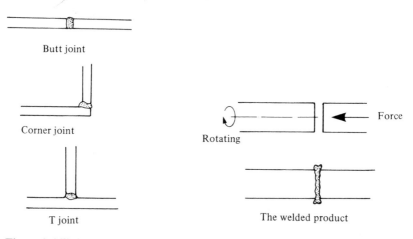

Figure 3.4 Weld joints   **Figure 3.5** Friction welding

Welding is capable of producing high-strength joints. However, because of the high temperatures involved, there may be detrimental changes in the materials being joined. There may also be local distortions due to uneven thermal expansion, residual stresses, or micro-structural changes. There are a large number of different welding processes, however they can be grouped into five main types.

Solid phase welding
Cold pressure welding uses mechanical deformation at room temperature to bring the two materials into intimate contact and so fuse them together. It is, however, only applicable to ductile metals such as aluminium or copper. Hot pressure welding is similar but uses heat to make the material more ductile.

Explosive welding uses explosive charges to impact the two materials together and fuse their interfaces. It is a useful technique for joining dissimilar materials. Friction welding has the two surfaces rubbing together (Figure 3.5) in order to clean them and to provide heat as a result of friction. Generally the parts joined by this process are cylindrical. With diffusion welding (or bonding) the workpieces are held together under light pressure and heated to a temperature which is sufficiently high to cause atoms to diffuse across the boundary between the two materials, the operation taking place in a vacuum, Dissimilar metals may be joined by this process.

Thermochemical welding

The heat to fuse the metals together is provided by a gas flame, usually acetylene and oxygen burnt in a torch (Figure 3.6), or by an exothermic chemical reaction producing liquid metal which is poured into the joint. The exothermic reaction is provided by iron oxide and aluminium being ignited and producing liquid iron. Oxyacetylene welding is a low-cost, portable operation usually used for quick machine repairs and small miscellaneous jobs with mainly ferrous alloys up to about 8 mm thick.

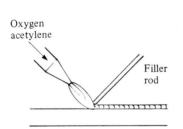

Figure 3.6 Oxyacetylene welding   **Figure 3.7** Spot welding

Electric-resistance welding

In this case the heat needed to fuse the metals together is provided either by the passage of an electric current across the interface of the joint, or by a current induced by electromagnetic induction in the metal near the joint. Widely used types of resistance welding are spot welding, seam welding and projection welding. Spot welding is used for joining sheet metal and involves squeezing the two sheets between a pair of electrodes (Figure 3.7). When a current is passed between the electrodes, a molten nugget of metal is produced at the interface between the two sheets; this rapidly solidifies when the current ceases. The process is widely used for high-speed mass production operations with steel sheets up to 3 mm thick.

Seam welding is similar to spot welding but, instead of cylindrical electrodes, copper alloy wheel electrodes are used. These rotate and drive the workpiece through the machine during the welding operation (Figure 3.8). The appearance of the weld is that of a series of overlapping spot welds or spots spaced at regular intervals, the motion of the wheels and the current pulsing being adjustable.

With spot and seam welding the current is made to flow through just one small area by the pressure from the electrodes forcing the sheets into intimate

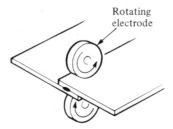

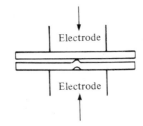

Figure 3.8 Seam welding   **Figure 3.9** Projection welding

contact at those points, hence it is only at these points that the welding occurs. With projection welding, the current is made to flow through just one small area by small projections raised on one of the sheets (Figure 3.9). During welding, the projections collapse as a result of the heat and pressure and the sheets are brought into close contact. This process is versatile and can be used for joining any two shapes, provided a contact point is possible where the weld is to be made. It is widely used for attaching small components, e.g. fasteners, to larger components.

Electric arc welding
Essentially three forms of electric arc welding are used, the distinction between them being the method adopted to protect the hot metal from attack by the atmosphere and the cleaning away of contaminating surface films on the materials being joined, the term *shielding* being used to describe the process. There is the unshielded arc, the flux-shielded arc and the gas-shielded arc. Basically the process of arc welding, in whatever form of shielding, involves producing an arc between a metal electrode and the workpiece by passing a high current between them. Flux-shielded arcs are probably the most widely used welding process.

With the metal arc version of flux-shielded arcs, the electric arc passes along a metal rod, called the core wire, which is covered with flux (Figure 3.10). The core wire melts and is transferred to the weld pool. Simultaneously the flux coating of the wire forms a molten slag which protects the liquid metal. In addition, protective gases are given off. This process is used mainly for welding steels, but can be used for other metals with the exception of aluminium alloys. It is used with materials from the thin to very thick. The equipment costs are fairly low.

The submerged arc version of flux-shielded arc welding uses a continuous bare electrode wire with the shielding flux being supplied separately in powdered or granule form in a layer over the work area, submerging it. The process is used with steels in shipbuilding, structural and general engineering, for metal thicknesses from about 8 mm upwards.

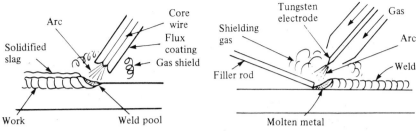

Figure 3.10 Metal arc welding   **Figure 3.11** Tungsten inert-gas welding

With gas-shielded arcs, the shielding is supplied by a continuous stream of gas over the work area. Tungsten inert-gas welding (TIG) uses an arc from a tungsten electrode in an atmosphere of argon of helium (Figure 3.11). Extra metal to fill the joint is supplied separately as the tungsten electrode is not consumed. High quality welds are produced, though the process is rather slow. Unlike flux-shielded arcs, no cleaning up of the weld area is required after the welding. The process is mainly used for the welding of sheet materials.

Plasma arc welding is similar to tungsten inert-gas welding with the arc being passed through a small orifice in a water-cooled nozzle before reaching the workpiece. Another version of gas-shielded arc welding is metal inert gas welding (MIG). In this version the arc is formed between a continuously fed electrode wire and the work, with a gas supply providing the shielding. The process is fairly costly and is used for the thicknesses from 0.5 mm upwards.

Radiant energy welding
In electron-beam welding a stream of electrons is produced and used to bombard the materials being welded, the result of the bombardment being the generation of heat. The entire operation is generally carried out in a vacuum. The process can produce very deep narrow welds. The equipment cost is fairly high.

Another radiant energy form of welding involves the use of a laser. Unlike electron-beam welding, the workpiece does not have to be in a vacuum. Both methods, however, are using focussed beams of energy. Because with lasers intense energy can be focussed down to a very small area, precision work is possible and high-melting-point metals can be welded, which would be virtually impossible by other means. Thicknesses up to about 3 mm can be handled.

Selecting a welding process
The choice of welding process for a particular situation is determined by the following factors:

(a) *The thickness of the materials being welded.* Thus, for instance, spot welding can be used for steels up to about 3 mm thick, and occasionally up to 6 mm. It cannot be used for thicker materials. Projection welding can also only be used for sheets up to about 3 mm thick. Submerged arc welding, however, is generally not suitable for materials thinner than about 8 mm.

(b) *The joint shape.* If the ends of two steel rods are to be joined then friction welding is a possible method; however, it would be unsuitable for a butt joint between the edges of two steel plates.

(c) *The need for inspection.* The design of the joint must take into account the need for an inspection of the weld on its completion. The accessibility of the joint also affects the choice of welding process, in that the welder must be able to get at the joint area in order to carry out the welding. Some processes are more easily adapted to awkward situations than others. For instance, the metal arc method only requires the electrode to be directed into the position, perhaps a corner.

(d) *Whether the process is to be automatic.* Where a larger number of welds of a similar type or a long length of weld is involved, automatic processes, rather than manual methods, are generally to be preferred. Submerged arc welding can be used in this way. Metal arc welding would not be suitable because of the operator skill that is required to maintain the arc and the metal feed.

Fastening systems with metals

The choice of fastener, from the very wide variety available, will depend on a number of factors:

(a) *Environmental.* The conditions under which the fastener is to be used, e.g. temperature and corrosive conditions.

(b) *Nature of the external loading on the fastener.* Different fasteners are appropriate for the different types of loading, e.g. tension, compression, shear, cyclic or impact, and the magnitude of the loading.

(c) *Life and service requirements.* Is the fastener to be permanent or demountable? Is there to be frequent assembly and disassembly?

(d) *The quantity of the fasteners required and their cost*

(e) *The method of assembly of the fastener.* This is often one of the prime factors determining choice in that over 60% of the cost of using a fastener in an assembly is likely to be the assembly operator's handling time. It can be as high as 90%. Thus a choice of fastener which permits a short handling time can have a very significant effect on the overall cost.

Fasteners can be classified into three types—threaded, non-threaded and special-purpose. The purpose of a fastener, in all forms, is however to provide a clamping force between two pieces of material. Of the metals used, steel is probably the most common, although aluminium alloys, brass and nickel are among other metals used. Aluminium alloy fasteners have the advantage over steel of being much lighter, non-magnetic and more corrosion-resistant. Nickel has the particular advantage of strength at high temperatures.

Threaded fasteners

With a threaded fastener, the clamping force holding the two pieces of material together is produced by a torque being applied to the fastener and being maintained during the service life of the fastener. Bolts mated with nuts, and screws and threads in the material, are examples of threaded fasteners.

The performance of a threaded fastener is affected by many factors, notably the thread form, the loading, the fastener material and the effect of coatings of the fastener material. Steel bolts, for instance, might be coated with zinc or cadmium to make them more resistant to corrosion.

Non-threaded fasteners

Examples of non-threaded fasteners are rivets, eyelets, pins and spring-retaining clips. Rivets can be used for joining dissimilar or similar materials, both metallic and non-metallic, to give permanent joints.

Figure 3.12 shows the basic stages that are typical of a riveting process. When the force applied to the rivet is sufficiently high, plastic deformation occurs and the shank of the rivet increases in diameter as its length decreases. That part of the shank within the hole increases in diameter until it fills the hole, and the unsupported part of the shank outside the hole continues to deform until a head is formed. A ductile material has to be used for the rivet material, e.g. mild steel, copper, brass, aluminium, aluminium alloy. For some materials the rivets are used in the cold state, in other cases they are used while hot.

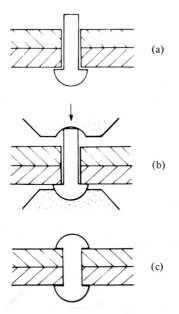

(a)

(b)

(c)

Figure 3.12 Typical stages in a riveting process

Figure 3.13 shows some of the forms of riveted joints. In some cases butt straps (thin sheets of material between the rivet heads and the materials being joined) are used. Other applications use multiple rows of rivets. In applications where the riveting force might be large enough to damage or distort the materials being joined, tubular or semi-tubular rivets might be used instead of solid rivets (Figure 3.14).

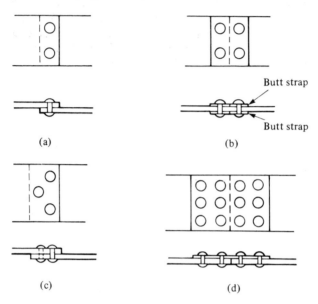

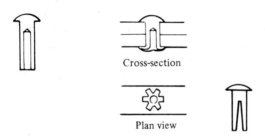

Figure 3.13 Riveted joints. (a) Single-riveted lap joint. (b) Single-riveted butt joint with two butt straps. (c) Double-riveted lap joint. (d) Double-riveted butt joint with a single butt strap

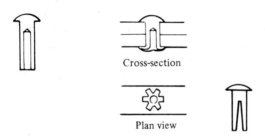

Cross-section

Plan view

Figure 3.14 A tubular rivet   **Figure 3.15** A bifurcated rivet

Tubular rivets can, in some circumstances, be self-piercing. This means that a hole does not have to be drilled prior to inserting the rivet. Tubular rivets are, however, limited to materials such as leather and fibre for self-piercing. Bifurcated rivets are used as self-piercing in a wider range of materials, e.g.

plywood, plastics and fibres. Figure 3.15 shows a type of form of such a rivet, such a rivet having just two prongs.

The term *blind rivet* is used for those rivets that are installed from just one side of the workpiece, requiring no other operator or holding tool on the other side. One form of such a rivet is the pop rivet. This consists of a hollow rivet assembled on a steel mandrel. When the rivet has been inserted into the hole in the workpiece a special tool is used to pull the mandrel and cause the hollow end of the rivet to expand on the blind side and so clamp the materials of the workpiece together (Figure 3.16). In some forms, the head of the mandrel breaks off and remains in the rivet to act as a plug, in other forms the head just breaks off and falls away to leave the centre of the rivet open.

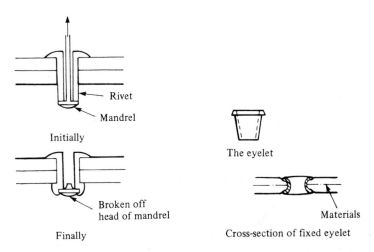

Initially

Finally

Figure 3.16 A pop rivet

The eyelet

Cross-section of fixed eyelet

Figure 3.17 One form of eyelet

The term eyelet is given to a small item used either as a hole strengthener or as a light load fastener. Figure 3.17 shows the basic form of an eyelet, consisting generally of a tubular body with a head at one end. It is inserted through the hole in the materials and pressure is applied to deform the non-head end and form a head at that end, hence clamping the materials.

Pins, either in the solid or tubular forms, are widely used for fastening, and also as hinges, pivots, etc. Thus taper pins (Figure 3.18) are used to join wheels on to the ends of shafts, the pin just being driven through holes in the two parts until it is fully home, giving a tight fit. Another form of pin is the split cotter pin (Figure 3.19). This is used where freedom of movement in the joint is required, or as a locking device for slotted nuts on bolts.

There is a wide variety of forms of spring-retaining clips. A simple form is a C-clip which is used to lock and retain components on shafts, the clip generally fitting into a groove on the shaft (Figure 3.20).

For more detailed information on fasteners, see D. M. Chaddock, *Introduction to Fastening Systems*, Oxford University Press, (1974).

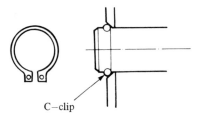

Figure 3.18 A taper pin   **Figure 3.19** A cotter-pin joint

C−clip

Figure 3.20 A spring-retaining clip

3.2 Assembly methods for plastics

The assembly processes that can be used with plastics can be considered to fall into four groups:

1 *Welding*. This process is similar to metal welding and involves using heat to fuse thermoplastic materials together.
2 *Adhesive bonding*. The types of adhesives used may be elastomers, thermoplastics, thermosets and two-polymer types.
3 *Riveting*. Both metal and thermoplastic rivets are used.
4 *Press and snap fits*. This is an important way of making both permanent and recoverable assemblies. Plastics can be distorted quite a considerable amount elastically and recover their original dimensions and shape afterwards. This is the basis of this type of assembly.
5 *Thread systems*. Screw threads and self-tapping screws are widely used.

Welding thermoplastics

There are nine methods of welding thermoplastics:

Spin welding	Ultrasonic welding
Vibration welding	Resistance-wire welding
Hot-gas welding	Dielectric welding
Hot-plate welding	Induction welding
Hot-wire welding	

Spin welding and vibration welding (sometimes referred to as oscillatory welding) are friction welding methods in which the weld is produced by the frictional heat developed at the interface between the two thermoplastic materials. Spin welding is only suitable for circular components and involves holding the lower part in a jig while the upper part is brought into contact with it while rotating at high speed (Figure 3.21). The friction at the interface rapidly causes the plastic to melt and the surfaces then fuse together. This is a reliable method of producing pressure and vacuum-tight joints having a bond which is almost as strong as the parent material. Vibration welding can be applied to non-circular shapes, the method involving an oscillatory motion rather than a rotation as with spin welding. The effects are the same.

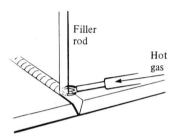

Figure 3.21 The basis of spin welding

Figure 3.22 Hot-gas welding

Hot-gas welding, hot-plate welding and hot-wire welding all involve the melting of the interface of the joint by direct heating. Hot-gas welding is similar to oxyacetylene welding of metals; heat is applied using a welding torch to blow hot gas on to the joint and a filler rod supplies molten plastic to fill the joint (Figure 3.22). The process requires a skilled operator if high-strength joints are to be produced: too little heating of the joint area, and the weld will be weak; too much heating, and the plastic will degrade and a poor weld will be produced. This process is used for the fabrication of large containers.

In hot-plate welding, the faces of the parts to be joined are pressed against a heated plate, coated with PTFE to prevent sticking; the plate is then withdrawn and the two surfaces to be joined are pressed together (Figure

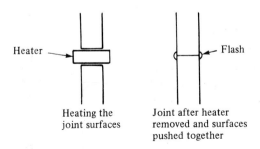

Figure 3.23 Hot-plate welding

3.23). By using specially designed heaters, three-dimensional shapes can be welded by this method. Nylons cannot be welded this way since they oxidise when the melted resin is exposed to air. Hot-plate welding is used for the on-site joining of thermoplastic pipes.

Hot-wire welding involves constant pressure being applied to the joint while an electrically heated wire passes through the joint, melting the plastic and thus forming a weld.

Ultrasonic welding essentially involves an input to the joint area of high frequency vibrations, of the order of 20 kHz. These cause the two surfaces of the joint to vibrate against each other and, as a result of friction, melt. The process is fast, some machines being capable of bonding some 30 parts per minute by this method, and is capable of being automated. Weld strength is consistently high and the process is very versatile. Some thermoplastics cannot, however, be welded by this method.

Resistance wire welding involves passing a current through a wire which has been inserted in the joint area. After the interfaces have melted and fused, the wire is left in position. This process has been used in the construction of the hulls of sailing craft.

Dielectric welding is used for the welding together of thin sheets of plastic. The materials are placed between the plates of a capacitor, acting as the dielectric. A high-frequency alternating voltage is applied to the capacitor and the resulting high-frequency electric field in the plastic causes heating and hence bonding. This process has been used for the production of upholstery, imitation leather, luggage and inflatables.

Induction welding involves inserting a strip of metal along the joint; in some cases this is a tape which is electrically conducting. No direct connections are made to this conductor but it is placed inside a coil through which an alternating current passes. Electromagnetic induction results in an e.m.f. being induced in the conductor and this causes local heating, hence melting of the plastic and so bonding.

Adhesive bonding of plastics

The main types of adhesives used with plastics are epoxy resins, acrylic acid diesters, two-part acrylics and the cyanocrylates.

Epoxy resins are thermosetting adhesives (see earlier in this chapter) and usually involve two components, the resin and the hardener, which have to be combined for the bond to be made. Generally they are used with thermosetting materials, producing good bonds with those materials, but are not so useful with thermoplastics in that poorer bonds are produced.

The acrylic acid diesters set when air is excluded, the term *anaerobic* often being used to indicate that they set without oxygen. The surfaces to be bonded are coated and then brought together under light pressure, this excluding air. They form good bonds with thermosets but are not suitable for use with the common engineering thermoplastics. Two-part acrylics involve a hardener being used with the acrylic resin, one surface of the joint being coated with hardener and the other with the resin. The two surfaces are then brought together for the bond to be made. These adhesives will bond to almost anything and are used with thermosets and many thermoplastics. The cyanocrylates, the 'super-glues', will form good bonds with thermosets and virtually all thermoplastics.

Some plastics can be bonded by the use of a solvent; nylons, polystyrene and PVC are examples of such plastics. The solvent is used to soften the interfaces of the joint and then light pressure is applied to bring the surfaces into close contact.

Riveting of plastics

Riveting using metal rivets can be used to make joints between plastics and between plastics and metals. Tubular or bifurcated rivets are generally used. The problem with riveting is that some plastics show a pronounced delayed response to deformation; they are said to exhibit a 'memory'. This means that though the rivet was initially fitted tightly, a less than tight fit might develop with time following the release of the load on the rivet. Only those plastics which have a good ductility and are not brittle, have high strength and good resistance to creep, are suitable for riveting.

A different version of riveting is known as ultrasonic staking. One of the parts is made with integral projecting rivets or 'stakes'. These, when mated with the other part, project through holes in it. The projecting part of the stake is deformed by application of ultrasonic energy into a mushroom-like head and so clamps the two parts together. This type of joining is frequently used when metal parts have to be attached to a plastic, the plastic being made with a stud-like projection and the metal part containing a hole in the relevant position.

Press and snap fits

One of the advantages of plastics is that they can be subjected to quite severe elastic distortion and still return to their original shape when the load is removed. Press and snap fits rely on this characteristic. Such joints may be designed for permanent or recoverable use. Snap fits are stronger and more dependable than press fits, relying on a mechanical interlocking of two components, as well as friction, whereas press fits rely only on friction.

A common form of snap fit is the hook joint. Figure 3.24 shows an example of such a snap fit. When the component is pushed into the hole, the end is deformed so that it can slide through the hole until it emerges from the other end. Then it expands and locks the component in position. The type of hook end shown in the figure is permanent in that it is not possible to disengage it. Figure 3.25 shows how the hook end varies when it is designed for use as a permanent joint and a recoverable joint. Figure 3.26 shows a press-fit. The component is a tight fit in the hole.

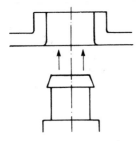

Figure 3.24 A snap-fit

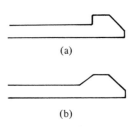

Figure 3.25 Hook end type of snap-fit design, (a) for permanent fixing, (b) for recoverable use

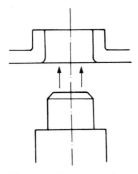

Figure 3.26 A press-fit

Thread systems with plastics

Screw threads are one means of joining plastic components. The main problem that can arise is the system coming loose due to the creep of the plastic when under load.

The thread is formed by material displacement when a thread-forming screw is screwed into a simple hole in the plastic. Self-tapping screws, in which the screw cuts its own thread when screwed into the plastic, can be used with a greater range of plastics.

3.3 Assembly operations

Consider a fairly simple assembly proposition, a bicycle pump. Figure 3.27 shows the various components that go together to make the pump. The parts are:

A barrel end, made of perhaps PVC

An insert in the barrel end, made of perhaps brass so that the connector can be screwed into it

A barrel, made perhaps from a PVC tube

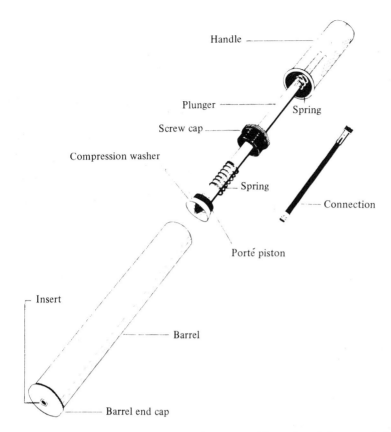

Figure 3.27 The parts of a bicycle pump. (Reproduced by permission of the Engineering Careers Information Service from leaflet *The bicycle pump project.*)

A handle, made perhaps from polypropylene
A plunger, made from possibly mild steel
Two springs, steel
A screw end cap, made perhaps from polypropylene
A porté piston, made from polypropylene
A compression washer, from possibly an artificial rubber

How should these parts be assembled, in what sequence, so that the entire operation is as quick as possible? You might like to try the exercise yourself with a bicycle pump. One possible assembly sequence is:

(a) Fit insert into barrel end-cap.
(b) Fit barrel end-cap onto barrel.
(c) Put the barrel with the end-cap to one side.
(d) Put the washer on the piston.
(e) Put the porté piston on the piston.

(f) Put one spring on the piston.
(g) Put the screw end-cap on the piston.
(h) Put a spring on the piston.
(i) Put the handle on the piston.
(j) Screw the end-cap down onto the barrel.

The final task would probably be to pack the pump, with the connector, in a suitable carton.

Some of the operations in the above sequence have to be in that order because it would be very difficult, if not impossible, to use a different sequence. Thus, for instance, the order which the items are put on the piston cannot easily be changed if the items are to be in the right places. However, it could be feasible to first fit the barrel end-cap onto the barrel and then fit the insert into the end-cap, instead of fitting the insert into the end-cap before it is fitted onto the barrel. These different sequences for fitting the insert might, however, require different fixtures to hold the workpiece during the fitting.

If the assembly is done manually, the times take for each of the operations can be measured and the overall time determined for the entire assembly.

Thus the times might be

Fit insert into barrel end-cap	5 s
Fit barrel end-cap onto barrel	3 s
Put the barrel with the end-cap to one side	1 s
Put the washer on the piston	2 s
Put the porté piston on the piston	2 s
Put one spring on the piston	2 s
Put the screw end-cap on the piston	2 s
Put a spring on the piston	2 s
Put the handle on the piston	3 s
Screw the end-cap down onto the barrel	2 s
Total assembly time	24 s

We can add to this time perhaps 5 s for picking up the pump and a connector, putting the two together and putting them in a carton. This would give a total basic time of 29 s. If we really want to find out how many pumps can be assembled by a worker in a day, we must add to the basic time allowances for relaxation (e.g. the worker pauses for a few seconds to blow his nose) and contingencies that cannot be easily predetermined (e.g. two springs are tangled together). Thus the standard time for the assembly might be 40 s.

By carrying out a prototype assembly it is possible to iron out some of the problems that could occur and perhaps change the sequence of the operations to give a faster assembly, or the design so that assembly is easier and faster, or the materials or the production processes used. For instance, it might be better to change the material, and the process, used for the end-cap so that there does not have to be separate threaded insert but the piece is made in metal in just one piece. This would then reduce the assembly time; however, it might cost more to produce the item this way and so the overall cost of the pump might be higher.

Designing for easier and faster assembly

In any assembly, one of the most significant costs is likely to be that of the labour involved in assembling the item. Such costs are likely to be a significant factor even when the assembly process is mechanised or automated. Thus designing components so that they can be easily and quickly assembled can have a significant effect on the costs.

With manual assembly, humans are involved rather than machines and humans are flexible enough to deal with irregular situations and make decisions as to how to proceed. However, if an assembly operation needs someone to stop, think, make a decision and then adapt to the new situation, there is a time factor involved and hence a cost factor. The more flexibility is needed, the slower will be the assembly process.

Thus, for instance, in the case of an item requiring electrical wiring, it makes assembly easier if the wires are coloured coded, and perhaps also the terminals so that a red wire has to be connected to a red terminal. If all the wires, or some of them, were the same colour and there was no help with a coding of the terminals, the process would be slower because the person carrying out the assembly would need more time to think about which wire was to be connected to which terminal. Colour coding could thus reduce the assembly time, it could also reduce the chance of errors and so the number of products that end up being rejected when inspected.

Another example of how assembly could be made easier and faster is where some component has to be inserted into another in perhaps a particular orientation. If a tapered tube has to be inserted just one way round into an assembly, marking the end that has to be inserted could reduce the time spent deciding which end is the right one. The marking might be a notch or a groove in this case. Designing the assembly so that either end of the tube could be inserted might, however, be better.

Figure 3.28 shows a number of situations where one component has to be inserted into another. With the circular cross-section there is complete symmetry and no particular orientation of the component is necessary for it to be inserted into the hole. With the square cross-section there are only a few orientations that allow insertion. For a rectangular cross-section the number of orientations is even smaller and with the irregular cross-section there is just one orientation. The fewer the ways in which a component can be orientated so that it can be inserted into the hole, the longer it will take to carry out the operation. Designing, where possible, so that there are as many orientations as possible for the component gives both easier and faster assembly.

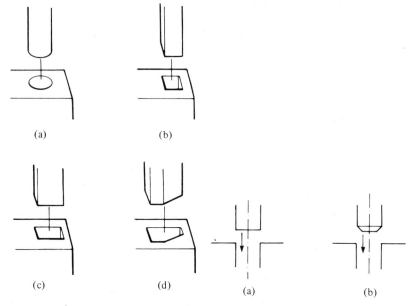

Figure 3.28 (a) Circular cross-section. (b) Square cross-section. (c) Rectangular cross-section. (d) Irregular cross-section

Figure 3.29 Design (b) is easier to assemble than (a)

Inserting a circular cross-section rod into a hole may take a significant time in getting the rod just correctly positioned to slide into the hole. The edges of the rod, and perhaps the hole, can be shaped so that they cause the rod to slide into the right position (Figure 3.29).

Jigs and fixtures

A jig is a device which determines the location dimensions required in some machining or assembly operation. A fixture is a device that holds the workpiece during machining or assembly. The terms are, however, not always precisely applied. Jigs and fixtures can make assembly both easier and faster.

Thus, for instance, in an assembly operation where two components have to be welded together, a jig can be used to ensure that the components are correctly located with respect to each other when the weld is made. In the bicycle pump assembly referred to earlier, an insert was put into the barrel end-cap. This insert has to be both at the right angle to the surface of the cap and central. This can be ensured by the use of a suitable jig, so designed that the insert can only be put into the cap at the right angle and in the right place.

Some of the basic factors that are considered in the design of jigs and fixtures are:

(a) Location of the workpiece, i.e. ensuring that it is in the right position

(b) Securing the workpiece while the operations are carried out, i.e. clamping the workpiece
(c) Location of the tool relative to the workpiece, i.e. ensuring that the operations occur in the right locations
(d) Support of the workpiece against any forces that are imposed by the process
(e) Rapid and easy operation

Mechanised assembly

The difference between mechanised and automated assembly is that an automated operation is one that not only is mechanised but has the built-in capacity to determine when corrective action is required and then take such action. A mechanised assembly does not have such capacity and ability to take corrective action, it is essentially just using a machine to carry out operations.

The major types of operations that are carried out by machines are workpiece orientation, workpiece transfer and placement, and inspection.

Workplace orientation

A variety of devices are used to ensure the correct orientation of a workpiece for some operations. One widely used method is the vibratory bowl feeder. This is a bowl having an internally inclined track climbing from the base of the bowl to

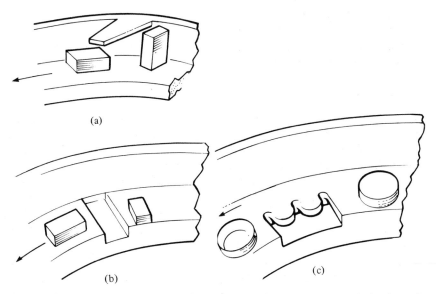

Figure 3.30 (a) A wiper, vertical components are deflected off the ramp while horizontal ones are not. (b) A dish-out, components not lying along the track fall out. (c) Components upside down are rejected

the station at the top of the bowl where the workpiece leaves in the right orientation. The bowl is made to vibrate in a twisting motion. This causes the workpieces to climb the ramp in a series of tiny hops. Along the track, the workpiece encounters devices that will orientate it or reject it back down to the bottom of the bowl because it is in an attitude which does not enable it to be suitably orientated. Figure 3.30 shows some of these devices.

When the components leave the vibratory bowl feeder they will all be in the same orientation. This orientation can be changed, if required, by changing the angle or the form of the track along which they are fed to the next operation.

Workpiece transfer and placement
Means have to be provided for transferring the workpiece from one work station to another and to place the workpiece correctly at each station so that the appropriate operations can be carried out. Feed tracks with perhaps an escapement mechanism might be used (Figure 3.31) or pick-and-place mechanisms (Figure 3.32). With the feed track and escapement mechanism, the track may be just gravity powered and the escapement a fairly simple device to allow a single component at a time to be released from the track. The rate at which the components are released can, with suitable choice of escapement mechanism, be pre-selected.

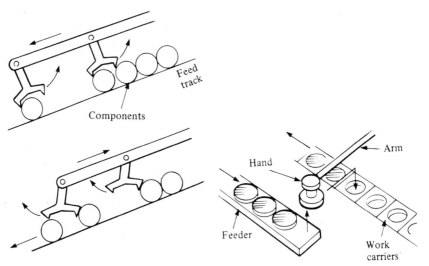

Figure 3.31 A feed track with an escape-   **Figure 3.32** A pick-and-place mechanism
ment mechanism

The pick-and-place mechanism essentially just involves an arm with a 'hand' for grasping the component. The arm moves the hand out to grasp the component, then pick it up and transfers it to the next work station. A wide variety of mechanisms are possible for workpiece transfer and placement and the above examples are only illustrative of the types of principles involved.

Inspection

Inspection can also be mechanised. Thus, for example, containers can be checked to see if they are filled by using, perhaps, a weighing mechanism or a means of determining level of contents within a container. The weighing method may employ a load cell, the level determination might employ a beam of light which becomes blocked if the container is full. In either case the measurement could be used to actuate an accept or reject mechanism.

There may be a considerable number of inspection points in a mechanised assembly process, checking that at each stage the required operation has been carried out properly. Thus if, at one work station, there is a need to spread adhesive on one component face before joining another to it, then an inspection mechanism might be used to check that the adhesive had actually been spread properly before the two components were brought together. Another inspection mechanism might then be used to check that the two components were actually brought together and bonded.

Automated assembly

In a completely automated system the raw materials are the input to the system and the output is the finished products. While many automated systems have been developed, they usefully operate on only part of the production cycle. Thus a single work station may be automated. With such a station a number of operations may be automatically carried out on the product.

A synchronous system, however, consists of the work being moved, perhaps on a rail or an overhead chain conveyor, from one work station to another. At each work station an operation or a number of operations may be carried out. All the workpieces are moved between stations at the same time and over the same distance. The whole line is, however, controlled by the greatest time required at any one work station.

A non-synchronous system has the workpiece allowed to spend differing amounts of time at the different work stations, generally by a system which allows the workpiece to be disengaged from the conveyor system at each work station. In order for this system to be effective and keep each work station fully occupied, buffer stocks need to be available.

The continuous system has the workpieces in constant motion and the work stations therefore have to move with the workpiece. This is not always feasible, e.g. with metal removal processes.

The advantages of automated assembly are a reduction in cost per part, increased productivity, more consistent quality, and also the removal of operators from awkward, dirty and perhaps hazardous work areas. However there is generally a high capital cost in setting up an automated system and the more work stations involved the more inefficient the system is likely to be.

Interchangeability

Interchangeability is the appropriate term to describe the assembly process where any one of a batch of components can be used with any one of the mating components. Thus if a shaft is required to fit in a hole in a component, it would not matter which shaft from a batch was chosen nor which component was chosen for it to be fitted in. In order to achieve this, the tolerances of the components must be specified to the standard required to suit the type of fit required.

An alternative to this is to use selective assembly. This involves the components in any batch being graded for size. Mating components can then be chosen from those size groups that will provide the type of fit required. This type of assembly is particularly feasible where mechanised or automated assembly and inspection is involved.

3.4 Limits and fits

A number of terms are associated with the specification of components that are to fit together.

Basic size

The basic size is the nominal size specified by the designer. This cannot however be produced exactly because of the inherent variations within the production process. It is the size by reference to which the size limits are specified, the same basic size being specified for both members of the fit.

Limits

The limits of the size are the maximum and minimum sizes permitted for the dimension concerned, the maximum limit of size being the greater of the two limits and the minimum limit the smaller of the two (Figure 3.33).

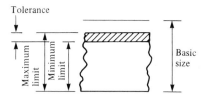

Figure 3.33 Limits for a shaft

Tolerance

The tolerance is the difference between the maximum and minimum limits of size.

Fit

Fit is the term used to describe the relationship resulting from the difference, before assembly, between the sizes of the two parts which are to be assembled. A clearance fit is one which always provides a clearance, i.e. the tolerance zone of the hole is entirely above that of the shaft (Figure 3.34a). An interference fit is one that always provides an interference, i.e. the difference between the sizes of the hole and the shaft before assembly is negative. This means that the tolerance zone of the hole is entirely below that of the shaft (Figure 3.34b). A transition fit is one which may provide either a clearance or an interference, the tolerance zones of the hole and the shaft overlap (Figure 3.34c).

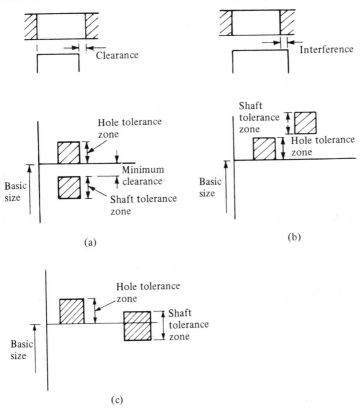

Figure 3.34 (a) A clearance fit, (b) an interference fit, (c) a transition fit

The hole basis of fits

The hole basis of fits is the system of fits in which the different clearances and interferences are all obtained by associating different shafts with a single hole. Figure 3.34 uses the principle in that all shaft fits are considered in relationship to a constant, single hole. This system is the one most commonly used, an alternative being to associate different holes with a single shaft.

Deviations

The deviation is the algebraic difference between a size and the corresponding basic size. The term upper deviation is used for the algebraic difference between the maximum limit of size and the corresponding basic size. The lower deviation is the algebraic difference between the minimum limit of the size and the corresponding basic size (Figure 3.35).

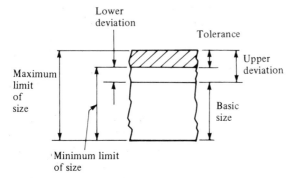

Figure 3.35 Deviations

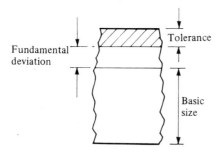

Figure 3.36 Fundamental deviation

Fundamental deviation

The fundamental deviation is that one of the upper or lower deviation, being the one nearest to the basic size, which can define the position of the tolerance zone. It is thus the deviation of the tolerance band away from the basic size (Figure 3.36). The fundamental deviation is positive when the deviation is added to the basic size to give the position of the tolerance band. Thus the shaft in Figure 3.36 has a positive fundamental deviation. A negative fundamental deviation means that the deviation has to be subtracted from the basic size to give the tolerance band. Thus an interference fit will have a positive fundamental deviation, a clearance fit will have a negative fundamental deviation. A transitional fit can have either a positive or a negative fundamental deviation.

The ISO system

Tolerance is a measure of the quality of the fit: the smaller the tolerance, the higher the quality. However, to maintain a constant quality over a range of basic sizes the tolerances need to be varied, a constant tolerance whatever the basic size would not give the same quality. For example, while a tolerance of 0.120 mm on a 50 mm diameter shaft would give a certain quality, the same quality on a 200 mm diameter shaft would require 0.160 mm tolerance. The larger the basic size, the larger the tolerance for the same quality.

Figure 3.37 shows the type of relationship that exists between tolerance and size for a particular quality. Over the diameter range 30–50 mm a tolerance of 0.100 mm, over the range 50–80 mm a tolerance of 0.120 mm, over the range 80–120 mm a tolerance of 0.140 mm—all these will give approximately the same quality. A range of these graphs can be drawn for different qualities. The ISO system provides a series of tolerance grades, each grade corresponding to a particular quality. The grade corresponding to the values given in Figure 3.37 is designated grade IT 10. By just specifying this tolerance grade, the tolerances for any diameter shaft can be found using tables, derived from the equation for the graph, and all will give the same quality. Tolerance grades are always prefixed by the letter IT and range from IT 01, IT 0, IT 1, IT 2, IT 3 to IT 16, a total of 18 grades; the earlier in this grade sequence, the finer the quality.

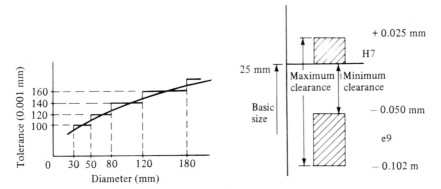

Figure 3.37 Tolerance values for the same quality

Figure 3.38 H7 hole with e9 shaft

When selecting a fit to be used for a given application it is necessary to choose an appropriate fundamental deviation and the appropriate tolerances for the mating parts. These choices will then determine the type of fit, i.e. the amount of clearance or interference. As with tolerances, fundamental deviations are a function of basic size. The ISO system provides 41 different fundamental deviation grades, within each grade the fundamental deviation varying with size to give the same fit characteristic.

The fundamental deviation grades are represented by a single letter or pair of letters, capital letters being used for holes and lower case letters for shafts. Thus for holes we have grades, A, B, C, CD, D, E, EF, F, FG, G, etc, while for shafts the grades are a, b, c, cd, d, e, ef, f, fg, g, etc. Letter grades below h or H represent negative deviations while the letters above h or H represent positive deviations. The letter a represents a large negative deviation, the letter z a large positive deviation. The letter h, or H, is zero fundamental deviation.

For each fundamental deviation grade, tables are available to give the deviations within each grade for different basic size ranges. Thus, for instance, the tables give, for a shaft in the size range over 30 mm to 40 mm with grade c, an upper deviation of −0.120 mm. For negative deviations the upper deviation is the fundamental deviation. For a shaft in the size range over 30 mm to 40 mm with grade y, the tables give a lower deviation of +0.094 mm. For positive deviations the lower deviation is the fundamental deviation.

In order to specify the fit it is necessary to quote both the tolerance grade and the fundamental deviation grade. The fit is thus specified by combining both the grade designations in the form H7 or e9, the letter indicating the fundamental deviation grade and the number the tolerance grade. For a shaft e9, the tables give for a shaft in the size range 30 mm to 50 mm:

Upper deviation	−0.050 mm
Tolerance	0.062 mm

Thus for a shaft of basic diameter 35 mm the upper limit is −0.050 mm and the lower limit is −(0.050 + 0.062) = −0.112 mm.

Working to the hole basis of fits means that the shafts are all considered in relation to a basic hole size. Because of this, the hole basis means that the holes are all designated in terms of the H grade as this is the grade with zero fundamental deviation. Thus the hole for which the e9 shaft might be used with could be, say, H7. The tables give for a hole in the size range 30 to 50 mm:

Lower deviation	0
Tolerance	0.025 mm

Thus for a hole of basic diameter 35 mm the lower limit is 0 and the upper limit +0.025 mm.

Figure 3.38 shows the data for H7 hole with the e9 shaft. The minimum clearance is 0.127 mm. The fit is a clearance fit.

A very wide range of tolerances and fundamental differences can be selected within the ISO system to give many different fit conditions. However the majority of fit conditions required for normal engineering can be provided by a small selection of grades:

Holes	H7, H8, H9, H11
Shafts	c11, d10, e9, f7, g6, h6, n6, p6, s6

Clearance fits are given by combining any of the above hole with shafts c11, d10, e9, f7, g6, h6. Transition fits are given by combining H7 with k6 or n6 and interference fits by combining H7 with p6 or s6.

Problems

1 What are the possible advantages of the use of adhesives over other methods of joining metals?
2 What factors can determine the time taken to make an adhesive bonded joint and hence the rate of production?
3 How does soldering differ from brazing?
4 How should the design of a brazed joint differ from that used for welding?
5 State two of the problems that may occur when welding is used to make a joint.
6 State four factors that should be taken into account in the choice of a welding process.
7 What factors determine the optimum fastening system to use?
8 Under what circumstances might a tubular rivet be used in preference to a solid rivet?
9 For what purposes might eyelets be used?
10 Suggest joining methods that might be used (a) to join two sheets of metal face-to-face together, (b) to attach a metal clip to a flat metal surface, (c) to join two thick steel bars, (d) to join two steel sheets edge-to-edge or with a slight overlap, (e) to attach leather to a metal frame.
11 Can (a) thermoplastics, (b) thermosets, be welded?
12 Explain how hot-plate welding is used to join two plastic components.
13 What are the characteristics of a plastic that can be successfully joined using metal rivets?
14 Explain the process of ultrasonic staking.
15 How do snap fits differ from press fits?
16 Suggest joining methods that might be used (a) to fix a metal clip on the face of a sheet of plastic, (b) to join two thermoplastic pipes, (c) to bond two sheets of thermoset materials together, face-to-face, (d) to join two sheets of thermoplastic materials along a line.
17 A small toy truck consists of the following parts: a wooden body, two metal axle rods to pass through holes in the body, four wheels with central holes to slide over axle, four caps to fit on the ends of the axle and prevent the wheels falling off, four washers to go on the axles between the wheels and the body and prevent the wheels binding against the body, two metal discs to be stuck on the front of the car as headlights, two pieces of red plastic to stick on the rear as tail lights, two strips of metal foil to be stuck on as bumpers, a strip of plastic to be stuck on as a windscreen. Suggest a possible sequence for assembling the toy truck.
18 List all the parts that constitute a bicycle and propose an assembly sequence. Assume that the wheel assemblies have been pre-assembled.

19 Select a product and plan the assembly of it from its constituent parts. Possible products could be a table lamp, a 13 A plug, a telephone handpiece, etc.

20 The bicycle pump shown in Figure 3.27 includes a barrel end made of PVC and an insert into that end, made of brass, so that the connector can be screwed into it. Suggest another possible way in which the end piece could be manufactured. What factors would need to be considered in comparing the costs of the two methods?

21 Give two examples of design changes that could make assembly easier and faster.

22 The three basic elements of mechanised assembly can be considered to be workpiece orientation, workpiece transfer and placement, and inspection. Explain briefly the type of operations involved in each.

23 State three advantages of automated assembly over manual assembly.

24 What are the basic characteristics of the automated systems termed (a) single work station, (b) synchronous system, (c) non-synchronous system, (d) continuous system.

25 An assembly process might be termed a selective assembly or an interchangeability system. Explain the difference.

26 Explain the following terms as used in limits and fits, (a) basic size, (b) limit, (c) tolerance, (d) fit, (e) fundamental deviation.

27 Distinguish between a clearance fit, an interference fit and a transitional fit.

28 Tolerance/0.001 mm

Basic size		IT 8	IT 9	IT 10
Over	*To*			
mm	mm			
10	18	33	52	70
18	30	39	62	84
30	50	46	74	100

Fundamental deviations for shafts/0.001 mm

Basic size			*Upper deviation*	
Over	*To*	d	e	
mm	mm			
10	18	−50	−32	
18	30	−65	−40	
30	50	−80	−50	

For the following shaft–hole combinations determine the tolerance, the lower and upper limits, the minimum and maximum clearances: (a) e9 with H8, basic size 15 mm, (b) d10 with H9, basic size 30 mm, (c) e9 with H9, basic size 35 mm.

4

Automation

4.1 Numerical control of machine tools

In the case of an operator-controlled machine tool the decisions as to how to operate the machine tool are taken by the operator, though assistance may be given in the form of digital readout of displacements and pre-set end-stops. In automatic operation of a machine tool, the decision-making role of the operator is replaced by some electronic or computerised system. This decides when the component has been machined to the right size.

The term *numerical control* (NC) is used to describe the control of machine tools by an electronic system responding to instructions expressed as a series of numbers. The term *computerised numerical control* (CNC) is used to describe the control of machine tools by a computer responding to instructions expressed as a series of numbers.

NC machine tools have been developed for a wide range of applications, e.g. metal cutting, welding and sheet metal forming. The most widely used are, however, those for metal cutting and are based on turning, milling and drilling.

The NC machine system

In essence, the system can be considered to consist of the following sequence:

1 A programmer translates information on drawings into a series of basic instructions that the machine has to follow.
2 The program is then put into a form that the control unit for the machine tool can follow. This might consist of holes punched in a length of tape or coded instruction fed into a computer by pressing keys on the computer keyboard.
3 The program then forms the input to the control unit for the machine tool and controls its operations in producing the part concerned.

The control system used can either be one where the drive motors of the machine tool follow precisely the instructions given to them without any modifications (open-loop control) or it can take the instructions and keep a check on what is happening, modifying the information in the light of this (closed loop control).

Closed-loop and open-loop systems
NC machines can have either closed-loop or open-loop control systems. In the closed-loop system there is a continual feedback of information as to how the machining is going; this is used to modify the inputs to the motors running the machining operations. With the open-loop system there is no feedback of information and so the machining takes place in accord with the initial input signal, without any later modifications.

Figure 4.1 illustrates the closed-loop system for a simple machining operation involving just the monitoring of a work table displacement. The sequence can be considered to be:

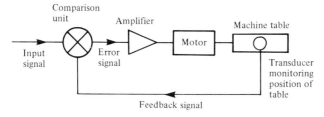

Figure 4.1 A closed loop system

1 The input signal to the amplifier results in the motor driving the work table and so machining occurring.
2 The displacement of the work table is continuously measured by a transducer and a signal is fed back to a comparison unit.
3 At the comparison unit the signal received is compared with the input signal. The signal received relates to the actual position of the work table at some instant and the input signal to the required position when the component has been machined to the required dimensions. The difference between these two signals, the error signal, is then fed to the amplifier and used to drive the motor and hence the work table. As long as there is an error signal, this sequence will continue. When the error signal becomes zero, i.e. when the actual displacement has become the required displacement, no signal is passed from the amplifier to the motor and the motion of the work table ceases.

With the open-loop system (Figure 4.2) there is no feedback signal from any monitoring of the work table position and hence no modification of the initial input signal to the amplifier and motor driving the work table. Unlike the closed-loop system, the open-loop system does not allow for any compensation to occur during the machining. Because of this, closed-loop systems are inherently more accurate than open-loop systems.

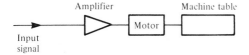

Figure 4.2 An open-loop system

With a closed-loop system there has to be one feedback loop for each of the following functions:

1 Each axis along which there can be relative movement between the tool and the work piece. Thus, for example, a vertical milling operation could require feedback loops for control of movements along the length of the workpiece, across its width and for cutting to different depths (Figure 4.3), i.e. the X, Y and Z axes.
2 Each type of rotary motion, i.e. rotation about the X, Y and Z axes (Figure 4.4).
3 Control of spindle rotation, speed and direction.
4 Supporting functions such as coolant and swarf removal.
5 Selection of tools when the machine has a number of cutting tools in a design turret or magazine.
6 Control of the holding of the workpiece.

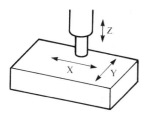

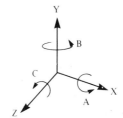

Figure 4.3 Movements requiring control in vertical milling. The Z axis always relates to a sliding motion parallel to the spindle axis

Figure 4.4 Rotary movements

Adaptive control
During machining cutting, conditions may change, perhaps as a result of tool wear. Such wear can have a number of effects. The torque needed to turn the machine spindle may increase, so that the cutting tool and workpiece temperature rises and the tool is likely to deflect. These effects will occur if the operating conditions are not adapted to take account of tool wear. With adaptive control the machine control unit responds to such changing conditions and might change feed rate or spindle speed. Adaptive control is a

closed-loop form of control; in the tool wear situation discussed above, the feedback is information on the spindle torque.

Adaptive control applied to the spindle speed and feed rate can mean that the programmer need not be concerned with these. The machine is simply programmed for optimum performance and the spindle speed and feed rate adjust to give this.

Data input for NC machines

The data input methods that can be used with NC machines are:

1 *Manual data input (MDI)*. An operator sets dials or position switches, presses keys on a keyboard, etc.

2 *Conversational manual data input*. The operator presses keys on a keyboard in response to questions which appear on the visual display unit (VDU). Thus, for example, there may be questions as to the material being machined and the surface finish required.

3 *Punched tape data input*. The data has been coded into perforations in a length of tape and a tape reader is used to transfer the data into the machine. The function of the tape reader is to detect the presence and position of the perforations in the tape.

4 *Magnetic tape data input*. The data is stored on magnetic tape in much the same way that a video recorder might be used in the home to store TV programmes.

5 *Magnetic disc data input*. This uses what is commonly referred to as a *floppy disc* and involves data storage in the same way as magnetic tape.

6 *Direct numerical control (DNC)*. This term is used when the input to the machine is from some master computer and can involve the computer inputting to a number of machine tools. The data input to the master computer can be via a keyboard, magnetic tape or disc.

Punched tape data input
The tape reader gets information by means of the presence or absence of a hole in the tape at a particular position. A photoelectric reader, detecting light passing through the holes, registers the light as either 'on' or 'off'. It is a binary code system whereby numerical information is presented in terms of 1 (on) or 0 (off). The system we traditionally use in counting is the denary system based on tens. In such a system the first digit, reading from right to left, is the units or 10^0, the second digit the tens or 10^1, the third digit the hundreds or 10^2, the fourth the thousands or 10^3. Thus the number 1472 is $(2 \times 10^0) + (7 \times 10^1) + (4 \times 10^2) + (1 \times 10^3)$.

In the binary system the first digit, reading from right to left, is 2^0, the second 2^1, the third 2^2, the fourth 2^3, etc. Thus the binary number 101 is $(1 \times 2^0) + (0 \times 2^1) + (1 \times 2^2)$ or, in the denary system, 5. The punched tap would show this number, binary 101, as a hole in the first column (reading from the

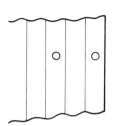

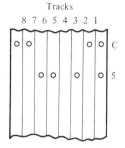

Figure 4.5 The punched tape entry for binary 101 (= denary 5)

Figure 4.6 The ISO tape code (even parity)

right), no hole in the second column and a hole in the third column (see Figure 4.5).

Not only numbers can be coded by this system, but also letters of the alphabet. Two different methods are employed, one according to the International Standards Organisation (ISO) and the other the Electrical Industries Association (EIA). On the ISO system the 26 letters of the alphabet are numbered from A to Z as 1 to 26 and the denary number is then converted into a binary number. Thus, for example, N is letter 14 and so its binary number is 1110. To distinguish between data inputs of numbers and letters, tracks on the punched tape are used only for this purpose. Thus the first five tracks, from the right, are used for either coded numbers or letters and then additional holes are punched in tracks five and six if the punching refers to a number or an additional hole in track seven if it is a letter.

An eighth track, used for a check on the accuracy of the tape punching, is referred to as a *parity track*. Each row across the width of the tape is made to contain an even number of holes. Thus, if tracks one to seven give an even number of holes, no punching is made in track eight. If tracks one to seven give an odd number of holes, a punching is made in track eight. The EIA system uses the eighth track for the same purpose, but requires an odd number of holes.

Figure 4.6 illustrates the ISO system applied to code the letter C (denary 3 and so binary 11) and the number 5 (binary 101). Thus an eight-track tape can be used to input all the letters of the alphabet and numbers 0 to 9. This does not use all the possible hole combinations and some of the others are used for such functions as decimal point, space, etc.

The term *character* is used for a horizontal row of holes in the tape. Characters may be grouped to give a *word*, e.g. G96 is a word involving three characters and so three horizontal rows of holes. A word can be used to describe a particular type of movement required, or perhaps a particular spindle speed. A set of words is called a *block*, blocks being identified by the letter N followed by digits indicating the number of the block. A block can thus contain information about the type of movement required, the spindle speed, the identity of the tool, etc.

Positioning

For NC machine tools, position is specified in terms of Cartesian coordinates. X, Y and Z axes are used (see Figure 4.4) with the X-axis generally involving motion parallel to the longest length of the machine table, the Y-axis motion parallel to the shortest length and the Z-axis motion that advances or retracts the spindle. In addition to these axes, letters A, B and C are used to describe rotary motions (see Figure 4.4).

Machining with an NC machine tool involves the control of the machine slides so that the position of the workpiece relative to the tool is controlled. In some cases the positioning may simply consist of moving the tool from operating in one position to operating in another, e.g. drilling holes in two places. This is called *point-to-point* positioning since the path followed by the tool between the positions is not specified, only the start and end points.

In other cases, e.g. profile milling, the path followed by the tool between positions is important, since over the entire path it is in contact with the workpiece and is cutting. Thus instead of point-to-point positioning we here use *continuous path* positioning. The term *interpolation* is used to describe the type of positioning control where there is intermediate information between the start and finish positions. *Linear interpolation* involves defining linear movements when one or more slides are moving at controlled rates. Since a cutting path (even a circular arc) can generally be considered to be made up of a number of straight-line paths, this method is widely used with NC machine tools.

Part programming

The term *part programming* is used for the preparation of the input instructions for NC or CNC machining of a part, i.e. a component,. The word 'part' refers to a component and not an incomplete program.

Before a part program can be compiled it is necessary to determine the correct sequence of operations needed, together with details of tooling requirements, work-holding requirements, cutting speeds, feed rates and spindle speeds. This information is then transformed into a suitable language for the NC or CNC machine control unit. A number of versions of such language exist.

For control systems using the system known as *word address* there is a common system described in BS 3635:1972. This employs preparatory functions to inform the machine control unit of the facilities required, and miscellaneous functions which inform the control unit of various changes of operation which may be required from time to time during machining, e.g. coolant on or off. The preparatory functions are specified by a G followed by two digits, e.g. G56 linear shift Z, and the miscellaneous functions by M followed by two digits, e.g. M09 coolant off. For further information on part programming using this system the reader is referred to more specialised texts, e.g. *Computer Numerical Control of Machine Tools* by G. E. Thyer (Heinemann).

Economics of numerical control

The following are the major economic factors that merit consideration in relation to NC or CNC machines.

1 There is a high capital outlay. Changing technology can mean that the capital investment in the machine has to be depreciated over a comparatively short time.
2 The cost of process planning, programming and tape preparation for an NC machine is higher than the process planning with traditional machines.
3 Productivity of NC machines is higher than that of traditional machines, because less operating time is taken up by non-cutting movements, taking measurements and tool-changing etc.
4 Fewer jigs and fixtures are required, those that are needed being relatively simple. This can result in significant cost-saving and a reduction in lead-time for getting work into production.
5 A consequence of the high capital cost and the rapid depreciation of NC machines is a need to consider multi-shift operation so that unit cost can be reduced.

4.2 Robotics

There are several definitions of what constitutes a robot. The British Robot Association gives the definition: 'The industrial robot is a reprogrammable device designed to both manipulate and transport parts, tools or specialised manufacturing implements through variable programmed motions for the performance of specified manufacturing tasks'. The definition used by the Robot Association of America is similar to this. The Japanese, however, use a definition which allows for the inclusion of a wider range of devices.

Essentially, present-day robots can be considered to be based on a mechanical arm attached to a machine that is either fixed or capable of motion. At the end of the arm can be something we could call a 'hand', in that it can grip objects or can take the form of a tool, perhaps a paint-sprayer. The arm and hand constitute what is called a *manipulator*. The manipulator is supplied with power, under the direction of a control system, so that it can move tools or workpieces to specific points and orientation. The position and motion of the arm and hand are monitored by sensors which provide feedback to control the motion of both arm and hand (the system is *closed-loop control*, explained earlier in this chapter).

Classification of robots

Robots can be classified in the following categories:

1 *Fixed-sequence robots* The movement of a fixed-sequence robot is deter-mined by mechanical end-stops, by means of limit switches which come into

operation when the end-stops are reached. Such a robot cannot be re-programmed to carry out other tasks; it is a machine that has to be reset for a new task. The term *pick-and-place device* is used for 'robots' which can carry out a fixed sequence of operations. By the British and American definition (but not the Japanese one) these are not true robots. The pick-and-place device is limited in its applications because the feedback from its sensors (the limit switches) is not continuous during the motion of the robot arm. Feedback occurs only at the limits of the arm's motion. It is this which prevents the device being reprogrammed.

2 *Servo-controlled robots*   The term *servomechanism* is used to describe a device that is driven by a signal representing the difference between the commanding and the actual device states, i.e., it is driven by the error signal (see Figure 4.1 for such a system). Thus, for example, an arm could have a servomechanism at the equivalent of the human shoulder to control the up-and-down movement of the arm. The servomechanism receives an input to move the arm to a particular position and the motion continues as long as there is a difference between the actual position of the arm and the required position. A servo-controlled robot will have servomechanisms at each of its joints. The motion at each joint is thus continuously controlled.

Servo-controlled robots can be also considered in two groups in terms of the way the arm moves between two points. In *point-to-point control* the movement is controlled by a simple specification of the two concerned. The path between the points is not specified. In *continuous path control*, however, the movement is controlled over the entire path. Between the end and start points the path is subdivided into a multitude of small paths, each defined by its own end and start points. Thus, instead of a specification of only the end and start points, a very large number of points along the intended path are specified.

Point-to-point robots are widely used, being perfectly adequate for jobs such as spot welding where a tool only has to be moved to specific points and the path followed between points is not important. Paint-spraying, however, requires a tool to be controlled over an entire path and so point-to-point control would be inadequate; continuous path control is required for such an operation.

Arm configurations

Robot arms can be classified according to the types of motion they are capable of. These enable the *work envelope* of the arm to be defined, this comprising all the points in the space around the robot which can be reached by the end of the robot arm. The main configurations are the following, though others are being developed.

1 *Cartesian-coordinate robot.* This type of robot is able to move along three mutually perpendicular axes, i.e., up-and-down, left-to-right, and forward-and-back (Figure 4.7a).

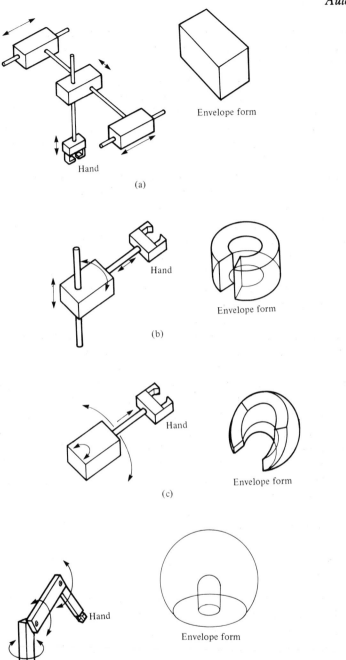

Figure 4.7 Arm configurations. (a) Cartesian coordinate robot. (b) Cylindrical coordinate robot. (c) Polar coordinate robot. (d) Jointed-arm robot

2 *Cylindrical-coordinate robot.* This type of robot has a horizontal arm which can move in-and-out, move round in a horizontal circle, and up-and-down about a vertical axis (Figure 4.7b).

3 *Polar-coordinate robot.* The horizontal arm can move in-and-out, move round in a horizontal circle and move round in a vertical circle about a horizontal axis (Figure 4.7c).

4 *Jointed-arm robot.* This has joints rather like those of the human arm. The entire arm can be rotated about a vertical axis. At the shoulder joint, the arm can be rotated in a vertical circle about a horizontal axis. At the elbow joint there is also rotation in a vertical circle about a horizontal axis (Figure 4.7d).

Degrees of freedom

Degrees of freedom are the number of ways a robot arm can be moved. *Transitional* degrees of freedom are motions in the right-to-left, forward-to-back, and up-and-down directions. *Rotational* degrees of freedom are pitch, roll and yaw, these being rotations about the three cartesian axes (X, Y and Z). Three translational and three rotational degrees of freedom are possible. While all six degrees are necessary to move an object to a point in space and correctly orientate it, for most applications four degrees of freedom are perfectly adequate.

The arm configurations described in Figure 4.7 each have three degrees of freedom, with the hand having further degrees of freedom.

Arm configuration	Degrees of freedom
Cartesian coordinate	3 translational
Cylindrical coordinate	2 translational, 1 rotational
Polar coordinate	1 translational, 2 rotational
Jointed arm	3 rotational

Drive systems

The drive system used is usually hydraulic or electrical. Hydraulic systems involve the conversion of high pressure of the hydraulic fluid into motion. Such a system is capable of moving large loads. Electrical systems use various forms of electrical motor to produce motion. Such systems are more efficient and quieter than hydraulic systems, but are not generally used for high payloads. In some environments electrical systems may pose a safety hazard with the possibility of a spark causing an explosion in a flammable atmosphere. Electrical systems tend to be more reliable than hydraulic systems.

Some hybrid systems exist, involving hydraulic systems for motion of the arm and electrical ones for the wrist.

Sensors

Sensors are used to provide data on the robot's surroundings and so give feedback signals which enable the motion of the robot to be controlled. Humans can determine the presence of some object by touching it, i.e. sensing by contact with the object, and by seeing it, i.e. sensing without making contact. Sensors used by robots can be grouped into the same two categories—contact sensors (the term *taction sensing* is used) and non-contact sensors.

Probably the simplest form of contact sensor is a microswitch which is activated when contact is made, e.g. 'fingers' with switches that are activated by contact with an object.

By using a small array of switches rather than just a single switch, more information than just contact at one point can be obtained. Microswitches are, however, just on-off devices and so no graded response to contact is possible, as occurs when a human hand touches or grasps an object. There is thus a requirement for transducers that give a response proportional to the force they experience and of which a large number can be packed into a small area. One way this is done is by utilizing the fact that the electrical resistance of a length of metal or semiconductor material depends on the forces stretching or compressing it (the principle of the electrical resistance strain gauge). By building an array of such transducers into the 'skin' of the robot a crude approximation to the touch-sensitive skin of humans can be realised.

The most common forms of non-contact sensor involve either infra-red radiation or light. The sensor consists of a transmitter which sends out radiation and a photoelectric sensor which detects any of this radiation that is reflected back by the object. Other more elaborate systems are being developed.

End-effectors

End-effector is the term used to describe the gripper, i.e. 'hand', or tool on the end of the robot arm. Grippers vary in terms of their gripping, sensory and manipulative capabilities. The choice, or design, of gripper will depend on the industrial application concerned. A common form of gripper employs two fingers to grasp objects, like a pair of pliers. These must be able to grip an object with sufficient force to be able to lift it but not so high a force as would deform it. To prevent objects slipping or pivoting, the gripper fingers are sometimes covered with resilient pads.

Where the objects being picked up have a smooth, clean and dry surface, a vacuum gripper can be used. This employs one or more suction cups with a pump to control the pressure in the cups and hold the object to them. The pressure might be increased to 'blow' the objects off the gripper. Ferrous materials might be lifted by a gripper incorporating an electromagnet.

Where grippers are required for more than one purpose, a three-finger gripper might be used or the arm could carry two independent grippers on one 'wrist'.

Wrist-mounted devices

For some applications, instead of the wrist terminating in a gripper, a tool might be mounted directly on the wrist. Typical tools are adhesive and sealant applicators, spot welding guns, arc welding torches, paint spray guns, water jets and light guides which pipe a laser beam to the wrist for cutting or welding.

Part orientation

A part that is presented to a robot must be positioned correctly to enable the robot to perform the appropriate action, e.g. picking it up. This may mean, for example, that all the screws that a robot picks up must be orientated in the same direction so that each is held by the head. If the robot has to pick up a spring there can be problems if several springs are tangled one with another. One way of avoiding such problems is to use a magazine in which all the parts are orientated in the same way and not tangled up. Parts can be delivered from the magazine by compressed air, gravity or perhaps a spring. Other methods were described in the previous chapter.

Programming

The term *manual programming* is used where, for a pick-and-place device, the sequence to be followed is determined by setting switches, stops, etc and is described in 'machine' language. Such a language consists essentially of instructions to open switch X, close switch Y, etc.

For other than simple pick-and-place devices, programming is either textual programming or teaching-by-showing. *Textual programming* involves giving the instructions to the robot through the keyboard of a computer terminal. *Teaching-by-showing* involves giving instructions to the robot by physically moving the robot arm through the required sequence of motions.

Applications of robots

The following are just a few of the possible applications of robots, illustrating their diversity.

Servicing a diecasting machine. This involves operations in close proximity to molten metal in hot and cramped conditions.

Carrying out heat treatment. The robot can pick up the hot metal, directly from inside the furnace, and give it the appropriate cooling treatment.

Machine loading and unloading. This could involve a single robot loading and unloading a number of CNC machines.

Assembly of components. This involves a robot taking components in the correct sequence and assembling them.

Robots thus have applications in environments which are uncomfortable or difficult for humans to operate in. They can carry out large numbers of simple transfer tasks, a type of operation that is extremely boring for humans. They can assemble parts from a diversity of components, doing the job more efficiently than humans.

Economics of robots

The cost factors involved in installing and using a robot system are:

1 Purchase price of the system. This includes the cost of the robot and any ancillary equipment required, e.g. conveyors or bowl feeders for part orientation.
2 Installation cost for the system. Some changes to factory layout may be required.
3 Special tooling costs. Special end-effectors may have to be designed and made.
4 Training costs. The work force required to operate the system will need training.
5 Operating cost. This is essentially the cost of power.
6 Maintenance costs.
7 Programming costs.
8 Depreciation.

A robot system can result in savings, for example:

1 Reduced labour costs, since a robot can displace labour and also operate more than one shift.
2 Increased throughput.
3 More efficient processing and materials handling.
4 A reduction in scrap, since inspection can be incorporated in the robot system and so automated.
5 More consistent quality.

4.3 Flexible manufacturing systems (FMS)

The term *flexible manufacturing system* (FMS) is used to describe a number of machining and associated work stations, e.g. inspection, controlled by a computer to provide automatic production of a range of components without manual intervention. Such a system is flexible in that it can switch from one type of component to another without interruption of the production process. Such a system uses CNC machine tools with some form of work transfer system. This can be a palletised transfer line, in which the work is moved around between machines on pallets carried by a conveyor, or a trolley system in which trolleys are guided around the production area on rails or by control cables buried in the factory floor. The automatic transfer of work to and from machines is an essential feature of such a system.

4.4 Computer-aided design (CAD)

The term *computer-aided design* (CAD) is used to describe the process by which designers use a computer as a means to create, evaluate, modify and finalise designs. The designer can use the computer to analyse data, carry out calculations and then transform the data into three-dimensional images on the computer visual display unit. These images can be rotated, viewed from different angles and sectioned. The design can be transformed into engineering drawings or used to generate a program for an NC machine tool.

4.5 Computer-aided manufacture (CAM)

Computer-aided manufacture (CAM) is the term used to describe manufacturing which is computer controlled. It is based on the use of:

1 CNC machine tools;
2 robotics;
3 production process planning to take account of the variety of products that have to be produced, possibly using group technology (see Section 11.2);
4 the computer to control also the entire management of production, e.g. inventory control and scheduling.

CAD and CAM can be integrated with, for example, the design system linked to the manufacturing processes via computer-generated part programming.

Problems

1 Explain the differences between closed loop and open-loop control.
2 What is meant by adaptive control?
3 List the ways by which data can be supplied as input to NC machines
4 Describe the ISO and EIA systems of coding.
5 Explain the difference between point-to-point and continuous path positioning.
6 What is meant by part programming?
7 Describe the difference between fixed sequence and servo-controlled robots.
8 Describe the types of motion possible, degrees of freedom and the work envelope for (a) cartesian-coordinate, (b) cylindrical-coordinate, (c) polar-coordinate, (d) jointed-arm robots.

9 Compare and contrast electrical and hydraulic drive systems for robots.

10 Give examples of contact and non-contact sensors used with robots.

11 Explain the significance of part orientation for the handling of parts by robots.

12 Explain the programming terms for robots of manual programming, textual programming and teaching-by-showing.

13 Give briefly the meaning of (a) FMS, (b) CAD, (c) CAM.

14 What major cost factors must be taken into account in considering the introduction of (a) numerical control machine tools, (b) robots?

PART TWO

Engineering Materials

This part of the book consists of six chapters dealing with the range of materials encountered in engineering, their properties and applications, together with a basic consideration of their microstructure.

Chapter 5 Properties of materials

Methods for tensile testing, impact tests, bend tests, hardness measurement and tests for fatigue and creep are described. The results of such tests are discussed in relation to their significance for engineering applications.

Chapter 6 Structure of metals

This is a basic consideration of the structure of metals in terms of crystal structure. The effects on crystal structure of cold and hot working and casting are considered, with their implications for properties.

Chapter 7 Structure of alloys

Alloy structures are introduced in terms of equilibrium diagrams. Consideration is given to non-equilibrium conditions and precipitation.

Chapter 8 Ferrous alloys

The iron-carbon system and critical change points are taken as the basis for a consideration of carbon steels, alloy steels and cast irons, with their heat treatment.

Chapter 9   Non-ferrous alloys

Aluminium, copper, magnesium, nickel, titanium and zinc alloys are reviewed in terms of their properties and structure.

Chapter 10   Non-metals

This includes a consideration of polymers—thermoplastics, thermo-sets and elastomers—in terms of their structure and properties. Ceramics and composites are also briefly discussed.

5

Properties of materials

5.1 The tensile test

In a tensile test, measurements are made of the force required to extend a standard size testpiece at a constant rate, the elongation of a specified gauge length of the test piece being measured by some form of extensometer. The measurements thus obtained are those of force applied and resulting extension. Figure 5.1 shows the form of a typical force–extension graph for a metal.

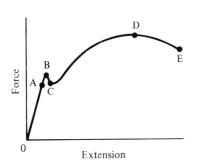

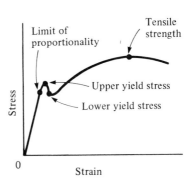

Figure 5.1 The form of a typical force–extension graph for a metal, e.g. a low carbon steel

Figure 5.2 Stress–strain graph for Figure 5.1

For the part of the force–extension graph 0A the extension is directly proportional to the force. *Hooke's law* is said to be obeyed. Point A is called the *limit of proportionality*. Up to a certain force, the material will return to its original dimensions when the force is removed, the region of the force–extension graph for which this occurs is called the *elastic region*. The maximum force for which this occurs is called the elastic limit. For many materials the elastic limit and limit of proportionality are almost identical.

Beyond the elastic limit the material will not return to its original dimensions when the force is removed, the material retaining a permanent change in dimensions. This is referred to as a *permanent set* or *plastic deformation*.

Beyond the limit of proportionality the extension is no longer proportional to the force. In some materials a situation may arise where the extension continues to increase without any increase in force; this point is known as the *yield point*. For the graph shown in Figure 5.1, typical of a low carbon steel, there are two distinct yield points B and C. Point B is called the *upper yield point*, point C the *lower yield point*. Data giving just a yield point, with no distinction between upper and lower points, refers to the lower yield point.

After the yield points have been passed, an increase in force is necessary for an increase in extension; however the two are not proportional to each other. At point D the force is a maximum, after that the force decreases though the extension increases until point E is reached when the material breaks.

In order that the data obtained from one particular cross-sectional area test piece can be applied to other cross-sectional area pieces of the same material, the force information is presented as the *stress*. This stress is the force divided by the initial cross-sectional area of the test piece. The extensions are also presented as *strain*, the extension divided by the initial gauge length. The force–extension graph can thus be used to obtain the more general stress–strain graph. For stress and strain defined as above the stress–strain graph will be of the same form as the force–extension graph with just the force values divided by the initial cross-sectional area and the extension values divided by the initial gauge length (Figure 5.2).

For a tensile stress-strain graph the term *tensile strength* is used for the maximum force (point D on Figure 5.1) divided by the initial cross-sectional area of the test piece. The slope of the stress–strain graph up to the limit of proportionality, i.e. stress/strain, is called the *tensile modulus* or *Young's modulus (Figure 5.3)*.

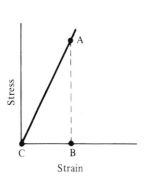

Figure 5.3 Young's modulus, or tensile modulus as it is often called, equals AB/BC

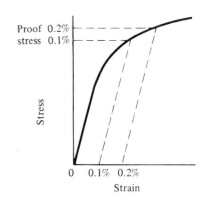

Figure 5.4 Determination of proof stress

Many materials do not have well-defined yield points, their stress–strain graph being of the form shown in Figure 5.4. In such instances a *proof stress* is specified rather than a yield stress. The 0.2% proof stress is defined as that stress which results in a 0.2% offset, i.e. the stress given by a line drawn on the stress–strain graph parallel to the linear part of the graph and passing through the 0.2% strain value. Similarly the 0.1% proof stress is given by drawing a line on the stress–strain graph parallel to the linear part of the graph and passing through the 0.1% strain value.

Other data is also often obtained during a tensile test. After the test piece has broken, the pieces are fitted together and the final gauge length measured. With the initial gauge length, this enables a quantity called the *percentage elongation* to be calculated.

$$\text{Percentage elongation} = \frac{(\text{final length}) - (\text{initial length})}{\text{initial length}} \times 100$$

Similarly, another quantity called the *percentage reduction in area* can be determined from the initial cross-sectional area and the smallest cross-sectional area at fracture.

Values for the tensile properties of metals depend on the temperature at which the data was obtained. In general the tensile modulus and the tensile strength both decrease with an increase in temperature. The percentage elongation tends to increase with an increase in temperature.

The data obtained from a tensile test are affected by the rate at which the test piece is strained, so in order to give standardised results, tensile tests are usually carried out at a constant strain rate. For metals tested at room temperature, the standard rate specified in BS 18 is a strain of 0.5 per minute or less.

Units

Stress, tensile strength, tensile modulus and proof stress all have the same basic unit, N m^{-2} or Pa (pascal), with $1 \text{ N m}^{-2} = 1 \text{ Pa}$. The stresses encountered in engineering are generally large and expressed as N mm^{-2}, MN m^{-2} or MPa, with $1 \text{ N mm}^{-2} = 1 \text{ MN m}^{-2} = 1 \text{ MPa}$. The tensile modulus is generally expressed as kN mm^{-2}, GN m^{-2} or GPa, with $1 \text{ kN mm}^{-2} = 1 \text{ GN m}^{-2} = 1 \text{ GPa}$.

The tensile test piece

In order to eliminate any variations in tensile test data due to differences in the shapes of test pieces, standard shapes are adopted. Figure 5.5 shows the forms of two standard test pieces, one being a flat test piece and the other a round test piece. The dimensions of standard test pieces are given in Table 5.1. For the tensile test data for the same material to give essentially the same stress–strain graph, regardless of the length of the test piece used, it is vital that these standard dimensions be adhered to. An important feature of the dimensions is the radius given for the shoulders of the test pieces. Variations in the radii can affect markedly the tensile test data. Very small radii can cause localised stress concentrations which may result in the test piece failing prematurely. The surface finish of the test piece is also important for the same reason.

Table 5.1   *Dimensions of standard test pieces (from BS 18)*

Flat test pieces b (mm)	L_o (mm)	L_c (mm)	L_t (mm)	r (mm)
25	100	125	300	25
12.5	50	63	200	25
6	24	30	100	12
3	12	15	50	6

Round test pieces A (mm^2)	d (mm)	L_o (mm)	L_c (mm)	r (mm) Wrought material	Cast material
200	15.96	80	88	15	30
150	13.82	69	76	13	26
100	11.28	56	62	10	20
50	7.98	40	44	8	16
25	5.64	28	31	5	10
12.5	3.99	20	21	4	8

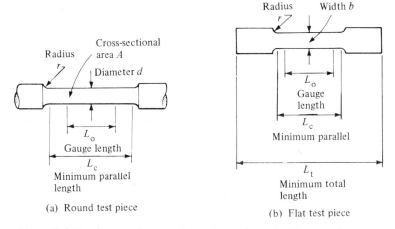

(a) Round test piece

(b) Flat test piece

Figure 5.5 Tensile test pieces (a) Round test piece, (b) Flat test piece

The round test pieces are said to be *proportional test pieces*, for which the relationship between the gauge length L_0 and the cross-sectional area of the piece A is specified in the relevant British Standard as being

$L_0 = 5.65 \sqrt{A}$

With circular cross-sections $A = \frac{1}{4}d\pi^2$ the relationship becomes, to a reasonable approximation,

$L_0 = 5d$

The reason for the specification of a relationship between the gauge length and the cross-sectional area of the test piece is in order to obtain reproducible test results for the same test material when different size test specimens are used. When a ductile material is being extended in the plastic region of the stress–strain relationship, the cross-sectional area of the piece does not reduce uniformly but necking occurs (Figure 5.6). The effect of this is to cause most of the further plastic deformation to occur in the necked region where the cross-sectional area is least. The percentage elongation can thus differ markedly for different gauge lengths encompassing this necked portion of the test piece (Figure 5.7). Doubling the gauge length does not double the elongation because most of the elongation is in such a small part of the gauge length. The same percentage elongation is, however, given if

$$\frac{\text{gauge length}}{\sqrt{(\text{cross-sectional area})}} = \text{a constant}$$

In the UK the constant is chosen to have the fixed value of 5.65.

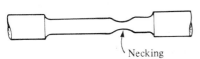

Figure 5.6 Necking of a tensile test piece

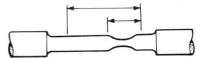

Figure 5.7 The percentage elongations for the two different gauge lengths differ considerably

The tensile test piece is usually chosen so that its properties are indicative of the properties of a component or components. This can present problems if the properties of a component are not the same in all parts of it. There can be a problem with a casting where the properties of the casting material may not be the same as that of a specially cast test piece, because of the different sizes of the two and hence the different cooling rates. Care has thus to be taken in interpreting the results of tensile test pieces.

The tensile test for plastics

Tensile tests can be used with plastic test pieces to obtain stress–strain data. The term 'tensile strength' has the same meaning as with metals. The tensile modulus (Young's modulus) can be determined in the same way as with metals, i.e. the slope of the stress–strain graph for stresses below the limit of proportionality. For many plastics there is, however, no straight-line part of the stress–strain graph and thus the tensile modulus cannot be determined in the way specified for metals. In such cases it is common practice to quote a modulus, termed the *secant modulus*, which is obtained by dividing the stress at a value of 0.2% strain by that strain (Figure 5.8).

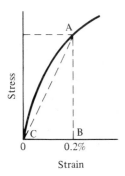

Figure 5.8 Part of the stress–strain graph for a plastic. The secant modulus is AB/BC, i.e. the slope of the line AC

The stress–strain properties of plastics are much more dependent on the rate at which the strain is applied than metals. Thus, for example, the stress–strain data may indicate a yield stress of 62 N mm^{-2} (MPa) when the rate of elongation is 12.5 mm/min but 74 N mm^{-2} (MPa) when it is 50 mm/min. Figure 5.9 shows the general forms of stress–strain graphs for plastics at different strain rates. Another factor that has a considerable effect on the stress–strain properties of plastics is temperature. Figure 5.10 shows how the tensile modulus tends to vary with temperature for plastics.

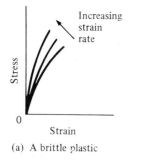
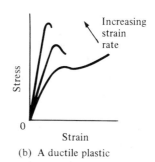

Figure 5.9 The effect of strain rate on the stress–strain graphs for plastics. (a) A brittle plastic, (b) a ductile plastic

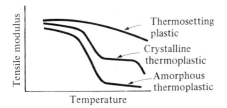

Figure 5.10 The effect of temperature on the tensile modulus of plastics

Interpreting tensile test data

Figure 5.11 shows the types of stress–strain graphs produced by brittle and ductile materials. *Brittle* materials show little plastic deformation before fracture, *ductile* materials show a considerable amount of plastic deformation. If you drop a china tea cup and it breaks you can stick the pieces together again and still have the same tea cup shape. The material used for the tea cup is a brittle material and little, if any, plastic deformation took place prior to fracture. If the wing of a motor car is involved in a collision it is likely to show considerable deformation rather than a fracture. The mild steel used for the car body work is a ductile material.

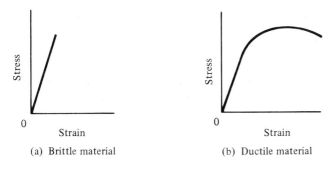

Figure 5.11 Stress–strain graphs to fracture. (a) Brittle material, (b) ductile material

A brittle material will show only a small percentage elongation, i.e. the length of the specimen is very different from the initial length as little plastic deformation has occurred (Figure 5.12). The material also shows little, if any, sign of necking. A ductile material, however, shows quite a large percentage elongation and quite significant necking.

Grey cast iron is a brittle material, it has a percentage elongation of about 0.5 to 0.7%. Mild steel is a reasonably ductile material and has a percentage elongation of the order of 30%.

Thermosetting plastics tend to behave as brittle materials. Thermoplastic materials can be either brittle or ductile depending on the temperature. Melamines are thermosetting materials and have percentage elongations of about 1% or less. High density polythene, a thermoplastic, can have percentage elongations as high as 800%.

The tensile modulus of a material can be taken as a measure of the stiffness of the material. The higher the value of the modulus the stiffer the material. i.e. the greater the force needed to produce a given strain within the limit of proportionality region (Figure 5.13). Mild steel has a tensile modulus of about 200 kN mm^{-2} (GPa) while an aluminium alloy may have a modulus of 70 kN mm^{-2} (GPa). A strip of mild steel is thus stiffer than a corresponding strip of aluminium alloy. Plastics have relatively low tensile modulus values when compared with metals, e.g. polythene has a modulus of about 0.1 to 1.2 kN mm^{-2} (GPa).

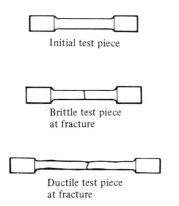

Initial test piece

Brittle test piece
at fracture

Ductile test piece
at fracture

Figure 5.12

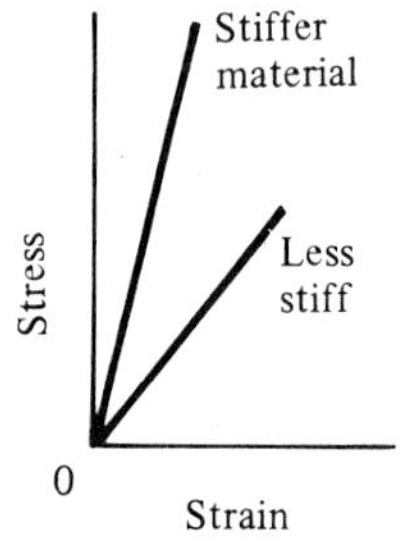

Stiffer
material

Less
stiff

Stress

0

Strain

Figure 5.13 Stress–strain graphs within the limit of proportionality region

The strength of a material is indicated by its tensile strength. An alloy steel may have a strength as high as 1500 Nmm^{-2} (MPa), while an aluminium alloy may only have a strength of 200 N mm^{-2} (MPa). The steel is obviously much stronger than the aluminium alloy. Plastics have tensile strengths lower than those of metals, though their strengths can be increased by incorporating fillers in the plastic. Polythene has a tensile strength of between 4 and 38 N mm^{-2} (MPa).

Table 5.2 gives typical tensile strength and tensile modulus values for metals and plastics.

5.2 Impact tests

Impact tests are designed to simulate the response of a material to a high rate of loading, and involve a test piece being struck a sudden blow. There are two main forms of test, the *Izod* and *Charpy* tests. Both tests involve the same type of measurement but differ in the form of the test pieces. Both involve a

Table 5.2 *Typical tensile test results*

Material	Tensile strength $(N\ mm^{-2}\ or\ MPa)$	Tensile modulus $(kN\ mm^{-2}\ or\ GPa)$
Aluminium alloys	100 to 550	70
Copper alloys	200 to 1300	110
Magnesium alloys	150 to 350	45
Nickel alloys	400 to 1600	200
Titanium alloys	400 to 1600	100
Zinc alloys	200 to 350	100
Grey cast iron	150 to 400	100
Mild steel	350 to 500	200
Ferritic stainless steel	500 to 600	200
Martensitic stainless steel	450 to 1300	200
Polythene, low-density	8–16	0.2
Polythene, high-density	22–38	0.9
PVC, no plasticiser	52–58	2.7
Polystyrene	35–60	3.3
Nylon 6	70–90	2.2

pendulum swinging down from a specified height to hit the test piece (Figure 5.14). The height to which the pendulum rises after striking and breaking the test piece is a measure of the energy used in the breaking. If no energy were used the pendulum would swing up to the same height as it started from. The greater the energy used in the breaking, the lower the height to which the pendulum rises.

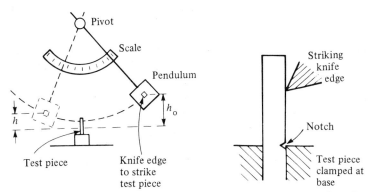

Figure 5.14 The principle of impact testing

Figure 5.15 Form of the Izod test piece (elevation view)

With the Izod test the energy absorbed in breaking a cantilevered test piece (Figure 5.15) is measured. The test piece is notched on one face and the blow is struck on the same face, at a fixed height above the notch. The test pieces used are, in the case of metals, either 10 mm square or 11.4 mm diameter if they conform to British Standards (BS 131: Part 1). Figure 5.16 shows the detail of a test piece conforming to British Standards. The British Standard test pieces for plastics (BS 2782: Part 3) are either 12.7 mm square or 12.7 mm by 6.4 to 12.7 mm, depending on the thickness of the material concerned (Figure 5.17). With metals the pendulum strikes the test piece at a speed between 3 and 4 m/s, with plastics the speed is 2.44 m/s.

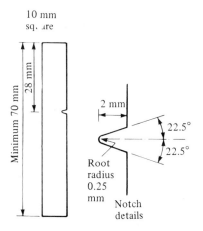

Figure 5.16 British Standard Izod test piece for a metal

With the Charpy test the energy absorbed in breaking a beam test piece (Figure 5.18) is measured. The test piece is supported at each end and is notched in the middle between the two supports. The notch is on the face directly opposite to where the pendulum strikes the test piece. For metals the British Standard test piece has a square cross-section of side 10 mm and a length of 55 mm (BS 131: Parts 2 and 3). Figure 5.19 shows the details of a standard test piece and the three forms of notch that are possible. The results obtained with the different forms of notch cannot be compared; for the purpose of comparison between metals, the same type of notch should be used. The test pieces for plastics are tested either in the notched or unnotched state. The notch is produced by milling a slot across one face, the slot of width 2 mm having a radius of less than 0.2 mm at the corners of the base and the walls of the slot. A standard test piece is 120 mm long, 15 mm wide and 10 mm thick in the case of moulded plastics. Different widths and thicknesses are used with sheet plastics. With metals the pendulum strikes the test piece at a speed between 3 and 5.5 m/s, with plastics the speed is between 2.9 and 3.8 m/s.

The results of impact tests need to specify not only the type of test, i.e. Izod or Charpy, but the form of the notch used. In the case of metals the results are expressed as the amount of energy absorbed by the test piece when it breaks. The results for plastics, however, are often given as absorbed energy divided

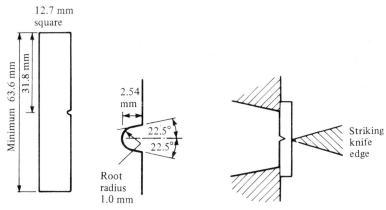

Figure 5.17 British Standard Izod test piece for a plastic

Figure 5.18 Form of the Charpy test piece (plan view)

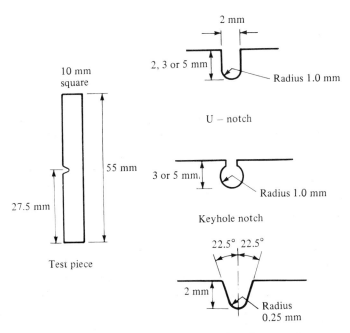

Figure 5.19 British Standard Charpy test piece for a metal

by either the cross-sectional area of the un-notched test piece or the cross-sectional area behind the notch in the case of notched test pieces.

Another test that is sometimes used with plastics involves a weight falling on to a disc of the material under test. The disc, for the British Standard test (BS

2782), is 57 to 64 mm in diameter, or a square of 57 to 64 mm side resting on an annular support of 50.8 mm inner diameter and maximum external diameter of 57.2 mm. The thickness of the test piece is 1.52 mm in the case of moulded plastic. The impact strength is the energy needed to fracture half of a large number of samples of the materials.

Interpreting impact test results

The fracture of materials can be classified roughly as either brittle or ductile fracture. With *brittle fracture* there is little, or no, plastic deformation prior to fracture. With *ductile fracture* the fracture is preceded by a considerable amount of plastic deformation. Less energy is absorbed with a brittle fracture than with a ductile fracture. Thus Izod and Charpy test results can give an indication of the brittleness of materials.

The appearance of the fractured surfaces after an impact test also gives information about the type of fracture that has occurred. With a brittle fracture of metals the surfaces are crystalline in appearance, with ductile fracture the surfaces are rough and fibrous in appearance. Also with a ductile failure there is a reduction in the cross-sectional area of the test piece, but with a brittle fracture there is virtually no change in the area. With plastics a brittle failure gives fracture surfaces which are smooth and glassy or somewhat splintered, with a ductile failure the surfaces often have a whitened appearance. With plastics the change in cross-sectional area can be considerable with a ductile failure.

One use of impact tests is to determine whether heat treatment has been successfully carried out. A comparatively small change in heat treatment can lead to quite noticeable changes in impact test results. The changes can be considerably more pronounced than changes in other mechanical properties,

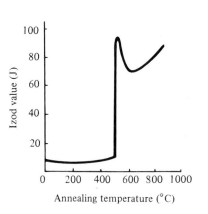

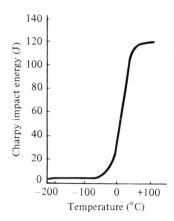

Figure 5.20 Effect of annealing temperature on Izod test values

Figure 5.21 Effect of temperature on the Charpy V-notch impact energies for a 0.2 per cent carbon steel

e.g. percentage elongation or tensile strength. Figure 5.20 shows Izod impact tests results for cold worked mild steel annealed to different temperatures. An impact test can thus indicate whether annealing has been carried out to the correct temperature.

The properties of metals change with temperature. For example, a 0.2% carbon steel changes from a brittle to a ductile material with a rise in temperature, undergoing a gradual transition at around normal room temperature (Figure 5.21): at about $-25°C$ it is a brittle material with a Charpy V-notch impact energy of only about 4 J, whereas at 100°C it is ductile with an impact energy of about 120 J. This type of change from a ductile to a brittle material can be charted by impact test results and the behaviour of the material at the various temperature can be predicted.

Table 5.3 gives typical impact test results for metals and plastics at 0°C.

Table 5.3 *Typical impact test results*

Material	Charpy V Impact strength (J)
Aluminium, commercial pure, annealed	30
Aluminium–1.5% Mn alloy, annealed	80
hard	34
Copper, oxygen free HC, annealed	70
Cartridge brass (70% Cu, 30% Zn), annealed	88
$\frac{3}{4}$ hard	21
Cupronickel (70% Cu, 30% Ni), annealed	157
Magnesium–3% Al, 1% Zn alloy, annealed	8
Nickel alloy, Monel, annealed	290
Titanium–5% Al, 2.5% Sn, annealed	24
Grey cast iron	3
Malleable cast iron, Blackheart, annealed	15
Austenitic stainless steel, annealed	217
Carbon steel, 0.2% carbon, as rolled	50

Material	Impact strength* ($kJ\ m^{-2}$)
Polythene, high-density	30
Nylon 6.6	5
PVC, unplasticised	3
Polystyrene	2
ABS	25

*Notch-tip radius 0.25 mm, depth 2.75 mm

5.3 Bend tests

The *bend test* is a simple test of ductility. It involves bending a sample of the material through some angle and determining whether the material is unbroken and free from cracks after such a bend. Figure 5.22 shows one way of

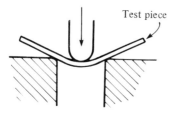

Figure 5.22 A bend test

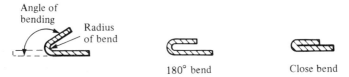

180° bend   Close bend

Figure 5.23 The angle of bend

conducting a bend test. The results of a bend test are specified in terms of the angle of bend (Figure 5.23).

5.4 Hardness measurements

The *hardness* of material may be specified in terms of some standard test involving indentation or scratching of the surface of the material. Hardness is essentially the resistance of the surface of a material to deformation. There is no absolute scale for hardness, each hardness form of test having its own scale. Though some relationships exist between results on one scale and those on another, care has to be taken in making comparisons because the tests associated with the scales are measuring different things.

The most common form of hardness tests for metals involves standard indentors being pressed into the surface of the material concerned. Measurements associated with the indentation are then taken as a measurement of the hardness of the surface. The Brinell test, the Vickers test and the Rockwell test are the main forms of such tests.

With the *Brinell test* (Figure 5.24) a hardened steel ball is pressed for a time of 10 to 15 s into the surface of the material by a standard force. After the load and the ball have been removed, the diameter of the indentation is measured. The Brinell hardness number (signified by HB) is obtained by dividing the size of the force applied by the spherical surface area of the indentation. This area can be obtained, either by calculation or the use of tables, from the values of the diameter of the ball used and the diameter of the indentation.

$$\text{Brinell hardness number} = \frac{\text{applied force}}{\text{spherical surface area of indentation}}$$

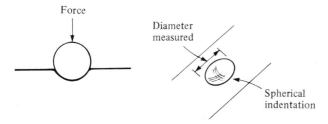

Figure 5.24 The basis of the Brinell hardness test

The units used for the area are mm^2 and for the force are kgf (1 kgf = 9.8 N). The British Standard for this test is BS 240.

The diameter D of the ball used and the size of the applied force F are chosen, for the British Standard, to give F/D^2 values of 1, 5, 10 or 30, the diameters of the balls being 1, 2, 5 or 10 mm. In principle, the same value of F/D^2 will give the same hardness value, regardless of the diameter of the ball used.

The Brinell test cannot be used with very soft or very hard materials. In the one case the indentation becomes equal to the diameter of the ball and in the other case there is either no or little indentation on which measurements can be based. The thickness of the material being tested should be at least ten times the depth of the indentation if the results are not to be affected by the thickness of the sample.

The *Vickers test* (Figure 5.25) uses a diamond indenter which is pressed for 10 to 15 s into the surface of the material under test. The result is a square-shaped impression. After the load and indenter are removed the diagonals of the indentation are measured. The Vickers hardness number (HV) is obtained by dividing the size of the force applied by the surface area of the indentation. The surface area can by calculated, the indentation is assumed to be a right pyramid with a square base (the vertex angle of the pyramid is assumed to be the same as the vertex angle of the diamond, i.e. 136°), or obtained by using tables and the diagonal values. The relevant British Standard is BS 427.

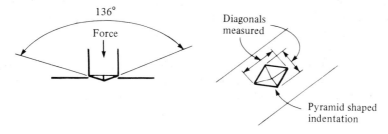

Figure 5.25 The basis of the Vickers hardness test

The Vickers test has the advantage over the Brinell test of the increased accuracy that is possible in determining the diagonals of a square as opposed to the diameter of a circle. Otherwise it has the same limitations as the Brinell test.

The *Rockwell test* (Figure 5.26) uses either a diamond cone or a hardened steel ball as the indenter. A force is applied to press the indenter in contact with the surface. A further force is then applied and causes an increase in depth of the indenter penetration into the material. The additional force is then removed and there is some reduction in the depth of the indenter due to the deformation of the material not being entirely plastic. The difference in the final depth of the indenter and the initial depth, before the additional force was applied, is determined. This is the permanent increase in penetration (e) due to the additional force.

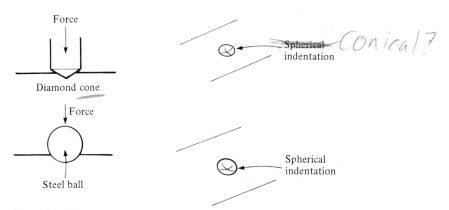

Figure 5.26 The basis of the Rockwell hardness test

Rockwell hardness number (HR) = $E - e$

where E is a constant determined by the form of the indenter. For the diamond cone indenter, E is 100; for the steel ball, E is 130.

There are a number of Rockwell scales, the scale being determined by the indenter and the additional force used. Table 5.4 indicates the scales and the types of materials for which each is typically used. The relevant British Standard for Rockwell tests is B.S. 891. In any reference to the results of a Rockwell test, the scale letter must be quoted. The B and C scales are probably the most commonly used for metals.

For the most commonly used indenters with the Rockwell test the size of the indentation is rather small. Localised variations of structure, composition and roughness can thus affect the results. The Rockwell test is, however, more suitable for workshop or 'on site' use as it is less affected by surface conditions than the Brinell or Vickers tests, which require flat and polished surfaces to permit accurate measurements. The test, as indicated above, cannot be used

Table 5.4   *Rockwell scales and typical applications*

Scale	Indenter	Additional force (kN)	Typical applications
A	Diamond	0.59	Thin steel and shallow case-hardened steel.
B	Ball 1.588 mm dia.	0.98	Copper alloys, aluminium alloys, soft steels.
C	Diamond	1.47	Steel, hard cast irons, deep case-hardened steel.
D	Diamond	0.98	Thin steel and medium case-hardened steel.
E	Ball 3.175 mm dia.	0.98	Cast iron, aluminium alloys, magnesium alloys, bearing metals.
F	Ball 1.588 mm dia.	0.59	Annealed copper alloys, thin soft sheet metals, brass.
G	Ball 1.588 mm dia.	1.47	Malleable irons, gun metals, bronzes, copper-nickel alloys.
H	Ball 3.175 mm dia.	0.59	Aluminium, lead, zinc.
K	Ball 3.175 mm dia.	1.47	Aluminium and magnesium alloys.
L	Ball 6.350 mm dia.	0.59	Plastics.
M	Ball 6.350 mm dia.	0.98	Plastics.
P	Ball 6.350 mm dia.	1.47	
R	Ball 12.70 mm dia.	0.59	Plastics
S	Ball 12.70 mm dia.	0.98	
V	Ball 12.70 mm dia.	1.47	

with thin sheet. A variation of the standard test, called the *Rockwell superficial hardness test*, can be used, however, with thin sheet. Smaller forces are used and the depth of the indentation is determined with a more sensitive device as much smaller indentations are used. The initial force is 29.4 N instead of 90.8 N. Table 5.5 lists the scales given by this test.

Table 5.5   *Rockwell scales for superficial hardness*

Scale	Indenter	Additional force (kN)
15–N	Diamond	0.14
30–N	Diamond	0.29
45–N	Diamond	0.44
15–T	Ball 1.588 mm dia.	0.14
30–T	Ball 1.588 mm dia.	0.29
45–T	Ball 1.588 mm dia.	0.44

Note: numbers with the scale letter refer to the additional force values when expressed in kgf units (1 kgf = 9.8 N).

Comparison of hardness scales

The Brinell and Vickers tests both involve measurements of the surface areas of indentations; the form of the indenter, however, is different. The Rockwell tests involve measurement of the depths of penetration of the indenter. Thus the various tests are concerned with different forms of material deformation as an indication of hardness. There are no simple theoretical relationships between the various hardness scales though some approximate, experimentally derived, relationships have been obtained. Different relationships hold for different metals.

There is an approximate relationship between hardness values and tensile strengths. Thus for annealed steels the tensile strength in N mm^{-2} (MPa) is about 3.54 times the Brinell hardness value, and for quenched and tempered steels 3.24 times the Brinell hardness value. For brass the factor is about 5.6, and for aluminium alloys about 4.2.

Hardness measurements with plastics

The Brinell, Vickers and Rockwell tests can be used with plastics. The Rockwell test, with its measurements of penetration depth rather than surface area of indentation, is more widely used. Scale R is a commonly used scale, e.g. Nylon 6 (Durethan 30S) has a Rockwell hardness on scale R of 120.

The British Standard test (B.S. 2782) involves an indenter, a ball of diameter 2.38 mm, being pressed against the plastic by an initial force of 0.294 N for 5 s and then an additional force of 5.25 N being applied for 30 s. The difference between the two penetration depths is measured and expressed as a *softness number*. This is just the depth expressed in units of 0.01 mm. Thus a penetration of 0.05 mm is a softness number of 5. The test is carried out at a temperature of 23 ± 1°C.

Another form of this test is the *Shore durometer*. This involves an indenter in the form of either a truncated cone with a flat end or spherically-ended cone.

The Moh scale of hardness

One form of hardness test is based on assessing the resistance of a material to being scratched. The *Moh scale* consists of ten materials arranged so that each one will scratch the one preceding it in the scale but not the one that succeeds it:

1. Talc
2. Gypsum
3. Calcspar
4. Fluorspar
5. Apatite
6. Felspar
7. Quartz
8. Topaz
9. Corundum
10. Diamond

Thus Felspar will scratch apatite but not quartz. Diamond will scratch all the materials, while talc will scratch none of them.

Ten styli of the materials in the scale are used for the test. The hardness number of a material under test is one number less than that of the substance that just scratches it. For instance, glass can just be scratched by felspar but not by apatite: glass thus has a hardness number of 5.

Typical hardness values

Table 5.6 gives typical hardness values for metals and plastics at room temperature. Figure 5.27 shows the general range of hardness values for different types of materials when related to Vickers, Brinell, Rockwell and Moh hardness scales.

Table 5.6 *Typical hardness values*

Material	Hardness value
Aluminium, commercial, annealed	21 HB
hard	40 HB
Aluminium-1.25% Mn alloy, annealed	30 HB
hard	30 HB
Copper, oxygen-free HC, annealed	45 HB
hard	105 HB
Cartridge brass, 70% Cu, 30% Zn, annealed	65 HB
hard	185 HB
Magnesium alloy–6% Al, 1% Zn, 0.3% Mn, forged	65 HB
Nickel alloy, Monel, annealed	110 HB
cold worked	240 HB
Titanium alloy–5% Al, 2.5% tin, annealed	360 HB
Zinc casting alloy A, as cast	83 HB
Grey cast iron	210 HB
White cast iron	400 HB
Malleable cast iron, Blackheart	130 HB
Carbon steel–0.2% carbon, normalised	150 HB
Stainless steel, austenitic (304), annealed	150 HB
cold worked	240 HB
PVC, no plasticiser	110 HRR
Polystyrene	80 HRM
ABS	70 HRM
Nylon 6	110 HRR

5.5 Fatigue

If you take a stiff piece of metal or plastic, and want to break it, then you will most likely flex the strip back and forth, as in Figure 5.28. This is generally an easier way of causing the material to fail than applying a direct pull.

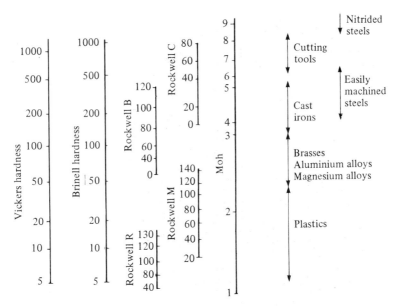

Figure 5.27 Hardness values

Figure 5.28

In service many components undergo thousands, often millions, of changes of stress. Some are repeatedly stressed and unstressed, while others undergo alternating stresses of compression and tension. For yet others, the stress may just fluctuate about some value. Many materials subject to such conditions fail, even though the maximum stress in any one stress change is less than the fracture stress as determined by a simple test. Such a failure, as a result of repeated stressing, is called a *fatigue failure* (Figure 5.29).

The source of the alternating stresses can be due to the conditions of use of a component. Thus, in the case of an aircraft, the changes of pressure between the cabin and the outside of the aircraft every time it flies subject the cabin skin to repeated stressing. Components such as a crown wheel and pinion are subject to repeated stressing by the very way in which they are used, while others receive their stressing 'accidentally'. Vibration of the component can occur as a result of the transmission of vibration from some machine nearby. Turbine blades may vibrate in use in such a way that they fail by fatigue. It has been said that fatigue causes at least 80 per cent of the failures in modern engineering components.

Figure 5.29 Fatigue failure of a large shaft. (From John, V. B., *Introduction to Engineering Materials*, by permission of Macmillan, London and Basingstoke)

A fatigue crack often starts at some point of stress concentration. This point of origin of the failure can be seen on the failed material as a smooth, flat, semicircular or elliptical region, often referred to as the nucleus. Surrounding the nucleus is a burnished zone with ribbed markings. This smooth zone is produced by the crack propagating relatively slowly through the material and the resulting fractured surfaces rubbing together during the alternating stressing of the component. When the component has become so weakened by

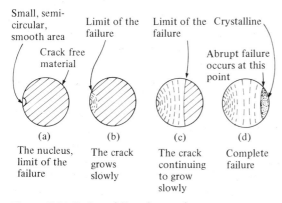

Figure 5.30 Fatigue failure in metal

the crack that it is no longer able to carry the load, the final, abrupt fracture occurs, which shows a typically crystalline appearance. Figure 5.30 shows the various stages in the growth of a fatigue crack failure.

Fatigue tests

Various fatigue tests have been devised to simulate the changes of stress to which the materials of different components are subjected when in service. Bending-stress machines bend a test piece of the material alternately one way and the other (Figure 5.31a), whereas torsional-fatigue machines twist it alternately in different directions (Figure 5.31b). Another type of machine can

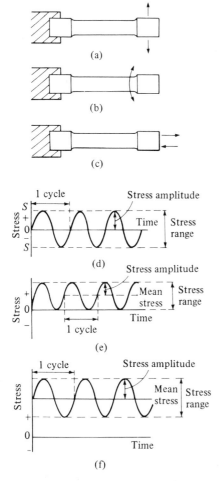

Figure 5.31 Fatigue testing. (a) Bending, (b) torsion, (c) direct stress, (d) alternating stress, (e) repeated stress, (f) fluctuating stress

be used to produce alternating tension and compression by direct stressing (Figure 5.31c).

The tests can be carried out with stresses which alternate about zero stress (Figure 5.31d), apply a repeated stress which varies from zero to some maximum stress (Figure 5.31e) or apply a stress which varies about some stress value and does not reach zero at all (Figure 5.31f).

In the case of the alternating stress (Figure 5.31d), the stress varies between $+S$ and $-S$. The tensile stress is denoted by a positive sign, the compressive stress by a negative sign; the stress range is thus $2S$. The mean stress is zero as the stress alternates equally about the zero stress. With the repeated stress (Figure 5.31e), the mean stress is half the stress range. With the fluctuating stress (Figure 5.31f), the mean stress is more than half the stress range.

In the fatigue tests, the machine is kept running, alternating the stress, until the specimen fails, the number of cycles of stressing up to failure being recorded by the machine. The test is repeated for the specimen subject to different stress ranges. Such tests enable graphs similar to those in Figure 5.32 to be plotted. The vertical axis is the *stress amplitude*, half the stress range. For a stress amplitude greater than the value given by the graph line, failure occurs for the number of cycles concerned. These graphs are known as *S/N graphs*, the S denoting the stress amplitude and the N the number of cycles.

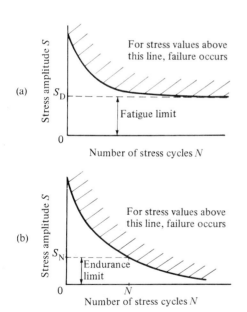

Figure 5.32 Typical S/N graphs for (a) a steel (b) a non-ferrous alloy

For the S/N graph in Figure 5.32a there is a stress amplitude for which the material will endure an indefinite number of stress cycles. The maximum value, S_D, is called the *fatigue limit*. For any stress amplitude greater than the fatigue limit, failure will occur if the material undergoes a sufficient number of stress cycles. With the *S/N* graph shown in Figure 5.32b there is no stress amplitude at which failure cannot occur; for such materials an *endurance limit* S_N is quoted. This is defined as the maximum stress amplitude which can be sustained for *N* cycles.

The number of reversals that a specimen can sustain before failure occurs depends on the stress amplitude; the bigger the stress amplitude, the smaller the number of cycles of stress reversals that can be sustained.

Some typical results for an aluminium alloy specimen are:

Stress amplitude (MN m^{-2} or MPa)	*Number of cycles before failure (millions)*
185	1
155	5
145	10
120	50
115	100

With a stress amplitude of 185 MN m^{-2} (MPa), e.g. a stress alternating from + 185 MN m^{-2} to −185 MN m^{-2}, one million cycles are needed before failure occurs. With a smaller stress amplitude of 115 MN m^{-2}, one hundred million cycles are needed before failure occurs. Figure 5.33 shows the *S/N* graph for the above data. Extrapolation of the graph seems to indicate that for a great number of cycles, failure will occur at even smaller stress amplitudes. There seems to be no stress amplitude for which failure will not occur; the material has no fatigue limit. If a component made of that material had a service life of 100 million stress cycles then we could specify that during

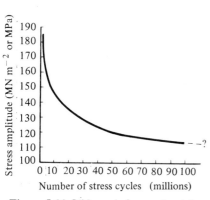

Figure 5.33 *S/N* graph for an aluminium alloy

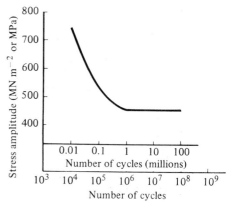

Figure 5.34 *S/N* graph for a steel. Note: number of cycles shown on logarithmic scale

the lifetime, failure should not occur for stress amplitudes less than 115 MN m^{-2}. The endurance limit for 100 million cycles is thus 115 MN m^{-2}.

The following are some typical results for a steel, the fatigue tests being bending stress (Figure 5.34).

Stress amplitude (MN m^{-2} or MPa)	*Number of cycles before failure (millions)*
750	0.01
550	0.1
450	1
450	10
450	100

With a stress amplitude of 750 MN m^{-2} (MPa), e.g. a stress alternating from +750 MN m^{-2} to −750 MN m^{-2}, 0.01 million cycles or ten thousand cycles are needed before failure occurs. For one million, ten million and one hundred million cycles the stress amplitude for failure is the same, 450 MN m^{-2}. For stress amplitudes below this value the material should not fail, however long the test continues. The fatigue limit is thus 450 MN m^{-2}.

The fatigue limit, or the endurance limit at about 500 million cycles, for metals tends to lie between about a third and a half of the static tensile strength. This applies to most steels, aluminium alloys, brass, nickel and magnesium alloys. For example, a steel with a tensile strength of 420 MN m^{-2} has a fatigue limit of 180 MN m^{-2}, just under half the tensile strength. If used in a situation where it was subject to alternating stresses, such a steel would need to be limited to stress amplitudes below 180 MN m^{-2} if it were not to fail at some time. A magnesium alloy with a tensile strength of 290 MN m^{-2} has an endurance limit of 120 MN m^{-2}, just under half the tensile strength. Such an alloy would need to be limited to stress amplitudes below 120 MN m^{-2} if it were to last to 500 million cycles.

Factors affecting the fatigue properties of metals

The main factors affecting the fatigue properties of a component are:

1. Stress concentrations caused by component design.
2. Corrosion.
3. Residual stresses.
4. Surface finish.
5. Temperature.

Fatigue of a component depends on the stress amplitude attained, the bigger the stress amplitude the fewer the stress cycles needed for failure. Stress concentrations caused by sudden changes in cross-section, keyways, holes or sharp corners can thus more easily lead to a fatigue failure. The presence of a countersunk hole was considered in one case to have led to a stress concentration which could have resulted in a fatigue failure. Figure 5.35 shows the effect on the fatigue properties of a steel of a small hole acting as a stress raiser. With the hole, at every stress amplitude value less cycles are needed to reach failure. There is also a lower fatigue limit with the hole present, 700 MN m^{-2} instead of over 1000 MN m^{-2}.

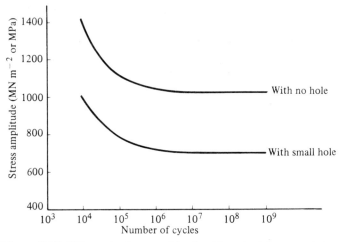

Figure 5.35 *S/N* graphs for a steel both with and without small hole acting as stress raiser. Note: number of cycles shown on logarithmic scale

Figure 5.36 shows the effect on the fatigue properties of a steel of exposure to salt solution. The effect of the corrosion resulting from the salt solution attack on the steel is to reduce the number of stress cycles needed to reach failure for every stress amplitude. The non-corroded steel has a fatigue limit of 450 MN

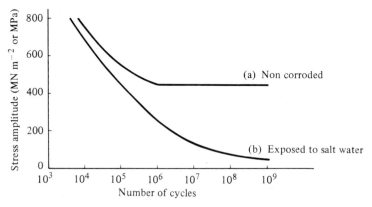

Figure 5.36 *S/N* graphs for a steel, (a) with no corrosion and (b) corroded by exposure to salt solution. Note: number of cycles shown on logarithmic scale

m^{-2}, the corroded steel has no fatigue limit. There is thus no stress amplitude below which failure will not occur. The steel can be protected against the corrosion by plating; for example, chromium or zinc plating of the steel can result in the same S/N graph as the non-corroded steel even though it is subject to a corrosive atmosphere.

Residual stresses can be produced by many fabrication and finishing processes. If the stresses produced are such that the surfaces have compressive residual stresses then the fatigue properties are improved, but if tensile residual stresses are produced at the surfaces then poorer fatigue properties result. The case-hardening of steels by carburising results in compressive residual stresses at the surface, hence carburising improves the fatigue properties. Figure 5.37 shows the effect of carburising a hardened steel. Many machining processes result in the production of surface tensile residual stresses and so result in poorer fatigue properties.

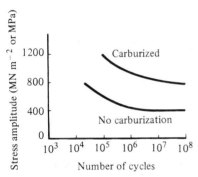

Figure 5.37 S/N graph for a steel, showing effect of carburisation. Note: number of cycles shown on logarithmic scale

The effect of surface finish on the fatigue properties of a component is very significant. Scratches, dents or even surface identification markings can act as stress raisers and so reduce the fatigue properties. Shot peening a surface produces surface compressive residual stresses and improves the fatigue performance.

An increase in temperature can lead to a reduction in fatigue properties as a consequence of greater oxidation or corrosion of the metal surface. For example, the nickel–chromium alloy Nimonic 90 undergoes surface degradation at temperatures around 700° to 800°C and there is a poorer fatigue performance as a result.

The fatigue properties of plastics

Fatigue tests can be carried out on plastics in the same way as with metals. A factor not present with metals is that when a plastic is subject to an alternating stress it becomes significantly warmer. The faster the stress is alternated, i.e., the higher the frequency of the alternating stress, the greater the temperature rise. Under very high frequency alternating stresses the temperature rise may be large enough to melt the plastic. To avoid this, fatigue tests are normally carried out with lower-frequency alternating stresses than is usual with metals. The results of such tests, however, are not entirely valid if the alternating stresses experienced by the plastic component in service are higher than those used for the test.

Figure 5.38 shows an *S/N* graph for a plastic, unplasticised PVC. The alternating stresses were applied with a square waveform at a frequency of 0.5 Hz, i.e. a change of stress every 2 s. The graph seems to indicate that there will be no stress amplitude for which failure will not occur; the material thus seems to have no fatigue limit.

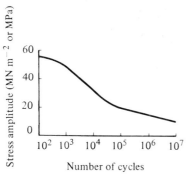

Figure 5.38 *S/N* graph for unplasticised PVC, alternating stress being a square waveform at frequency 0.5 Hz. Note: number of cycles shown on logarithmic scale

5.6 Creep

There are many situations where a piece of material is exposed to a stress for a protracted period of time. The stress/strain data obtained from the conventional tensile test refer generally to a situation where the stresses are applied for quite short intervals of time and so the strain results refer only to the immediate values resulting from stresses. What is the result of applying stress to a material over a long period of time? If you tried such an experiment with a strip of lead you would find that the strain would increase with time—the material would increase in length with time even though the stress remained constant. This phenomenon is called *creep*, which can be defined as the continuing deformation of a material with the passage of time when the material is subject to a constant stress.

For metals, other than the very soft metals like lead, creep effects are negligible at ordinary temperatures, but become significant at higher temperatures. For plastics, creep is often quite significant at ordinary temperatures and even more noticeable at higher temperatures.

Figure 5.39 shows the essential features of a creep test. A constant stress is applied to the specimen, sometimes by the simple method of suspending loads from it. Because creep tests with metals are usually performed at high temperatures, a furnace surrounds the specimen, the temperature of the furnace being held constant by a thermostat. The temperature of the specimen is generally measured by a thermocouple attached to it.

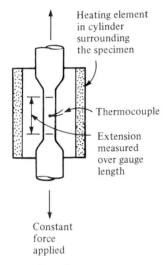

Figure 5.39 A creep test

Figure 5.40 shows the general form of results from a creep test. The curve generally has three parts. During the *primary creep* period the strain is changing

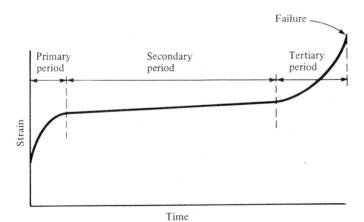

Figure 5.40 Typical creep curve for a metal

but the rate at which it is changing with time decreases. During the *secondary creep* period the strain increases steadily with time at a constant rate. During the *tertiary creep* period the rate at which the strain is changing increases and eventually causes failure. Thus the initial stress, which did not produce early failure, will result in a failure after some period of time. Such an initial stress is referred to as the *stress to rupture* in some particular time. Thus an acrylic plastic may have a rupture stress of 50 N mm^{-2} at room temperature for failure in one week.

Factors affecting creep behaviour of metals

For a particular material, the creep behaviour depends on both the temperature and the initial stress; the higher the temperature the greater the creep, also the higher the stress the greater the creep. Figure 5.41 shows both these effects. Thus, to minimise creep, stress and temperature must be controlled.

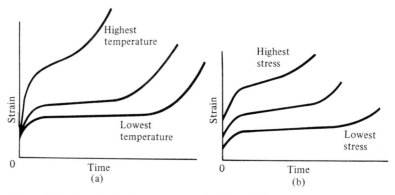

Figure 5.41 Creep behaviour for a material (a) at different temperatures but subject to constant stress (b) at different stresses but subject to constant temperature.

Figure 5.42 shows one way of presenting creep data, indicating the design stress that can be permitted at any temperature if the creep is to be kept within specified limits. In the example given, the limit is 1% creep in 10 000 hours. Thus for the Pireks 25/20 nickel–chrome alloy a stress of 58.6 N mm^{-2} at a temperature of 800°C will produce the 1% creep in 10 000 h. At 1050° a stress of only 10.3 N mm^{-2} will produce the same creep in 10 000 h. Figure 5.43 shows how stress to rupture the material in 10 000 h varies with temperature. At 800°C a stress of 65.0 N mm^{-2} will result in the Pireks 25/20 alloy failing in 10 000 h; at 1050°C a stress of only 14.5 N mm^{-2} will result in failure in the same time.

Another factor that determines the creep behaviour of a metal is its composition. Figure 5.44 shows how the stress to rupture different materials in 1000 h varies with temperature. Aluminium alloys fail at quite low stresses when the temperature rises over 200°C. Titanium alloys can be used at higher

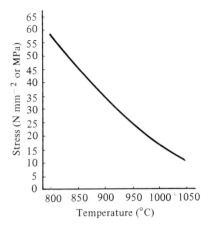

Figure 5.42 Data to give 1% creep in 10 000 h for Pireks 25/20 alloy (0.45% C, 0.8% Mn, 1.2% Si, 20% Cr, 25% Ni). (Courtesy of Darwins Alloy Castings Ltd)

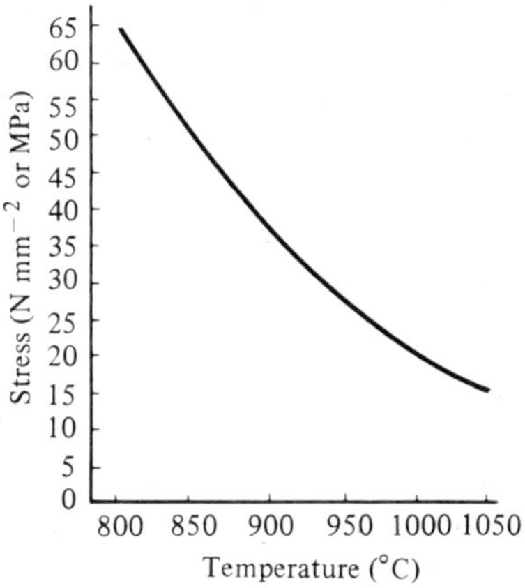

Figure 5.43 Data showing stress to rupture at 10 000 h for Pireks 25/20 alloy. (See also Figure 5.42)

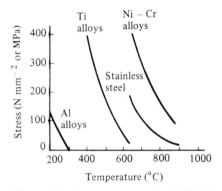

Figure 5.44 Stress to rupture in 1000 h for different materials

temperatures before the stress to rupture drops to very low values, while stainless steel is even better and nickel–chromium alloys offer yet better resistance to creep.

Factors affecting creep behaviour of plastics

Creep can be significant in plastics at normal temperatures. The creep behaviour of a plastic depends on temperature and stress, just as in metals. It also depends on the type of plastic involved—flexible plastics show more creep than stiff ones.

Figure 5.45 shows how the strain on a sample of polyacetal at 20°C varies with time for different stresses. The higher the stress, the greater the creep. As

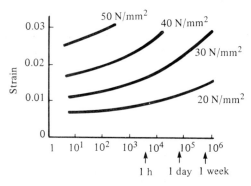

Figure 5.45 Creep behaviour of polyacetal at different stresses. Note logarithmic time scale

can be seen from the graph, the plastic creeps quite substantially in a period of just over a week, even at relatively low stresses.

Figure 5.45 is the form of graph obtained by plotting of results derived from creep tests. Based on the Figure 5.45 graph of strain against time at different stresses, a graph of stress against strain for different times can be produced. Thus, for a time of 10^2 s, a vertical line drawn on Figure 5.45 enables the stresses needed for different strains after this time to be read from the graph (Figure 5.46a). The resulting stress/strain graph is shown in Figure 5.46b and is known as an *isochronous stress/strain graph*. For a specific time the quantity obtained by dividing the stress by the strain for the isochronous stress/strain graph can be calculated and is known as the *creep modulus*. It is not the same as Young's modulus though can it be used to compare the stiffness of plastics. The creep modulus varies both with time and strain and Figure 5.47 shows

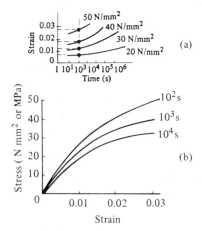

Figure 5.46 (a) Obtaining stress/strain data (b) Isochronous stress/strain graph

how, at 0.5% strain and 20°C, it varies with time for the polyacetal described in Figures 5.45 and 5.46.

Figure 5.48 shows how the stress to rupture a plastic, Durethan (a polyamide), varies with time at different temperatures. The higher the temperature, the lower the stress needed to rupture the material after any particular time.

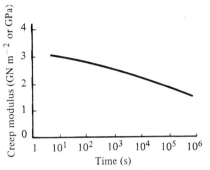

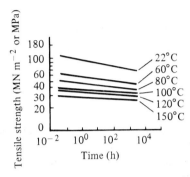

Figure 5.47 Variation of creep modulus with time for polyacetal at 0.5% strain and 20°C

Figure 5.48 Creep rupture graph for Durethan BKV 30 (Courtesy Bayer UK Ltd)

5.7 Corrosion

The car owner can rightly be concerned about rust patches appearing on the bodywork of the car, as the rust not only makes the bodywork look shoddy but indicates a mechanical weakening of the material. The steel used for the car bodywork has thus changed with time due to an interaction between it and the environment. The possibility of such corrosion is therefore a factor that has to be taken into account when a material is selected for a particular purpose.

Most metals react, at moderate temperatures, only slowly with the oxygen in the air. The result is the build-up of a layer of oxide on the surface of the metal, which can insulate the metal from further reaction with the oxygen. Aluminium is an example of a metal that builds up a protective oxide layer which is a very effective barrier against further oxidation. If oxidation were the only reaction which tended to occur between a metal and its environment then probably corrosion would present few problems. The presence of moisture in the environment can very markedly affect corrosion, as can the presence of chemically active pollutants.

Iron rusts when the environment contains both oxygen and moisture, but not with oxygen alone or moisture alone. Iron nails kept in a container with dry air do not rust, nor if they are kept in oxygen-free, e.g. boiled, water. But in moist air they rust readily.

Chemically-active pollutants in the environment can have a marked effect on corrosion, especially those that are soluble in water. Man-made pollutants such as the oxides of carbon and sulphur, produced in the combustion of fuels, dissolve in water to give acids, which readily attack metals and many other materials. Marine environments also are particularly corrosive, due to the high concentrations of salt from the sea. The sodium chloride reacts with metals to produce chlorides of the metals which are soluble in water and thus cannot act as a protective layer on the surface of a metal as a non-soluble oxide may do. The salt may also destroy the protective oxide layer that has been acquired by a metal.

Galvanic corrosion

A simple electrical cell could be just a plate of copper and one of zinc (Figure 5.49) dipping into an *electrolyte*, a solution that conducts electricity. Such a cell gives a potential difference between the two metals – it can be measured with a voltmeter or used to light a lamp. Different pairs of metals give different potential differences. Thus a zinc–copper cell gives a potential difference of about 1.1 V, and an iron–copper cell about 0.8 V. A zinc–iron cell gives a potential difference of about 0.3 V; this value is, however, the difference between the potential differences of the zinc–copper and iron–copper cells. It is as though we had a cell made up of zinc–copper–iron.

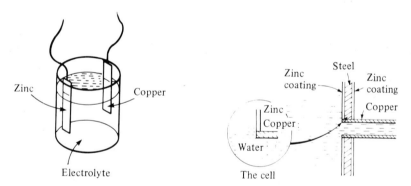

Figure 5.49 A simple cell

Figure 5.50 Copper–zinc cell for copper pipe connection to galvanised steel water tank

By tabulating values of the potential differences between the various metals and a standard, a table can be produced from which the potential differences between any pair of metals can be forecast. The standard used is hydrogen and Table 5.7 shows the potential differences relative to hydrogen for a number of metals. The table gives, for a silver–aluminium cell, a potential difference of 2.47 V, that is $+0.80 - (-1.67)$ V. The metal highest up the table behaves as the negative electrode of the cell while the metal lowest in the table is the positive electrode. The term *cathode* is used for the negative electrode and *anode* for the positive electrode. Thus for the silver–aluminium cell the silver is the negative electrode and the aluminium the positive electrode.

Table 5.7   *Potential differences relative to hydrogen*

Metal	Potential difference (V)
Gold	+1.7
Silver	+0.80
Copper	+0.34
Hydrogen	0.00
Lead	−0.13
Tin	−0.14
Nickel	−0.25
Iron	−0.44
Zinc	−0.76
Aluminium	−1.67
Magnesium	−2.34
Sodium	−2.71

With a copper–zinc cell it is found that the copper electrode remains unchanged after a period of cell use but the zinc electrode is badly corroded. With any cell the anode is corroded and the cathode not affected. The greater the cell potential difference, the greater the corrosion of the anode.

Table 5.6 lists what is called the *electromotive series* or *galvanic series*. Tables of such series are available for metals in various environments. Such tables are of use in assessing the possibilities of corrosion when two different metals are in electrical contact, either directly or through a common electrolyte.

If a copper pipe is connected to a galvanised steel tank, perhaps the cold water storage tank in your home, a cell is created (Figure 5.50) and corrosion follows. Galvanised steel is zinc-coated steel; there is thus a copper–zinc cell. With such a cell the copper is the cathode and the zinc the anode; the zinc thus corrodes and so exposes the iron. Iron–copper is also a good cell with the iron as the anode. The result is corrosion of the iron and hence the overall result of connecting the copper pipe to the tank is likely to be a leaking tank.

Stainless steel and mild steel form a cell, the mild steel being the anode. Thus if a stainless steel trim, on say a car, is in electrical contact with mild steel bodywork, then the bodywork will corrode more rapidly than if no stainless steel trim were used. The electrical connection between the stainless steel and the mild steel may be through water gathering at the junction between the two. With oxygen-free clean water the cell potential difference may be only 0.15 V, but if the water were sea water containing oxygen the potential difference could become as high as 0.75 V. So the use of salt to melt ice on the roads leads to greater corrosion of cars.

Galvanic cells can be produced in a number of other ways. An alloy or a metal containing impurities can give rise to galvanic cells within itself. For example, brass is an alloy of copper and zinc, a copper–zinc cell has the zinc as the anode and thus the zinc corrodes. The effect is called *dezincification*, one example of *demetallification*. After such corrosion the remaining metal is likely to be porous and lacking in mechanical strength.

A similar type of corrosion takes place for carbon steels in the pearlitic condition, the cementite in the steel acting as the cathode and the ferrite as the anode. The ferrite is thus corroded.

Cast-iron is a mixture of iron and graphite, the graphite acting as the cathode and the iron as the anode. The iron is thus corroded, the effect being known as *graphitisation*.

Variations in concentration of the electrolyte in contact with a metal can lead to corrosion. That part of the metal in contact with the more concentrated electrolyte acts as a cathode while the part in contact with the more dilute electrolyte acts as an anode and so corrodes most. Such a cell is known as a *concentration cell*.

Another type of concentration cell is produced if there are variations in the amount of oxygen dissolved in the water in contact with a metal. That part of the metal in contact with the water having the greatest concentration of oxygen acts as a cathode, while the part of the metal in contact with the water having the least concentration acts as an anode and so corrodes most. A drop of water on a steel surface is likely to have a higher concentration of dissolved oxygen near its surface where it is in contact with air than in the centre of the drop (Figure 5.51). The metal at the centre of the drop acts as an anode and so corrodes most.

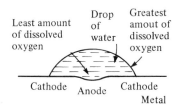

Figure 5.51 Example of a concentration cell

Variations in stress within a metal or a component can lead to the production of cells and hence corrosion. A component which is partly heavily cold-worked and partly less-worked will contain internal stresses which can result in the heavily-worked part acting as an anode and the less-worked part as a cathode. Therefore the heavily cold-worked part corrodes most.

Corrosion prevention

Methods of preventing corrosion, or reducing it, can be summarised as:

(1) Selection of appropriate materials.

(2) Selection of appropriate design.
(3) Modification of the environment.
(4) Use of protective coatings.

In selecting materials, care should be taken not to use two different materials in close proximity, particularly if they are far apart in the galvanic series, giving a high potential difference between them. The material which acts as the anode will be corroded in the appropriate environment. It is not desirable to connect copper pipes to steel water tanks. Steel pipes to a steel tank would be better.

However there are situations where the introduction of a dissimilar metal can be used for protection of another metal. Pieces of magnesium or zinc placed close to buried iron pipes can protect the pipes in that the magnesium or zinc acts as the anode relative to the iron which becomes the cathode. The result is corrosion of the magnesium or zinc and not the pipe. Such a method is known as *galvanic protection*.

The steel hull of a ship can be protected below the water line by fixing pieces of magnesium or zinc to it. The steel then behaves as the cathode, the magnesium or zinc becoming the anode and so corroding. Another way of making a piece of metal act as a cathode and so not corrode, is to connect it to a source of e.m.f. in such a way that the externally applied potential difference makes the metal the cathode in an electrical circuit (Figure 5.52).

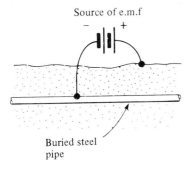

Figure 5.52 Corrosion prevention using an applied e.m.f.

Selection of an appropriate metal for a specific environment can do much to reduce corrosion.

Copper. When exposed to the atmosphere copper develops an adherent protective layer which then insulates it from further corrosion. Copper is used for water pipes, e.g. domestic water supply pipes, as it offers a high resistance to corrosion in such situations.

Copper alloys. Demetallification can occur. Some of the alloying metals, however, can improve the corrosion resistance of the copper by improving the development of the adherent protective surface layer.

Iron and steel. This is corroded readily in many environments and exposure to sea water can result in graphitisation. Stress corrosion can occur in certain environments.

Alloy steels. The addition of chromium to steel can improve considerably the corrosion resistance by modifying the surface protective film produced under oxidising conditions. The addition of nickel to an iron–chromium alloy can further improve corrosion resistance and such a material can be used in sea water with little corrosion resulting. Austenitic steels, however, are susceptible to stress corrosion in certain environments. Iron–nickel alloys have good corrosion resistance and such alloys with 20% nickel and 2 to 3% carbon are particularly good for marine environments.

Zinc. Zinc can develop a durable oxide layer in the atmosphere and then becomes resistant to corrosion.

Aluminium. Readily developing a durable protective surface film, aluminium is then resistant to corrosion. Aluminium alloys are subject to stress corrosion.

The avoidance of potential crevices (Figure 5.53) which can hold water or some other electrolyte and so permit a cell to function, perhaps by bringing two dissimilar metals into electrical contact or by producing a concentration cell, should be avoided in design. Suitable design can also do much to reduce the incidence of stress corrosion.

Corrosion can be prevented or reduced by modification of the environment in which a metal is situated. Thus in the case of a packaged item an impervious packaging can be used so that water vapour cannot come into contact with the metal. Residual water vapour within the package can be removed by including a dessicant such as silica gel, within the package.

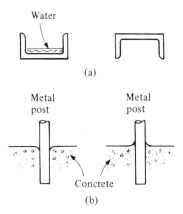

Figure 5.53 Ways of reducing corrosion by preventing water from collecting. (a) Simple inversion, (b) using a fillet to eliminate crevice

Where the environment adjacent to the metal is a liquid it is possible to add certain compounds to the liquid so that corrosion is inhibited, such additives being known as *inhibitors*. In the case of water-in-steel radiators or boilers, compounds which provide chromate or phosphate ions may be used as inhibitors, as they help to maintain the protective surface films on the steel.

One way of isolating a metal surface from the environment is to cover the metal with a protective coating, which can be impervious to oxygen, water or other pollutants. Coatings of grease, with perhaps the inclusion of specific corrosion inhibitors, can be used to give a temporary protective coating. Plasticised bitumens or resins can be used to give harder but still temporary coatings, while organic polymers or rubber latex can be applied to give coatings which can be stripped off when required.

One of the most common coatings applied to surfaces in order to prevent corrosion is paint, different types having different resistances to corrosion environments. Thus some paints have a good resistance to acids while others are good for water.

Steel components can be protected by dipping them in molten zinc to form a thin surface coating of zinc on the steel surface. This method is known as *galvanising*. The zinc acts as an anode with the steel being the cathode (Figure 5.54), and zinc corrodes, rather than the steel, when the surface layer is broken. Small components can be coated with zinc by heating them in a closed container with zinc dust. This process is called *sherardising*.

Other metals can be used to coat steel, often by means of electroplating; thus nickel-plated steel offers some protection from the environment. Often a layer of chromium is applied to a steel surface over a base coat of nickel; the chromium layer is quite resistant to corrosion. Such coatings, though offering protection, are not so effective as zinc when the layer is broken, the zinc being sacrified in place of the steel.

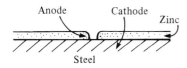

Figure 5.54 Galvanising steel

Steel surfaces are often treated with phosphoric acid or solutions containing phosphate ions, resulting in the formation of a phosphate coating, the process being known as *phosphating*. The coating bonds well with the surface and, though giving some corrosion protection, is generally used as a precursor for othere coatings, perhaps paint.

Several metals can have corrosion-resistant surface layers produced by the application of solutions of chromates to their surfaces. A steel surface which has had the phosphating treatment may then be subject to a chromating treatment, the result being a good corrosion-resistant surface layer on the metal.

Aluminium exposed to the atmosphere has generally an oxide surface layer which offers some corrosion resistance and can be thickened by an electrolyte process. The treatment is known as *anodising*. For more details the reader is referred to T. K. Ross, *Metal Corrosion* (Oxford University Press).

The stability of non-metals

Polymers are largely unaffected by weak acids or alkalis, oils and greases, but may be dissolved by some organic solvents. Exposure to air and sunlight can result in some deterioration of properties. This may show as a loss of colour of the surface, a loss in elasticity and possibly a loss in strength. This is discussed in more detail in section 10.5.

Ceramics are relatively stable when exposed to the atmosphere, though the presence of sulphur dioxide in the atmosphere and its subsequent change to sulphuric acid can result in deterioration of ceramics, e.g. damage to stone and brick of building. Damage to ceramics may also result from the freezing of water which has become absorbed into the pores of the material. Thermal shock is often a problem with ceramics, because of their low thermal conductivity. This can lead to flaking of surfaces.

5.8 Thermal properties

For engineers, the three important effects of the heating of a solid are that its temperature rises, it transmits heat, and it expands. The amount by which the temperature rises depends on the *specific heat capacity* of the material: the higher the specific heat capacity for a given mass, the greater the temperature rise.

$$\text{Specific heat capacity} = \frac{\text{heat input}}{\text{mass} \times \text{change in temperature}}$$

The transmission of heat through a solid depends on the property called *thermal conductivity*. The higher the conductivity, the greater the rate at which heat is conducted.

$$\text{Thermal conductivity} = \frac{\text{rate of transfer of heat}}{\text{cross-sectional area} \times \text{temperature gradient}}$$

The expansion of a solid is described by the *thermal expansivity*, or coefficient of thermal expansion. The bigger this expansivity, the more a solid will expand.

$$\text{Thermal expansivity} = \frac{\text{change in length}}{\text{original length} \times \text{temperature change}}$$

Table 5.8 gives some typical values.

Table 5.8 *Typical values of thermal properties*

		Thermal expansivity $(10^{-5} K^{-1})$	Specific heat capacity $(kJ\ kg^{-1}\ K^{-1})$	Thermal conductivity $(W\ m^{-1}\ K^{-1})$
Metals	Aluminium	2.3	0.88	230
	Copper	1.7	0.38	380
Ceramics	Alumina	0.8	0.8	20
	Glass, Pyrex	0.3	0.8	1.1
Polymers	PVC	5–18	1.1	0.19
	Polystyrene	6–8	1.2	0.16

Problems

1 Explain the terms: tensile strength, tensile modulus, limit of proportionality, yield stress, proof stress, percentage elongation, percentage reduction in area.

2 Explain why tensile test pieces have a standard relationship between gauge length and cross-sectional area.

3 The following results were obtained from a tensile test of an aluminium alloy. The test specimen had a diameter of 11.28 mm and a gauge length of 56 mm. Determine (a) the stress–strain graph, (b) the tensile modulus, (c) the 0.1% proof stress.

Load (kN)	0	2.5	5.0	7.5	10.0	12.5	15.0	17.5
Extension (10^{-2} mm)	0	1.8	4.0	6.2	8.4	10.0	12.5	14.6

Load (kN)	20.0	22.5	25.0	27.5	30.0	32.5	35.0
Extension (10^{-2} mm)	16.3	19.0	21.2	23.5	25.7	28.1	31.5

Load (kN)	37.5	38.5	39.0	39.0	(broke)
Extension (10^{-2} mm)	35.0	40.0	61.0	86.0	

4 A flat tensile test piece of steel had a gauge length of 100.0 mm. After fracture, the gauge length was 131.1 mm. What was the percentage elongation?

5 The following data was obtained from a tensile test on a stainless steel test piece. Determine (a) the limit of proportionality stress, (b) the tensile modulus, (c) the 0.2% proof stress.

Stress (N mm^{-2})	0	90	170	255	345	495	605
Strain (10^{-4})	0	5	10	15	20	30	40

Stress (N mm^{-2})	700	760	805	845	880	895
Strain (10^{-4})	50	60	70	80	90	100

6 Determine from the stress–strain graph for cast iron given in Figure 5.55, the tensile strength and the limit of proportionality.

7 Determine from the stress–strain graph for an aluminium alloy given in Figure 5.56, the tensile modulus.

8 What is meant by a 'proportional test piece?
9 Figure 5.57 shows part of the stress–strain graph for a sample of nylon 6. Estimate (a) the tensile modulus and (b) the tensile strength for the sample.
10 What is the secant modulus of elasticity?

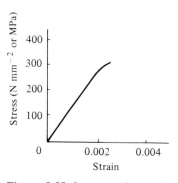

Figure 5.55 Stress–strain graph for a sample of cast iron

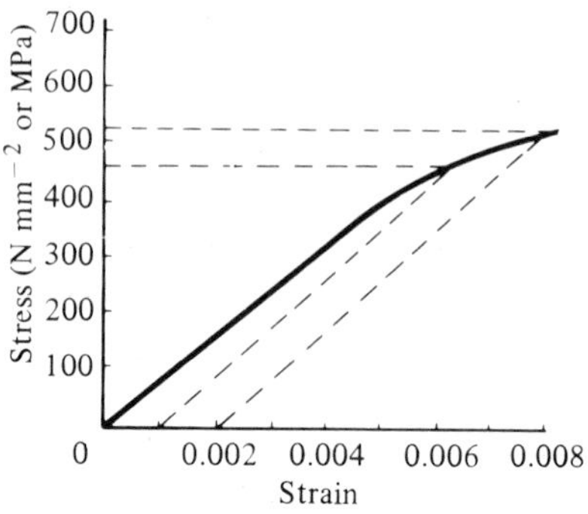

Figure 5.56 Part of the stress–strain graph for a sample of an aluminium alloy

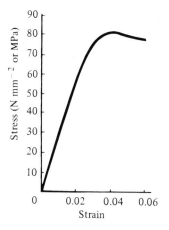

Figure 5.57 Part of the stress–strain graph for Nylon 6 (Durethan SK)

11 What is the effect of strain rate on the data obtained from tensile tests for plastics?
12 Sketch the form of the stress–strain graphs for (a) brittle stiff materials, (b) brittle non-stiff materials, (c) ductile stiff materials, (d) ductile non-stiff materials.
13 The effect of working an aluminium alloy (1.25% manganese) is to change the tensile strength from 110 N mm^{-2} to 180 N mm^{-2} and the percentage elongation from 30% to 3%. Is the effect of the working to (a) make the material stronger, (b) make the material more ductile?

14 A sand-cast aluminium alloy (12% silicon) is found to have a percentage elongation of 5%. Would you expect this material to be brittle or ductile?

15 An annealed titanium alloy has a tensile strength of 880 N mm^{-2} (MPa) and a percentage elongation of 16%. A nickel alloy, also in the annealed condition, has a tensile strength of 700 N mm^{-2} (MPa) and a percentage elongation of 35%. Which alloy is (a) the stronger, (b) the more ductile in the annealed condition?

16 Cellulose acetate has a tensile modulus of 1.5 kN mm^{-2} (GPa) while polythene has a tensile modulus of 0.6 kN mm^{-2} (GPa). Which of the two plastics will be the stiffer?

17 Describe the Izod or the Charpy impact test.

18 Describe the difference between the appearance of a brittle and a ductile fracture in an impact test piece.

19 Explain how impact tests can be used to determine whether a heat treatment process has been carried out successfully.

20 The following are Izod impact energies at different temperatures for samples of annealed cartridge brass (70% copper – 30% zinc). What can be deduced from the results?

Temperature (°C)	+27	−78	−197
Impact energy (J)	88	92	108

21 The following are Charpy V-notch impact energies for annealed titanium at different temperatures. What can be deduced from the results?

Temperature (°C)	+27	−78	−196
Impact energy (J)	24	19	15

22 The following are Charpy impact strengths for nylon 6.6 at different temperatures. What can be deduced from the results?

Temperature (°C)	−23	−33	−43	−63
Impact strength (kJ m^{-2})	24	13	11	8

23 The impact strength of samples of nylon 6, at a temperature of 22°C, is found to be 3 kJ m^{-2} in the as moulded condition, but 25 kJ m^{-2} when the sample has gained 2.5% in weight through water absorption. What can be deduced from the results?

24 Describe a bend test and explain the significance of the results of such a test.

25 Describe the principles of the Brinell, Vickers and Rockwell hardness measurement methods.

26 How can a hardness measurement give an indication of the tensile strength of a material?

27 Outline the limitations of the Brinell hardness test.

28 With Rockwell test results, a letter A, B, C, etc. is always given with the results. What is the significance of the letter?

29 Which hardness test could be used with thin steel sheet?

30 Explain what is meant by the Moh scale of hardness.
31 A sample of brass can just be scratched by calcite but not by gypsum. What would be its Moh hardness number?
32 Specify the type of test that could be used in the following instances.
   (a) A large casting is to be produced and a check is required as to whether the correct cooling rate occurs.
   (b) The storekeeper has mixed up two batches of steel, one batch having been surface hardened and the other not. How could the two be distinguished?
   (c) What test could be used to check whether tempering has been correctly carried out for a steel?
   (d) A plastic is modified by the inclusion of glass fibres. What test can be used to determine whether this has made the plastic stiffer?
   (e) What test could be used to determine whether a metal has been correctly heat treated?
   (f) What test could be used to determine whether a metal is in a suitable condition for forming by bending?
33 Explain what is meant by fatigue failure.
34 List the types of test available for the determination of the fatigue properties of specimens.
35 Describe the various stages in the failure of a component by fatigue.
36 Explain the terms 'fatigue limit' and 'endurance limit'.
37 Figure 5.33 shows the S/N graph for an aluminium alloy.
   (a) For how many stress cycles could a stress amplitude of 140 MN m^{-2} be sustained before failure occurs?
   (b) What would be the maximum stress amplitude that should be applied if the component made of the material is to last for 50 million stress cycles?
   (c) The alloy has a tensile strength of 400 MN m^{-2} and a yield stress of 280 MN m^{-2}. What should be the limiting stress when such an alloy is used for static conditions? What should be the limiting stress when the alloy is used for dynamic conditions where the number of cycles is not likely to exceed 10 million?
38 Explain an S/N graph and state the information that can be extracted from the graph.
39 What is the fatigue limit for the uncarburised steel giving the S/N graph in Figure 5.37?
40 What is the endurance limit for the unplasticized p.v.c. at 10^6 cycles that gave the S/N graph in Figure 5.38?
41 Plot the S/N graph for the nickel-chromium alloy Nimonic 90, which gave the following fatigue test results. Determine from the graph the fatigue limit.

Stress amplitude (MN m^{-2} or MPa)	Number of cycles before failure
750	10^5
480	10^6
350	10^7
320	10^8
320	10^9
320	10^{10}

42 Plot the *S/N* graph for the plastic (cast acrylic) which gave the following fatigue test results when tested with a square waveform at 0.5 Hz. Why specify this frequency? What is the endurance limit at 10^6 cycles?

Stress amplitude (MN m^{-2} or MPa)	Number of cycles before failure
70	10^2
62	10^3
58	10^4
55	10^5
41	10^6
31	10^7

43 Figure 5.58 shows the *S/N* graph for a nickel-based alloy Iconel 718.
(a) What is the fatigue limit?
(b) What is the significance of the constant stress amplitude part of the graph from 10^0 to 10^4 cycles?
(c) The graph is for the material at 600°C. The tensile strength at that temperature is 1000 MN m^{-2}. What can be added to your answer to part (b)?

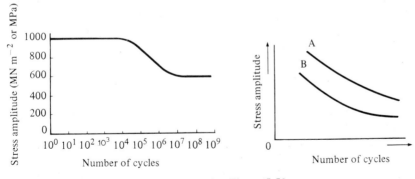

Figure 5.58 *S/N* graph for a nickel-based alloy, Iconel 718

Figure 5.59

44 Figure 5.59 shows two *S/N* graphs, one for the material in an un-notched state, the other for the material with a notch. Which of the graphs would you expect to represent each condition? Give reasons for your answer.

45 List factors that contribute to the onset of fatigue failure and those which tend to resist fatigue.

46 Explain what is meant by 'creep'.

47 Describe the form of a typical strain/time graph resulting from a creep test and explain the significance of the slope of the graph at its various stages.

48 Describe the effect of (a) increased stress and (b) increased temperature on the creep behaviour of materials.

49 For the alloy described in Figure 5.42 estimate the stress that will result in a 1% creep at a temperature of 900°C after 10 000 h.

50 For the alloy described in Figure 5.43, estimate the stress to rupture at 10 000 h for a temperature of 900°C.

51 Under what circumstances would you consider that a metal like that described in Figures 5.42 and 5.43, would be necessary? What would you estimate the limiting temperature for use of such an alloy?

52 Explain how an isochronous stress/strain graph for a polymer can be obtained from creep test results.

53 Explain the significance of the graph shown in Figure 5.48 for the creep rupture behaviours of Durethan.

54 Durethan, as described by Figure 5.48, is used for car fan blades, fuse box covers, door handles and plastic seats. How would the behaviour of the material change when the temperature or stress rises?

55 Figure 5.60 shows how the strain changes with time for two different polymers when they are subjected to a constant stress. Describe how the materials will creep with time. Which material will creep the most?

56 Figure 5.61 shows the stress rupture properties of two alloys, one 50% chromium and 50% nickel, the other (IN 657) 48–52% chromium, 1.4–1.7% niobium, 0.1% carbon, 0.16% nitrogen, 0.2% carbon + nitrogen, 0.5% maximum silicon, 1.0% iron, 0.3% maximum manganese and the remainder nickel. The creep rupture data is presented for two different times, 1000 h and 10 000 h.

(a) What is the significance of the difference between the 1000 h and 10 000 h graphs?

(b) What is the difference in behaviour of the 50 Cr–50 Ni alloy and the IN 657 alloy when temperatures increase?

(c) The IN 657 alloy is said to show 'improved hot strength' when compared with the 50 Cr–50 Ni alloy. Explain this statement.

57 It has been observed that cars in a dry desert part of a country remain remarkably free of rust when compared with cars in a damp climate such as the UK. Offer an explanation for this.

58 Why does the de-aeration of water in a boiler reduce corrosion?

59 What criteria should be used if corrosion is to be kept to a low value when two dissimilar metals are joined together?

60 Aluminium pipes are to be used to carry water into a water tank. Possible materials for the tank are copper or galvanised steel. Which material would you advocate if corrosion is to be minimised?

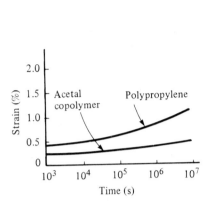

Figure 5.60

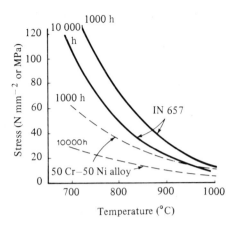

Figure 5.61 Creep rupture data from 'High chromium Cr–Ni alloys to resist residual fuel oil ash corrosion'. (Courtesy of Inco Europe Ltd)

61 It is found that for a junction between mild steel and copper in a sea water environment that the mild steel corrodes rather than the copper. With a mild steel–aluminium junction in the same environment it is found that the aluminium corrodes more than the mild steel. Explain the above observations.

62 Pieces of magnesium placed close to buried iron pipes are used to reduce the corrosion of the iron. Explain.

63 Corrosion of a stainless steel flange is found to occur when a lead gasket is used. Explain.

64 Propose a method to give galvanic protection for a steel water storage tank.

65 Why should copper piping not be used to supply water to a galvanised steel water storage tank?

66 What are the main ways galvanic cells can be set up in metals?

67 Compare the use of zinc and tin as protective coatings for steel.

6

Structure of metals

6.1 Crystals

The form of crystals and the way in which they grow can be explained if matter is considered to be made up of small particles which are packed together in a regular manner. Figure 6.1 shows how a simple cube can be made by stacking four spheres. The cube can 'grow' if further spheres are added equally to all faces of the four-sphere cube (Figure 6.2). The result is a bigger cube which can be considered to be made up of a larger number of the basic four-sphere cube.

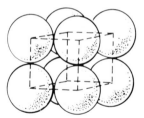

Figure 6.1 A simple cubic structure

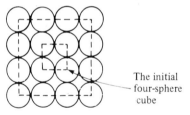

The initial four-sphere cube

Figure 6.2 A two-dimensional view of the 'growing' four-sphere cube

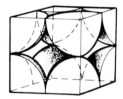

Figure 6.3 The simple cubic structure

The dotted lines in Figure 6.1 enclose what is called the *unit cell*. In this case the unit cell is a cube. The unit cell is the geometric figure which illustrates the

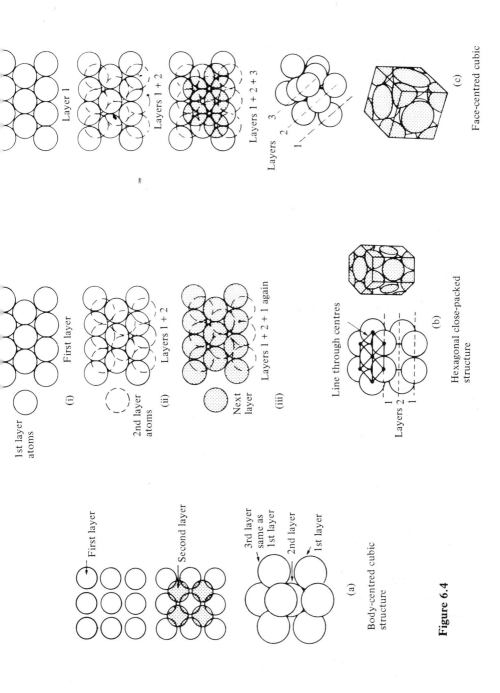

Layer 1

Layers 1 + 2

Layers 1 + 2 + 3

Layers
3
2
1

(c)

Face-centred cubic

1st layer atoms ◯

(i)

First layer

2nd layer atoms ◯

(ii)

Layers 1 + 2

Next layer ◯

(iii)

Layers 1 + 2 + 1 again

Line through centres

Layers 2
1
1

(b)

Hexagonal close-packed
structure

First layer

Second layer

3rd layer
same as
1st layer

2nd layer

1st layer

(a)

Body-centred cubic
structure

Figure 6.4

grouping of the particles in the solid. This group is repeated many times in space within a crystal, which can be considered to be made up, in the case of a *simple cubic crystal*, of a large number of these unit cells stacked together. Figure 6.3 shows that portion of the stacked spheres that is within the unit cell. The crystal is considered to consist of large numbers of particles arranged in a regular, repetitive pattern, known as the *space lattice*, as in Figure 6.2.

It is this regular, repetitive pattern of particles that characterises crystalline material. A solid having no such order in the arrangement of its constituent particles is said to be *amorphous*.

The simple cubic crystal shape is arrived at by stacking spheres in one particular way. By stacking spheres in different ways, other crystal shapes can be produced (Figure 6.4a, b and c). With the simple cubic unit cell the centres of the spheres lie at the corners of a cube. With the *body-centred cubic* unit cell the cell is slightly more complex than the simple cubic cell in having an extra sphere in the centre of the cell. The *face-centred cubic* cell is another modification of the simple cubic cell, having spheres at the centre of each face of the cube. Another common arrangement is the *hexagonal close-packed* structure.

Metals as crystalline substances

Metals are crystalline substances. The term *grain* is used to describe the crystals within the metal. A grain is simply a crystal that lacks its characteristic geometrical shape and flat faces because its growth has been impeded by contact with other crystals. Within a grain the arrangement of particles is just as regular and repetitive as within a crystal with smooth faces. A simple model of a metal with its grains is given by the raft of bubbles on the surface of a liquid (Figure 6.5). The bubbles pack together in an orderly and repetitive manner but if 'growth' is started at a number of centres then 'grains' are produced. At the boundaries between the 'grains' the regular pattern breaks down as the pattern changes from the orderly pattern of one 'grain' to that of the next 'grain'.

The grains in the surface of a metal are not generally visible. They can be made visible by careful etching of the surface with a suitable chemical. The chemical preferentially attacks the grain boundaries.

Here are some examples of the different forms of crystal structure adopted by metallic elements.

Body-centred cubic	Face-centred cubic	Hexagonal close-packed
Chromium	Aluminium	Beryllium
Molybdenum	Copper	Cadmium
Niobium	Lead	Magnesium
Tungsten	Nickel	Zinc

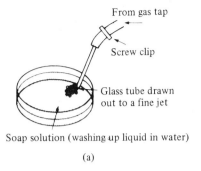

From gas tap

Screw clip

Glass tube drawn
out to a fine jet

Soap solution (washing up liquid in water)

(a)

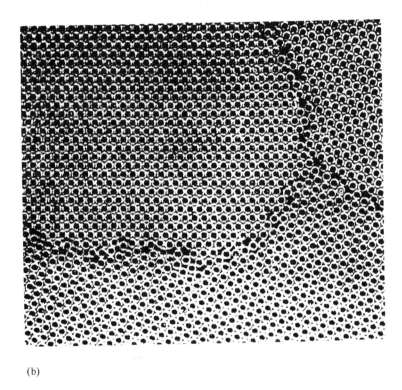

(b)

Figure 6.5 (a) Simple arrangement for producing bubbles, (b) 'grains' in a bubble raft.
(Courtesy of the Royal Society)

Growth of metal crystals

How do metal crystals grow in liquid metal? Figure 6.6 shows the various stages that can occur when a metal solidifies. Crystallisation occurs round small nuclei, which may be impurity particles. The initial crystals that form have the shape of the crystal pattern into which the metal normally solidifies, e.g. face-centred cubic in the case of copper. However, as the crystal grows it tends to develop spikes. The shape of the growing crystal thus changes into a 'tree-like' growth called a *dendrite* (Figure 6.7). As the dendrite grows, so the spaces between the arms of the dendrite fill up. Outward growth of the dendrite ceases when the growing arms meet other dendrite arms. Eventually the entire liquid solidifies. When this happens there is little trace of the dendrite structure, only the grains into which the dendrites have grown.

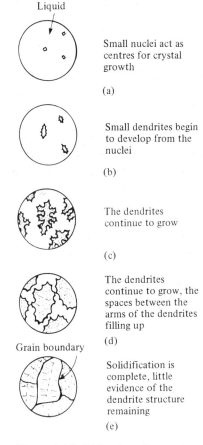

Liquid

Small nuclei act as centres for crystal growth

(a)

Small dendrites begin to develop from the nuclei

(b)

The dendrites continue to grow

(c)

The dendrites continue to grow, the spaces between the arms of the dendrites filling up

(d)

Grain boundary

Solidification is complete, little evidence of the dendrite structure remaining

(e)

Figure 6.6 Solidification of a metal

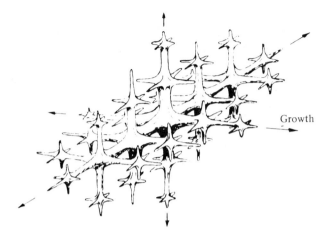

Figure 6.7 Growth of a metallic dendrite (From Higgins, R. A., *Properties of Engineering Materials*. Hodder & Stoughton)

Why do metals tend to grow from the melt as dendrites? Energy is needed to change a solid, at its melting point, to a liquid without any change in temperature occurring; this energy is called *latent heat*. Similarly, when a liquid at the fusion point (i.e. the melting point) changes to a solid, energy has to be removed, no change in temperature occurring during the change of state; this is the latent heat. Thus, when the liquid metal in the immediate vicinity of the metal crystal face solidifies, energy is released which warms up the liquid in front of that advancing crystal face. This slows, or stops, further growth in that direction. The result of this action is that spikes develop as the crystal grows in the directions in which the liquid is coolest. As these warm up in turn, so secondary, and then tertiary, spikes develop as the growth continues in those directions in which the liquid is coolest. This type of growth can be considered in terms of the 'crystal shapes' generated by stacking spheres in an orderly manner (Figure 6.4). Instead of stacking the spheres over the entire 'crystal' surface, the spheres are only stacked on parts of that surface (Figure 6.8). The result is that, although the material is still growing into a crystal with the same unit cell arrangement, the growth is not even in all directions; spurs develop.

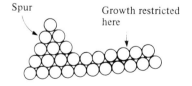

Figure 6.8 Dendritic growth

6.2 Plastic deformation of metals

When a material is stretched beyond its elastic limit, plastic deformation can occur. With plastic deformation a material does not return to its original dimensions when the applied stress is removed but is permanently deformed. An important group of forming processes used with metals involves shaping the metals by plastic deformation. These processes include rolling, drawing, pressing, spinning, forging and extrusion.

What happens to a metal when plastic deformation occurs? Metals are crystalline. This means that the metal atoms are arranged in a regular manner within the crystal. This regular array of atoms enables us to consider the atoms in a crystal as lying in a number of planes (Figure 6.9). The different planes have different densities of atoms and different spacings between parallel, like, planes. When plastic deformation occurs we can consider planes of atoms to be sliding past each other (Figure 6.10). The plane along which this movement occurs is called a *slip plane*. This slip of planes tends to occur between those parallel planes with the highest atomic density and the greatest separation (Figure 6.11).

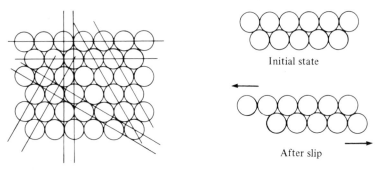

Figure 6.9 Some of the possible planes in a regularly packed array of atoms

Initial state

After slip

Figure 6.10 Slip of atom planes

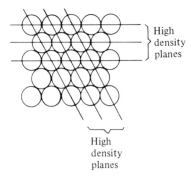

High density planes

High density planes

Figure 6.11

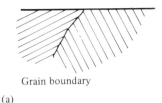

Grain boundary

(a)

(b)

Figure 6.12 (a) Before the application of stress. (b) After plastic deformation

Metals are composed of many crystals. A crystal within a metal is just a region of orderly packed atoms. Such a region is generally referred to as a *grain*. The surfaces that divide the different regions or orderly packed atoms are termed *grain boundaries*. When plastic deformation occurs in a metal, movement occurs along slip planes and the result is rather like Figure 6.12. Slip occurs only in those planes which are at suitable angles to the applied stress. The result is that the surface of the metal shows a series of steps due to the different movements of the various planes of atoms. These can be seen

Figure 6.13 Slip steps in polycrystalline aluminium. (Courtesy of the Open University, TS251/6, © 1973 Open University Press)

under a microscope (Figure 6.13). As will be seen from the figure, the slip lines do not cross over from one grain to another; the grain boundaries restrict the slip to within a grain. Thus the bigger the grains the more slippage that can occur; this would show itself as a greater plastic deformation. A fine grain structure should therefore have less slippage and so show less plastic deformation, i.e. be less ductile. A brittle material is thus one in which each little slip process is confined to a short run in the metal and not allowed to spread; a ductile material is one in which the slip process is not confined to a short run and does spread over a large part of the metal.

6.3 Cold working

Suppose you were to take a carbon steel test piece and perform a tensile test on it. You might, for instance, find that the material showed a yield stress of 430 N mm^{-2} (MPa). If the test is continued beyond this point but the stress released before the tensile strength is reached a permanent deformation of the test piece will be found to have occurred. Figure 6.14 illustrates this sequence of events. Suppose you now repeat the test. This time the yield stress is not 430 N mm^{-2} (MPa) but 550 N mm^{-2} (MPa). The material has a much higher yield stress (Figure 6.15). This phenomenon is called *cold working*.

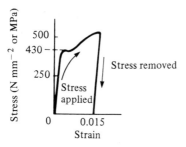

Figure 6.14

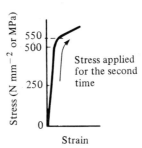

Figure 6.15

It is not only the yield stress that changes during cold working and it is not only carbon steel that shows such changes. The term 'cold working' is applied to any process that results in plastic deformation at a temperature which does not alter the structural changes produced by the working. Table 6.1 shows some of the changes that take place when a sheet of annealed aluminium is rolled and its thickness reduced (Figure 6.16). As the amount of plastic deformation is increased, so the tensile strength increases, the hardness increases and the elongation decreases. The material is becoming harder as a result of the cold working, which is sometimes known as *work hardening*. The more the material is worked, the harder it becomes.

Also, as the percentage elongation results in Table 6.1 indicate, the more a material is worked the more brittle it becomes. A stage can, however, be reached when the strength and hardness are at a maximum and the elongation is at a minium, so that further plastic deformation is not possible since the material is too brittle. Table 6.1 shows the rolled aluminium sheet reaching this condition at about a 60% reduction in thickness. The material is then said to be *fully work hardened*.

Table 6.1 *Effect of rolling on annealed aluminium sheet*

Reduction in sheet thickness (%)	Tensile strength ($N mm^{-2}$ or MPa)	Elongation (%)	Hardness (HV)
0	92	40	20
15	107	15	28
30	125	8	33
40	140	5	38
60	155	3	43

(Based on a table in John, V. B., *Introduction to Engineering Materials*, Macmillan)

Figure 6.17 shows how, for a number of materials, the hardness depends on the amount of cold working. The percentage reduction in thickness of a sheet or the percentage reduction in cross-sectional area of other forms of material is taken as a measure of the amount of working. As will be seen from the graph, as the amount of cold working increases the hardness increases until at some value of cold working a maxium hardness is reached. The material is then fully work hardened.

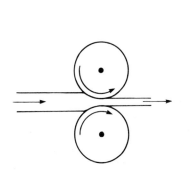

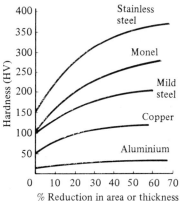

Figure 6.16 Sheet passing through rollers and being plastically deformed

Figure 6.17 The effect of cold working on hardness

The structure of cold worked metals

When stress is applied to a metal grain, deformation starts along the slip planes most suitably orientated. The effect of this is to cause the grains to become elongated and distorted. Figure 6.18 shows the results of heavy rolling of a tin bronze ingot (4% tin). The grains have become elongated into fibre-like structures, which has the effect of giving the material different mechanical properties in different directions, a greater strength along the grain than at right angles to the grain. This effect can be used to advantage by the designer.

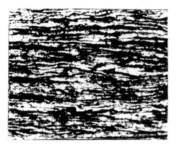

Figure 6.18 The grains in a heavily rolled tin bronze ingot. (From Rollason, E. C. *Metallurgy for Engineers*, Edward Arnold)

The effect of heat on cold-worked metals

Cold-worked metals generally have deformed grains and have often become rather brittle due to the working. In this process of deforming the grains, internal stresses build up.

When a cold-worked metal is heated to temperatures up to about 0.3 T_m,

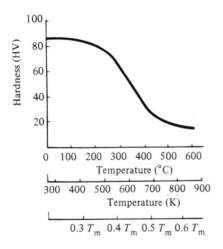

Figure 6.19 The effect of heat treatment on cold worked copper

where T_m is the melting point of the metal concerned on the Kelvin scale of temperature, then the internal stresses start to become relieved. This does not involve changes in grain structure but just some slight rearrangement of atoms in order that the stresses diminish. This process is known as *recovery*. Copper has a melting point of 1083°C, or 1356 K. Hence stress relief with copper requires heating to above about 407 K, i.e. 134°C.

If the heating is continued to a temperature of about 0.3 to 0.5 T_m there is a very large change in hardness. The strength and also the structure of the metal change. Table 6.2 shows how the hardness of copper changes, the copper having been subject to a 30% cold working. Figure 6.19 shows the same results graphically. Between 0.3 T_m and 0.5 T_m there is a very large change in hardness. The strength also decreases while the elongation increases. What is happening is that the metal is recrystallising.

Table 6.2 *Effect of temperature on the hardness of cold worked copper*

Temperature (°C)	(K)		Hardness (HV)
Initally			86
150	423	(0.3 T_m)	85
200	473		80
250	523	(0.4 T_m)	74
300	573		61
350	623	(0.5 T_m)	46
450	723		24
600	873	(0.6 T_m)	15

With *recrystallisation* crystals begin to grow from nuclei in the most heavily deformed parts of the metal. The temperature at which recrystallisation just starts is called the *recrystallisation temperature*. This is, for pure metals, about 0.3 to 0.5 T_m.

Material	Melting point (°C)	(K)	Recrystallisation temperature (°C)	(K)	
Aluminium	660	933	150	423	0.5 T_m
Copper	1083	1356	200	473	0.3 T_m
Iron	1535	1808	450	723	0.4 T_m
Nickel	1452	1725	620	893	0.5 T_m

As the temperature is increased from the recrystallisation temperature so the crystals grow until they have completely replaced the original distorted cold worked structure. Figure 6.20 illustrates this sequence and its relationship to the changes in physical properties.

The sequence of events that occur when a cold worked metal is heated can be broken down into three phases:

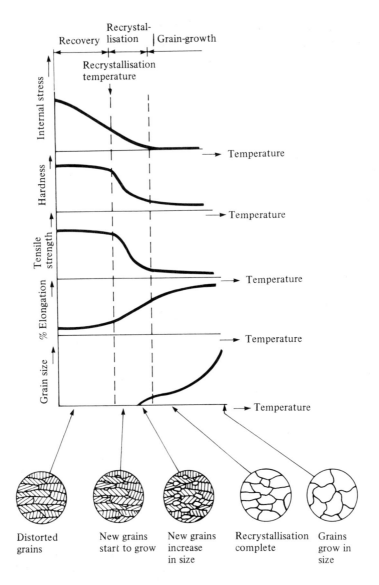

Figure 6.20 The effect of an increase in temperature on cold worked materials

(1) Recovery. The only significant change during this phase is the relief of internal stresses.

(2) Recrystallisation. The hardness, tensile strength and percentage elongation all change noticeably during this phase.

(3) Grain growth. The hardness, tensile strength and percentage elongation change little during this phase. The only change is that the grains grow and the materials become large-grained.

During the *grain-growth* phase the newly formed grains grow by absorbing other neighbouring grains. The amount of grain growth depends on the temperature and the time for which the material is at that temperature (Figure 6.21).

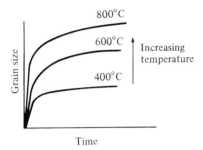

Figure 6.21 The effect of time and temperature on grain growth

The term *annealing* is used for the heating process used to change the properties of a material. Thus, in the case of aluminium that has been cold worked and has become too brittle to work further, heating to above the recrystallisation temperature of 150°C enables new grains to grow and the material to become more ductile. The aluminium can then be worked further. This sequence of events, cold working followed by annealing and then further cold working, is used in many manufacturing processes.

Factors affecting recrystallisation

(1) A minimum amount of deformation is necessary before recrystallisation can occur. The permanent deformation necessary depends on the metal concerned.

(2) The greater the amount of cold work, the lower the crystallisation temperature for a particular metal.

(3) Alloying increases the recrystallisation temperature.

(4) No recrystallisation takes place below the recrystallisation temperature. The higher the temperature above the recrystallisation temperature, the shorter the time needed at that temperature for a given crystal condition to be attained.

(5) The resulting grain size depends on the annealing temperature. The higher the temperature, the larger the grain size.

(6) The amount of cold work prior to the annealing affects the size of the grains. Figure 6.22 shows the effect on the grain size of annealing aluminium that had been subject to different amounts of cold work. The greater the amount of cold work, the smaller the resulting grain size. The greater the amount of cold work, the more centres are produced for crystal growth.

Figure 6.22 The effect on the grain size of annealing aluminium that has been subject to different amounts of cold work. Annealing involved, in all cases, 5 hours at 640°C. (From Rollason, E. C., *Metallurgy for Engineers*, Edward Arnold)

6.4 Hot working

Cold working involves plastically deforming materials below the recrystallisation temperature. The result is a harder, less ductile, material with deformed grains. *Hot working* involves deforming a material at a temperature greater than the recrystallisation temperature. As soon as a grain becomes deformed it recrystallises, no hardening occurs and the working can be continued without any difficulty. No interruption of working is needed to anneal the material, as is the case with cold working.

The grain structure of hot-worked material depends on the temperature at which the working occurs, the type of working process involved and the cooling rate of the material after the hot working. If the temperature is just above the recrystallisation temperature, a fine grain structure is produced; if it is well above this temperature then large grains will be produced.

A disadvantage of hot working is that oxidisation of the metal surfaces occurs; cold working does not have this problem. Another disadvantage is that the material will have comparatively low values of hardness and tensile strength, with high elongation. Hot and cold working processes are often both used in a particular shaping operation. The first operation involving large amounts of plastic deformation is carried out by hot working. After cleaning, the material is then cold worked to increase the strength and hardness and give a good surface finish.

6.5 Cast and wrought products

Casting involves the shaping of a product by the pouring of liquid metal into a mould. The grain structure within the product is determined by the rate of cooling. Thus the metal in contact with the mould cools faster than that in the centre of the casting. This gives rise to small crystals, termed *chill crystals*, near the surfaces. These are small because the metal has cooled too rapidly for the crystals to grow to any size. The cooling rate nearer the centre is, however, much less and so some chill crystals can develop in an inward direction. This results in large elongated crystals perpendicular the mould walls, these being called *columnar crystals*. In the centre of the mould the cooling rate is the lowest. While growth of the columnar crystals is taking place, small crystals are growing in this central region. These grow in the liquid metal which is constantly on the move due to convection currents. The final result is a central region of medium-sized, almost spherical, crystals called *equiaxed crystals*. Figure 6.23 shows all these types of crystals in a casting section.

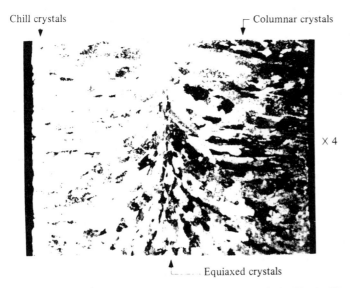

Chill crystals

Columnar crystals

X 4

Equiaxed crystals

Figure 6.23 Grain structure in an aluminium casting. (From Monks, H. A. and Rochester, D. C., *Technician Structure and Properties of Metals 2*, Cassell & Co.)

In general, a casting structure having entirely small equiaxed crystals is preferred. This type of structure can be promoted by a more rapid rate of cooling for the casting. Castings in which the mould is made of sand tend to have a slow cooling rate as sand has a low thermal conductivity. Thus sand castings tend to have large columnar grains and hence relatively low strength. Die casting involving metal moulds has a much faster rate of cooling and so gives castings having a bigger zone of equiaxed crystals. As these are smaller than columnar crystals, the casting has better properties. The following values show the types of differences that can occur with aluminium casting alloys:

Material	Tensile strength (N mm^{-2} or MPa)		Percentage elongation	
	Sand cast	Die cast	Sand cast	Die cast
5% Si, 3% Cu	140	150	2	2
12% Si	160	185	5	7

Castings do not show directionality of properties, the properties being the same in all directions. They do, however, have the problems, produced by working from a liquid metal, of blowholes and other voids occurring during solidification.

Manipulative processes involve the shaping of a component by plastic deformation. The products given by such methods are said to be *wrought*. Hot working processes involve rolling, forging or extruding at temperatures in excess of the recrystallisation temperature. Cold working processes such as rolling, drawing, pressing, spinning and impact extrusion involve temperatures less than the recrystallisation temperature.

Rolling is a continuous process in which the material is passed between a pair of rotating rollers and emerges with a reduced thickness (as in Figure 6.16). Hot rolling is often at a temperature of about $0.6\ T_m$. At this temperature work hardening does not occur. The product is a relatively soft material with low tensile strength, the surfaces of the material are also oxidised. A ductile material can be cold rolled. Such a process work hardens the material and can lead to a useful gain in tensile strength and hardness. A good surface finish can be produced. Rolling does, however, give a directionality to the properties of the finished material due to the deformation of the grains in the direction of the rolling. The following values were obtained for rolled brass strip (70% copper, 30% zinc):

Angle to rolling direction	Tensile strength (N mm^{-2} or MPa)	Percentage elongation
0°	740	3
45°	770	3
90°	850	2

Forging involves squeezing a ductile material between a pair of dies. The term 'forging' is generally applied only to the hot working process, though cold forging can be carried out with some of the very ductile non-ferrous materials. The flow of the material during the squeezing operation does give a directionality to the properties of the material.

Extrusion is rather similar to the squeezing of toothpaste out of the tube. The form of the ejected toothpaste is determined by the nozzle through which it is ejected. Extrusion involves forcing a metal through a die. Hot extrusion involves temperatures of the order of $0.65\ T_m$ to $0.9\ T_m$. The result of hot extrusion is a product with comparatively low tensile strength and soft. Cold extrusion gives a cold-worked product with higher strength and hardness. The process is not, however, possible where the cross-sectional area of the material is considerably reduced as a result of too much work hardening. The process of extrusion does give a product with directionality of properties.

In general, wrought products have a directionality of properties due to the process used giving rise to grain deformation in some particular direction.

6.6 Effect of grain size on properties

Grain size is an important factor in determining the mechanical properties of a material. It is not so much the size of the grain that is important as the length of grain boundary. Fine-grain material contains a greater length of grain boundaries than a coarse-grained material. Grain boundaries restrict the amount of slip that can occur, in that slip within one grain is not easily transmitted across a grain boundary to cause slip in a neighbouring grain. Thus the more grain boundaries there are the more slip is restricted. Coarse-grained materials thus have lower yield stresses than fine-grained materials. This also means that a coarse-grained material will have a lower tensile strength than a fine-grained material. Hardness is related to tensile strength and thus coarse-grained material will have a lower hardness than fine-grained material. Ductile materials are those which have a large plastic region in their stress–strain relationship (see page 127), thus we would expect the large-grain materials to be the more ductile as plastic deformation is easier.

The mechanical properties of a metal can be changed by changing the size or the shape of the grains. Cold-working distorts the shape of the grains and so increases tensile strength and hardness, while decreasing ductility. Annealing of a cold-worked material can increase ductility, while decreasing tensile strength and hardness, by reforming the grains without the distortion. It also enables the grain size to be controlled.

Problems

1 Distinguish between cold-working and hot-working processes.
2 Describe the effect on the mechanical properties of a metal of cold working.
3 Explain what is meant by the recrystallisation temperature.
4 What is meant by 'work hardening'?
5 What are the properties of a fully work hardened material in comparison with its properties before any working occurs?
6 How does cold working change the structure of a metal?
7 Describe how the mechanical properties of a cold-worked material change as its temperature is raised from room temperature to about $0.6\,T_m$, where T_m is the melting point temperature in degrees Kelvin.

8 What factors affect the recrystallisation temperature of a metal?

9 How is the recrystallisation temperature of a pure metal related to its melting point?

10 Zinc has a melting point of 419°C. Estimate the recrystallisation temperature for zinc.

11 Magnesium has a melting point of 651°C. What order of temperatures would be required to (a) stress relieve, (b) anneal a cold-worked piece of magnesium?

12 What is the effect on the grain size of a metal after annealing of the amount of cold work the material had originally been subject to?

13 How does the temperature at which hot working is carried out determine the grain size and so the mechanical properties?

14 Describe the grain structure of a typical casting.

15 Why are the properties of a material dependent on whether it is sand-cast or die-cast?

16 What is meant by directionality in wrought products?

17 Why are the mechanical properties of a rolled metal different in the direction of rolling from those at right-angles to this direction?

18 How does a cold-rolled product differ from a hot-rolled product?

19 Brasses have recrystallisation temperatures of the order of 400°C. Roughly what temperature should be used for hot extrusion of brass?

20 Describe how grain size and shape affect the mechanical properties of a metal.

21 A brass, 65% copper and 35% zinc, has a recrystallisation temperature of 300°C after having been cold-worked so that the cross-sectional area has been reduced by 40%.
  (a) How will further cold working change the structure and the properties of the brass?
  (b) To what temperature should the brass be heated to give stress relief?
  (c) To what temperature should the brass be heated to anneal it and give a relatively small grain size?
  (d) How would the grain size, and the mechanical properties, change if the annealing temperature used for (c) was exceeded by 100°C?

22 Use Figure 6.17 for this question.
  (a) What is the maximum hardness possible with cold-rolled copper?
  (b) Copper plate, already cold worked 10% is further cold worked 20%. By approximately how much will the hardness change?
  (c) Mild steel is to be rolled to give thin sheet. This involves a 70% reduction in sheet thickness. What treatment would be suitable to give this reduction and a final product which was no harder than 150 HV?

23 As a fine-grained product is generally desirable after some process, it is common to lower the temperature of a hot-working process to a working temperature just above the recrystallisation temperature for the final part of the process. Why should this lead to fine grains?

7

Structure of alloys

7.1 Alloys

An *alloy* is a metallic material consisting of an intimate association of two or more elements: brass is an alloy composed of copper and zinc; bronze is an alloy of copper and tin. The everyday metallic objects around you will be made almost invariably from alloys rather than the pure metals themselves. Pure metals do not always have the appropriate combination of properties needed; alloys can, however, be designed to have them.

The coins in your pocket are made of alloys. Coins need to be made of a relatively hard material which does not wear away rapidly, i.e. the coins have to have a 'life' of many years. Coins made of, say, pure copper would be very soft; not only would they suffer considerable wear but they would bend in your pocket.

Coins (British)	Percentage by mass			
	Copper	Tin	Zinc	Nickel
1p, 2p	97	0.5	2.5	—
5p, 10p, 50p	75	—	—	25

If you put sand in water, the sand does not react with the water but retains its identity, as does the water. The sand in water is said to be a *mixture*. In a *mixture*, each component retains its own physical structure and properties. Sodium is a very reactive substance, which has to be stored under oil to stop it interacting with the oxygen in the air, and chlorine is a poisonous gas. Yet when these two substances interact, the product, sodium chloride, is eaten by you and me every day. The product is common salt. Sodium chloride is a *compound*. In a *compound* the components have interacted and the product has none of the properties of its constituents. Alloys are generally mixtures though some of the components in the mixture may interact to give compounds as well.

7.2 Solutions

If you drop a pinch of common salt, sodium chloride, into cold water it will dissolve. The sodium chloride is said to be *soluble* in the cold water. Up to 36 g of sodium chloride can be dissolved in 100 g of cold water; more than that amount will not dissolve. With the 36 g dissolved in 100 g the resulting solution is said to be *saturated*. The *solubility* of sodium chloride in cold water is said to be 36 g per 100 g of water. If 40 g of sodium chloride is put into 100 g of cold water only 36 g of it will dissolve, the remaining 4 g remaining as solid. The solubility of sodium chloride in water depends on the temperature, hot water dissolving more sodium chloride than cold water. The solubility of sodium chloride in water does not, however, vary to a considerable extent with temperature, only slightly increasing as the temperature increases (Figure 7.1).

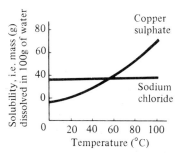

Figure 7.1 The solubility variation with temperature for sodium chloride (common salt) and copper sulphate. The solubility of copper sulphate increases considerably more with temperature than does that of sodium chloride

The solubility of copper sulphate in water does, however, increase quite significantly with temperature, as Figure 7.1 shows. Figure 7.2 shows this solubility variation with temperature plotted in a different way, the temperature axis being vertical rather than horizontal. The reason for this will become apparent later in this chapter when phase diagrams are considered. Suppose we have 40 g of copper sulphate dissolved in 100 g of hot water, say at 90°C. At this temperature the solution will not be saturated. If now the temperature decreases, then at a temperature of just under 60°C the solution will become saturated. Further cooling will result in the excess copper sulphate coming out of solution as a precipitate. At 0°C there will only be about 17 g still in solution.

Some substances, e.g. sand, are not soluble in water. When sand is mixed with water the result is always that none of the sand enters into a solution. The sand is said to be *insoluble* in water.

The terms 'soluble' and 'insoluble' can also be used when we mix two liquids. Oil and water are insoluble in each other; when the two are mixed the result is just a mixture with the oil and the water retaining their separate

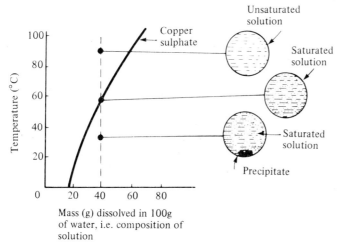

Figure 7.2 An alternative way of plotting in the data given in Figure 7.1 for the solubility of copper sulphate in water

identities. When alcohol is mixed with water, the result is a solution; the water and the alcohol are said to be soluble in each other. It is not possible to identify the water and the alcohol as separate entities in the solution. When two liquid metals are mixed, say copper and nickel, a solution is produced in that it is not possible to identify in the liquid either the copper or the nickel. The copper and the nickel are said to be soluble in each other in the liquid state.

When a salt solution is cooled to such an extent that it solidifies, the resulting solid is a mixture of sodium chloride crystals and ice crystals. In the solid state the salt and the water retain their separate identities; in the solid state they are insoluble in each other. Cadmium and bismuth are soluble in each other in the liquid state but insoluble in each other in the solid state. Copper and nickel, however, are soluble in each other in the solid state. The resulting copper-nickel alloy has, in the solid state, a structure in which it is impossible to distinguish the copper from the nickel. The copper and nickel are thus said to form a *solid solution*.

Solid solutions

Copper forms face-centred cubic crystals (Figure 7.3a), as also does nickel (Figure 7.3b). When copper and nickel are in solid solutions a single face-centred lattice is formed (Figure 7.3c). Such a solid solution is said to be *substitutional* in that when, say, nickel is added to the copper, the nickel atoms substitute for the copper atoms in the copper face-centred lattice. This substitution may be ordered, the atoms of the added metal always taking up the same fixed places in the lattice, or it can be disordered. In the disordered solid

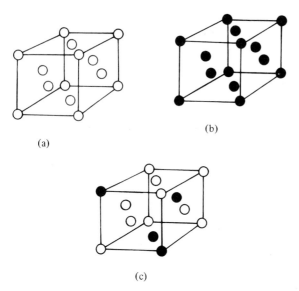

Figure 7.3 (a) The copper lattice, (b) The nickel lattice, (c) The copper-nickel solid solution lattice

solution the added atoms appear virtually at random throughout the lattice. Most solid solutions are disordered lattices.

In a copper–nickel solid solution the copper and nickel atoms are virtually the same size. This is necessary for such a solid solution. Another form of solid solution can, however, occur when the sizes of the two atoms are considerably different. With an *interstitial* solid solution the added atoms are small enough to fit in between the atoms in the lattice. Carbon can form an interstitial solid solution with the face-centred cubic form of iron. Figure 7.4 shows the lattice of the solid solution.

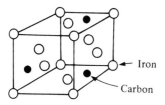

Figure 7.4 An interstitial solid solution. The iron atoms form the face-centred cubic lattice with the carbon atoms fitting in between the iron atoms

Alloy types

When an alloy is in a liquid state the atoms of the constituents are distributed at random through the liquid. When solidification occurs a number of possibilities exist.

(1) The two components separate out with each in the solid state maintaining its own separate identity and structure. The two components are said to be *insoluble* in each other in the solid state.

(2) The two components remain completely mixed in the solid state. The two components are said to be *soluble* in each other in the solid state, the components forming a solid solution.

(3) On solidifying, the two components may show *limited solubility* in each other.

(4) In solidifying, the elements may combine to form *intermetallic compounds*.

7.3 Solidification

When pure water is cooled to 0°C it changes state from liquid to solid, i.e. ice is formed. Figure 7.5 shows the type of graph that is produced if the temperature of the water is plotted against time during a temperature change from above 0°C to one below 0°C. Down to 0°C the water only exists in the liquid state. At 0°C solidification starts to occur and while solidification is occurring the temperature remains constant. Energy is still being extracted from the water but there is no change in temperature during this change of state. This energy is called *latent heat*. The *specific latent heat of fusion* is defined as the energy taken from, or given to, 1 kg of a substance when it changes from liquid to solid, or solid to liquid, without any change in temperature occurring.

All pure substances show the same type of behaviour as the water when they change state. Figure 7.6 shows the cooling graph for copper when it changes state from liquid to solid.

During the transition of a pure substance from liquid to solid, or vice versa, the liquid and solid are both in existence. Thus for the water, while the latent heat is being extracted there is both liquid and ice present. Only when all the latent heat has been extracted is there only ice. Similarly with the copper,

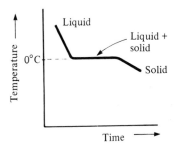

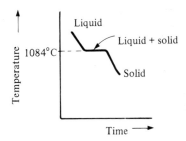

Figure 7.5 Cooling curve for water during solidification

Figure 7.6 Cooling curve for copper during solidification

during the transition from liquid to solid at 1084°C, while the latent heat is being extracted both liquid and solid exist together.

The cooling curves for an alloy do not show a constant temperature occurring during the change of state. Figure 7.7 shows cooling curves for two copper–nickel alloys. With an alloy, the temperature is not constant during solidification. The temperature range over which this solidification occurs depends on the relative proportions of the elements in the alloy. If the cooling curves are obtained for the entire range of copper–nickel alloys a composite diagram can be produced which shows the effect the relative proportions of the constituents have upon the temperatures at which solidification starts and that at which it is complete. Figure 7.8 shows such a diagram for copper–nickel alloys.

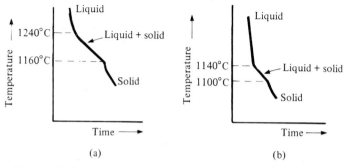

(a)   (b)

Figure 7.7 Cooling curves for copper–nickel alloys, (a) 70% copper – 30% nickel, (b) 90% copper – 10% nickel

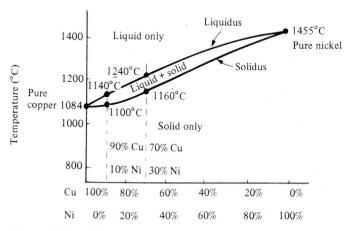

Figure 7.8 Equilibrium diagram for copper–nickel alloys

Thus for pure copper there is a single temperature point of 1084°C, indicating that the transition between liquid and solid takes place at a constant temperature. For 90% copper – 10% nickel the transition between liquid and solid starts at 1140°C and terminates at 1100°C when all the alloy is solid. For 70% copper – 30% nickel the transition between liquid and solid starts at 1240°C and terminates at 1160°C when all the alloy is solid.

The line drawn through the points at which each alloy in the group of alloys ceases to be in the liquid state and starts to solidify is called the *liquidus line*. The line drawn through the points at which each alloy in the group of alloys becomes completely solid is called the *solidus line*. These liquidus and solidus lines indicate the behaviour of each of the alloys in the group during solidification. The diagram in which these lines are shown is called the *thermal equilibrium diagram*.

The thermal equilibrium diagram is constructed from the results of a large number of experiments in which the cooling curves are determined for the whole range of alloys in the group. The diagram provides a forecast of the states that will be present when an alloy of a specific composition is heated or cooled to a specific temperature. The diagrams are obtained from cooling curves produced by very slow cooling of the alloys concerned. They are slow because time is required for equilibrium conditions to obtain at any particular temperature, hence the term thermal equilibrium diagram.

Phase

A *phase* is defined as a region in a material which has the same chemical composition and structure throughout. A piece of pure copper which throughout is the face-centred cubic structure, has but a single phase at that temperature. Molten copper does, however, represent a different phase in that the arrangement of the atoms in the liquid copper is different from that in the solid copper. A completely homogeneous substance at a particular temperature has only one phase at that temperature. If you take any piece of that homogeneous substance it will show the same composition and structure.

Common salt, in limited quantities, can be dissolved completely in water at given temperature; the salt is said to be soluble in the water. The solution is completely homogeneous throughout, so there is thus just one phase present.

Liquid copper and liquid nickel are completely miscible, as are most liquid metals. The copper–nickel solution is completely homogeneous and thus at the temperature at which the two are liquid there is but one phase present. When the liquid alloy is cooled it solidifies. In the solid state the two metals are completely soluble in each other and so the solid state for this alloy has but one phase. In the case of the 70% copper–30% nickel alloy (Figure 7.7a), the liquid phase exists above 1240°C; between 1240°C and 1160°C there are two phases when both liquid and solid are present. Below 1160°C the copper–nickel alloy exists in just one phase as the two metals are completely soluble in each other and give a solid solution. The thermal equilibrium diagram given in Figure 7.8 for the range of copper–nickel alloys is thus a diagram showing the phase or phases present at any particular temperature of any composition of copper–nickel alloy.

7.4 Equilibrium diagrams and solubility

The equilibrium diagram for the copper–nickel alloy is typical of that given when two components are soluble in each other, both in the liquid and solid states. Figure 7.9 shows the general form of such an equilibrium diagram.

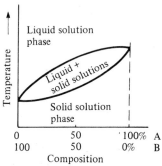

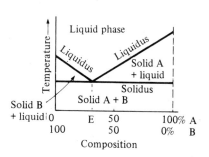

Figure 7.9 Equilibrium diagram for two metals that are completely soluble in each other in the liquid and solid states

Figure 7.10 Equilibrium diagram for two metals that are completely soluble in each other in the liquid state and completely insoluble in each other in the solid state

Figure 7.10 shows the type of equilibrium diagrams produced when the two alloy components are completely soluble in each other in the liquid state but completely insoluble in each other in the solid state. The solid alloy shows a mixture of crystals of the two metals concerned. Each of the two metals in the solid alloy retains its independent identity. At one particular composition, called the *eutectic composition* (marked as E in Figure 7.10), the temperature at which solidification starts to occur is a minimum. At this temperature, called

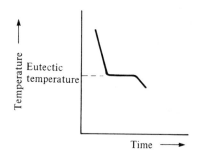

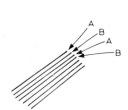

Figure 7.11 Cooling curve for the eutectic composition

Figure 7.12 The laminar structure of the eutectic

the *euctectic temperature*, the liquid changes to the solid state without any change in temperature (Figure 7.11). The solidification at the eutectic temperature, for the eutectic composition, has both the metals simultaneously coming out of the liquid. Both metals crystallise together. The resulting structure, known as the *eutectic structure*, is generally a laminar structure with layers of metal A alternating with layers of metal B (Figure 7.12).

The properties of the eutectic can be summarised as:

(1) Solidification takes place at a single fixed temperature.
(2) The solidification takes place at the lowest temperature in that group of alloys.
(3) The composition of the eutectic composition is a constant for that group of alloys.
(4) It is a mixture, for an alloy made up from just two metals, of the two phases.
(5) The solidified eutectic structure is generally a laminar structure.

Figure 7.13 illustrates the sequence of events that occur when the 80% A–20% B liquid alloy is cooled. In the liquid state both metals are completely soluble in each other and the liquid alloy is thus completely homogeneous. When the liquid alloy is cooled to the liquidus temperature, crystals of metal A start to grow. This means that a metal A is being withdrawn from the liquid, the composition of the liquid must change to a lower concentration of A and a higher concentration of B. As the cooling proceeds and the crystals of A continue to grow so the liquid further decreases in concentration of A and increases in concentration of B. This continues until the concentrations in the liquid reach that of the eutectic composition. When this happens solidification of the liquid gives the eutectic structure. The resulting alloy has therefore crystals of A embedded in a structure having the composition and structure of the eutectic. Figure 7.14 shows the cooling curve for this sequence of events.

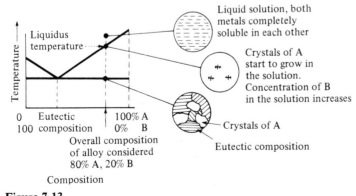

Figure 7.13

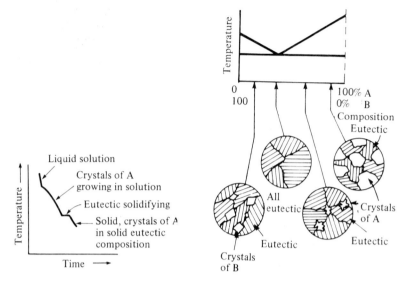

Figure 7.14 Cooling curve for an 80% A–2% B alloy   **Figure 7.15**

Apart from an alloy having the eutectic composition and structure when the alloy is entirely of eutectic composition, all the other alloy compositions in the alloy group shows crystals of either metal A or B embedded in eutectic structure material (Figure 7.15). Thus for two metals that are completely insoluble in each other in the solid state:

(1) The structure prior to the eutectic composition is of crystals of B in material of eutectic composition and structure.

(2) At the eutectic structure the material is entirely eutectic in composition and structure.

(3) The structure after the eutectic composition is of crystals of A in eutectic composition and structure material.

Many metals are neither completely soluble in each other in the solid state nor completely insoluble; each of the metals is soluble in the other to some limited extent. Lead–tin alloys are of this type. Figure 7.16 shows the equilibrium diagram for lead–tin alloys. The solidus line is that line, started at 0% tin – 100% lead, between the (liquid + α) and the α areas, between the (liquid + α) and the (α + β) areas, between the (liquid + β) and the (α + β) areas, and between the (liquid + β) and the β areas. The α, the β, and the (α + β) areas all represent solid forms of the alloy. The transition across the line between α and (α + β) is thus a transition from one solid form to another solid form. Such a line is called the *solvus*. Figure 7.17 shows the early part of Figure 7.16 and the liquidus, solidus and solvus lines.

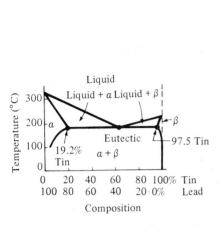

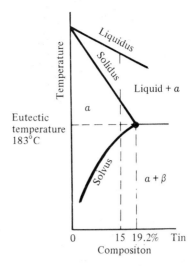

Figure 7.16 Equilibrium diagram for lead–tin alloys

Figure 7.17

Consider an alloy with the composition 15% tin–85% lead (Figure 7.17). When this cools from the liquid state, where both metals are soluble in each other, to a temperature below the liquidus then crystals of the α phase start to grow. The α phase is a solid solution. Solidification becomes complete when the temperature has fallen to that of the solidus. At that point the solid consists entirely of crystals of the α phase. This solid solution consists of 15% tin completely soluble in the 85% lead at the temperature concerned. Further cooling results in no further change in the crystalline structure until the temperature has fallen to that of the solvus. At this temperature the solid solution is saturated with tin. Cooling below this temperature results in tin coming out of solution in another solid solution β. The more the alloy is cooled the greater the amount of tin that comes out of solution, until at room temperature most of the tin has come out of the solid solution. The result is largely solid solution crystals, the α phase, having a low concentration of tin in lead, mixed with small solid solution crystals, the β phase, having a high concentration of tin in lead.

At the eutectic temperature the maximum amount of tin that can be dissolved in lead in the solid state is 19.2% (see Figure 7.17) Similarly the maximum amount of lead that can be dissolved in tin, at the eutectic temperature, is 2.5%.

The eutectic composition is 61.9% tin–38.1% lead. When an alloy with this composition is cooled to the eutectic temperature the behaviour is the same as when cooling to the eutectic occurred for the two metals insoluble in each other in solid state (Figure 7.13) except that, instead of pure metals separating out to give a laminar mixture of the metal crystals, there is a laminar mixture of crystals of the two solid solutions α and β. The α phase has the composition of 19.2% tin–80.8% lead, the β phase has the composition 97.5% tin–2.5% lead. Cooling below the eutectic temperature results in the α solid solution giving up tin, due to the decreasing solubility of the tin in the lead, and the β solid solution giving up lead, due to the decreasing solubility of the lead in the tin. The result at room temperature is a structure having a mixture of alpha and beta solid solution, the alpha solid solution having a high concentration of lead and the beta a high concentration of tin.

For alloys having a composition with between 19.2% and 61.9% tin, cooling from the liquid results in crystals of the α phase separating out when the temperature falls below that of the liquidus. When the temperature reaches that of the solidus, solidification is complete and the structure is that of crystals of the α solid solution in eutectic structure material. Further cooling results in α solid solution losing tin. The eutectic mixture has the α part of it losing tin and the β part losing lead. The result at room temperature is a structure having the α solid solution crystals with a high concentration of lead and very little tin and some β precipitate, and the eutectic structure a mixture of α with high lead concentration and β with high tin concentration.

For alloys having a composition with between 61.9% and 97.5% tin, cooling from the liquid results in crystals of the β phase separating out when the temperature falls below that of the liquidus. Otherwise the events are the same as those occurring for compositions between 19.2% and 61.9% tin. The result at room temperature is a structure having the β solid solution crystals with a high concentration of tin and very little lead and some α precipitate, and the eutectic structure a mixture of α with high lead concentration and β with high tin concentration.

For alloys having a composition with more than 97.5% tin present, crystals of the β phase begin to grow when the temperature falls below that of the liquidus. When the temperature falls to that of the solidus, solidification becomes complete and the solid consists entirely of β solid solution crystals. Further cooling results in no further change in the structure until the temperature reaches that of the solvus. At this temperature the solid solution is saturated with lead and cooling below this temperature results in the lead coming out of solution. The result at room temperature, when most of the lead has come out of the solid solution, is β phase crystals having a high concentration of tin mixed with α phase crystals with a high concentration of lead.

Figure 7.18 shows the types of structure that might be expected at room temperature for lead–tin alloys of different compositions.

For many alloys, the phase diagrams are more complex than those already considered in this chapter. The complexity occurs because of the formation of further phases. These can be due to the formation of intermetallic compounds. Figure 7.19 shows the magnesium–tin thermal equilibrium diagram. This can

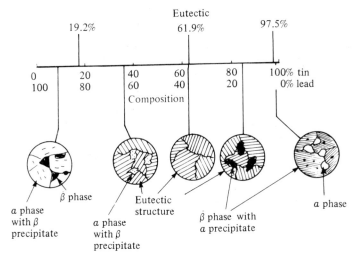

Figure 7.18 Lead–tin alloys

be considered to be essentially two equilibrium diagrams stuck together, the dividing line between the two occurring at 29.1% magnesium–70.9% tin. This is the composition of the intermetallic compound between magnesium and tin (Mg_2Sn).

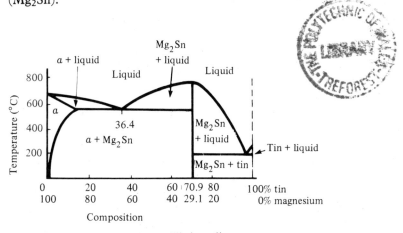

Figure 7.19 Magnesium–tin equilibrium diagram

Because it is possible to consider each part of the magnesium–tin thermal equilibrium diagram separately, the diagram up to 70.9% tin can be considered as being that due to two materials partially soluble in each other in the solid state, the substances between magnesium and the intermetallic compound. Between 70.9% and 100% tin, the equilibrium diagram is like that of two materials insoluble in each other in the solid state, the materials being tin and the intermetallic compound. Thus an alloy of composition 20% tin–80% magnesium will, at room temperature, have α phase and eutectic structure, this being a mixture of α phase and intermetallic compound. An alloy of 36.4% tin–63.6% magnesium will have purely the eutectic structure, a mixture of α phase and intermetallic compound. An alloy of 80% tin–20% magnesium will consist of intermetallic compound crystals in a eutectic composed of a mixture of intermetallic compound and tin.

Other forms of thermal equilibrium diagrams are produced when, during the cooling process from the liquid, a reaction occurs between the solid that is first produced and the liquid in which it is forming. The reaction is called a *peritectic reaction*. Figure 7.20 shows part of the thermal equilibrium diagram for copper–zinc alloys, peritectic reactions occurring during the solidification of such alloys.

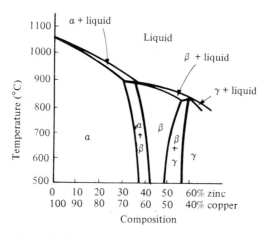

Figure 7.20 Part of the copper–zinc thermal equilibrium diagram

7.5 Non-equilibrium conditions

The term 'equilibrium' has been used in connection with the thermal equilibrium diagrams used to predict the structure of alloys. This has been considered to mean a very slow cooling of the alloy from the liquid state. Why should the cooling be slow?

Consider the solidification of a copper–nickel alloy, e.g. 70% copper–30% nickel. Figure 7.21 shows the relevant part of the thermal equilibrium diagram. When the liquid copper–nickel alloy cools to the liquidus temperature, small dendrites of copper-nickel solid solution form. Each dendrite will have the composition of 53% copper–47% nickel. This is the composition of the solid that can be in equilibrium with the liquid at the temperature concerned, the composition being obtained from the thermal equilibrium diagram by drawing a constant temperature line at this temperature and finding the intersection of the line with the solidus. As the overall composition of the liquid plus solid is 70% copper–30% nickel, the formation of dendrites, having a greater percentage of nickel, means that the remaining liquid must have a lower concentration of nickel than 30%.

As the alloy cools further, so the dendrites grow. At 1200°C the expected composition of the solid material would be 62% copper–38% nickel, the liquid having the composition 78% copper–32% nickel. So the percentage of copper has increased from the 53% at the liquidus temperature, while the percentage of nickel has decreased. If the dendrite is to have a constant composition, movement of atoms within the solid will have to occur. The term *diffusion* is used for the migration of atoms. The nickel atoms will have to move outwards from the initial dendrite core and copper atoms will have to move inwards to the core; Figure 7.22 illustrates this process. This diffusion takes time, in fact the process of diffusion in a solid is very slow.

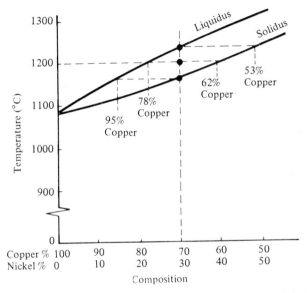

Figure 7.21 Thermal equilibrium diagram for copper-rich alloys of copper–nickel

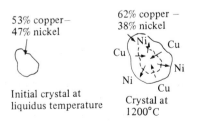

Figure 7.22 Diffusion during crystal growth

As the alloy cools further so the expected composition of the solid material changes until at the solidus temperature the composition becomes 70% copper–30% nickel, the last drop of the liquid having the composition 95% copper–5% nickel. For the solid to have this uniform composition there must have been a diffusion of copper atoms inwards to the core of the dendrite and a corresponding movement of nickel atoms outwards. For this to happen, the entire process of cooling from the liquid must take place very slowly. This is what is meant by equilibrium conditions.

In the normal cooling of an alloy, in perhaps the production of a casting, the time taken for the transition from liquid to solid is relatively short and inadequate for sufficient diffusion to have occurred for constant composition solid to be achieved.

The outcome is that the earlier parts of the crystal growth have a higher percentage of nickel than the later growth parts. The earlier growth parts have, however, a lower percentage of copper than the later growth parts (Figure 7.23). This effect is called *coring*; the more rapid the cooling from the liquid, the more pronounced the coring, i.e. the greater the difference in composition between the earlier and later growth parts of crystals. Figure 7.24 shows the cored structure of the 70% copper–30% nickel alloy. The photograph shows the etched surface of the alloy; as the amount of etching that takes place is determined by the composition of the metal, the earlier and later parts of the dendritic growth are etched to different degrees and so show on the photograph as different degrees of light and dark. The effect of this is to show clearly the earlier parts of the dendritic growth within a crystal grain.

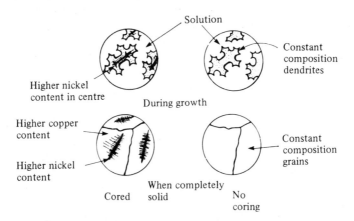

Figure 7.23 Coring with a 70% copper–30% nickel alloy

Coring can be eliminated after an alloy has solidified by heating it to a temperature just below that of the solidus and then holding it at that temperature for a sufficient time to allow diffusion to occur and a uniform composition to be achieved.

Magnification × 50

Figure 7.24 The cored structure in a 70% copper–30% nickel alloy. (Courtesy of the Open University, TS251/5, © 1973 Open University Press)

7.6 Precipitation

If a solution of sodium chloride in water is cooled sufficiently, sodium chloride precipitates out of the solution. This occurs because the solubility of sodium chloride in water decreases as the temperature decreases. Thus very hot water may contain 37 g per 100 g of water. Cold water is saturated with about 36 g. When the hot solution cools down the surplus salt is precipitated out of the solution. Similar events can occur with solid solutions.

Figure 7.25 shows part of the copper–silver thermal equilibrium diagram. When the 5% copper–95% silver alloy is cooled from the liquid state to 800°C a solid solution is produced. At this temperature the solid solution is not saturated but cooling to the solvus temperature makes the solid solution saturated. If the cooling is continued, slowly, precipitation occurs. The result at room temperature is a solid solution containing a coarse precipitate.

The above discussion assumes that the cooling occurs very slowly. The formation of a precipitate requires the grouping together of atoms. This requires atoms to diffuse through the solid solution. Diffusion is a slow process; if the solid solution is cooled rapidly from 800°C, i.e. quenched, the precipitation may not occur. The solution becomes *supersaturated*, i.e. it contains more of the α phase than the equilibrium diagram predicts. The result of this rapid cooling is a solid solution, the α phase, at room temperature.

The supersaturated solid solution may be retained in this form at room temperature, but the situation is not very stable and a very fine precipitation may occur with the elapse of time. This precipitation may be increased if the solid is heated for some time (the temperature being significantly below the solvus temperature). The precipitate tends to be very minute particles dispersed throughout the solid. Such a fine dispersion gives a much stronger and harder alloy than when the alloy is cooled slowly from the α solid solution. This hardening process is called *precipitation hardening*. The term *natural ageing* is sometimes used for the hardening process that occurs due to precipitation at room temperature and the term *artificial ageing* when the precipitation occurs as a result of heating.

Figure 7.26 shows part of the thermal equilibrium diagram for aluminium-copper alloys. If the alloy with 4% copper is heated to about 500°C and held at that temperature for a while, diffusion will occur and a homogeneous α solid solution will form. If the alloy is then quenched to about room temperature, supersaturation occurs. This quenched alloy is relatively soft. If now the alloy is heated to a temperature of about 165°C and held at this temperature for about ten hours, a fine precipitate is formed. Figure 7.27 shows the effects on the alloy structure and properties of these processes. The effect is to give an alloy with a higher tensile strength and harder.

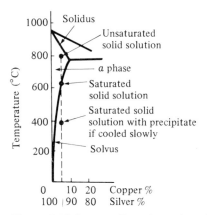

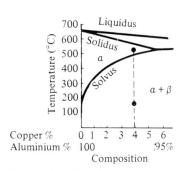

Figure 7.25 Copper–silver thermal equilibrium diagram

Figure 7.26 Aluminium–copper thermal equilibrium diagram

Not all alloys can be treated in this way. Precipitation hardening can only occur, in a two-metal alloy, if one of the alloying elements has a high solubility at high temperatures and a low solubility at low temperatures, i.e. the solubility decreases as the temperature decreases. This means that the solvus line must slope as shown in Figure 7.28. Also the structure of the alloy at temperatures above the solvus line must be a single-phase solid solution. The alloy systems that have some alloy compositions that can be treated in this way are mainly non-ferrous, e.g. copper–aluminium and magnesium–aluminium.

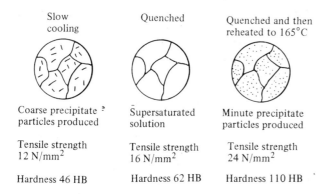

Slow cooling | Quenched | Quenched and then reheated to 165°C

Coarse precipitate ? particles produced

Supersaturated solution

Minute precipitate particles produced

Tensile strength 12 N/mm²

Tensile strength 16 N/mm²

Tensile strength 24 N/mm²

Hardness 46 HB

Hardness 62 HB

Hardness 110 HB

Figure 7.27 The effect of heat treatment for a 96% aluminium–4% copper alloy

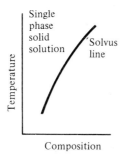

Figure 7.28 The required form of solvus line for precipitation hardening

Problems

1 What is the difference between a mixture of two substances and a compound of the two?
2 Explain what is meant by the terms soluble, insoluble, liquid solution, solid solution.
3 Sketch the form of cooling curve you would expect when a sample of pure iron is cooled from the liquid to solid state.
4 Figure 7.29 shows the cooling curves for copper–nickel alloys. Use these to plot the copper–nickel thermal equilibrium diagram.
5 Use either Figure 7.8 or your answer to Problem 4 to determine the liquidus and solidus temperatures for a 50% copper–50% nickel alloy.
6 Germanium and silicon are completely soluble in each other in both the liquid and solid states. Plot the thermal equilibrium diagram for germanium–silicon alloys from the following data.

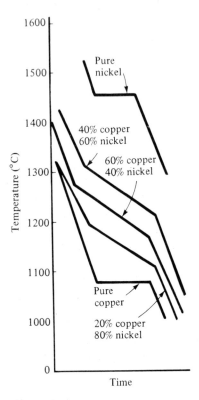

Figure 7.29

Alloy		Liquidus temperature	Solidus temperature
Germanium %	Silicon %	(°C)	(°C)
100	0	958	958
80	20	1115	990
60	40	1227	1050
40	60	1315	1126
20	80	1370	1230
0	100	1430	1430

7 Explain what is meant by the liquidus, solidus and solvus lines on a thermal equilibrium diagram.

8 Describe the form of the thermal equilibrium diagrams that would be expected for alloys of two metals that are completely soluble in each other in the liquid state but in the solid state are (a) soluble, (b) completely insoluble, (c) partially soluble in each other.

9 The lead–tin thermal equilibrium diagram is given in Figure 7.16
   (a) What is the composition of the eutectic?

(b) What is the eutectic temperature?

(c) What will be the expected structure of a solid 40% tin–60% copper alloy?

(d) What will be the expected structure of a solid 10% tin–90% copper alloy?

(e) What will be the expected structure of a solid 90%–10% copper alloy?

10 For (a) a 40% tin–60% lead alloy, (b) an 80% tin–20% lead alloy, what are the phases present at temperatures of (i) 250°C, (ii) 200°C? (See Figure 7.16).

11 Explain what is meant by coring and the conditions under which it occurs.

12 Explain how precipitation hardening is produced.

13 The relevant part of the aluminium–copper thermal equilibrium diagram is given in Figure 7.26. What type of microstructure would you expect for a 2% copper–98% aluminium alloy after it has been heated to 500°C, held at that temperature for a while, and then cooled (a) very slowly to room temperature, (b) very rapidly to room temperature?

14 Use the information given in the thermal equilibrium diagram for lead–tin alloys, Figure 7.16, for this question.

(a) Sketch the cooling curves for liquid to solid transitions for (i) 20% tin–80% lead, (ii) 40% tin–60% lead, (iii) 60% tin–40% lead (iv) 80% tin–20% tin alloys.

(b) Solder used for electrical work in the making of joints between wires has about 67% lead–33% tin. Why is this alloy composition chosen?

(c) What is the lowest temperature at which lead–tin alloys are liquid?

8

Ferrous alloys

8.1 Iron alloys

Pure iron is a relatively soft material and is hardly of any commercial use in that state. Alloys of iron with carbon are, however, very widely used and are classified according to their carbon content:

Material	Percentage carbon
Wrought iron	0 to 0.05
Steel	0.05 to 2
Cast iron	2 to 4.3

The percentage of carbon alloyed with iron has a profound effect on the properties of the alloy. The term *carbon steel* is used for those steels in which essentially just iron and carbon are present. The term *alloy steel* is used where other elements are included. The term *ferrous alloy* is used for all iron alloys.

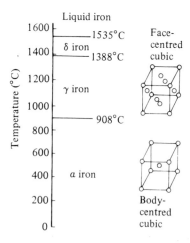

Figure 8.1 The forms of pure iron

8.2 The iron–carbon system

Pure iron at ordinary temperatures has the body-centred cubic structure. This form is generally referred to as α *iron*. The iron will retain this structure up to a temperature of 908°C. At this temperature the structure of the iron changes to face-centred cubic. This is referred to as γ *iron*. At 1388°C this structure changes again to give a body-centred cubic structure known as δ *iron*. The iron retains this form up to the melting point of 1535°C. Figure 8.1 summarises these various changes.

The name *ferrite* is given to the two body-centred cubic forms of iron, i.e. the α and δ forms. The name *austenite* is given to the face-centred cubic form, i.e. the γ form.

Figure 8.2 shows the iron–carbon system thermal equilibrium diagram. The α iron will accept up to about 0.02% of carbon in solid solution. The γ iron will accept up to 2.0% of carbon in solid solution. With these amounts of carbon in solution the α iron still retains its body-centred cubic structure, and the name ferrite, and the γ iron its face-centred structure, and the name austenite (Figure 8.3). The solubility of carbon in iron, in both the austenite and ferrite forms, varies with temperature. With slow cooling, carbon in excess of that

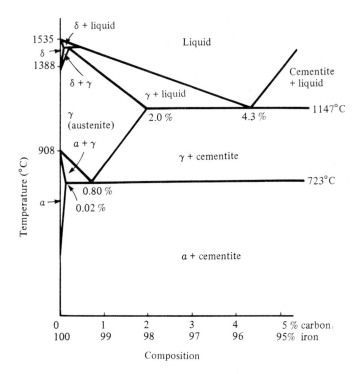

Figure 8.2 The iron–carbon system. (Note: This is really the iron–iron carbide system)

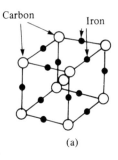

(a)

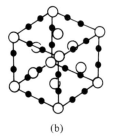

(b)

Figure 8.3 (a) Ferrite, the body centred cubic form, can only accept up to 0.02% carbon in solid solution. (b) Austenite, the face-centred cubic form, can accept up to 2.0% carbon in solid solution

which the α or γ solid solutions can hold at a particular temperature will precipitate. The precipitate is not, however, carbon but as iron carbide (Fe_3C), a compound formed between the iron and the carbon. This iron carbide is known as *cementite*. Cementite is hard and brittle.

Consider the cooling from the liquid of an alloy with 0.80% carbon (Figure 8.4). For temperatures above 723°C the solid formed is γ iron, i.e. austenite. This is a solid solution of carbon in iron. At 723°C there is a sudden change to give a laminated structure of ferrite plus cementite. This structure is called *pearlite* (Figure 8.5). This change at 723°C is rather like the change that occurs at a eutectic, but there the change is from a liquid to a solid, here the change is from one solid structure to another. This type of change is said to give a *eutectoid*. The eutectoid structure has the composition of 0.8% carbon – 99.2% iron in this case.

Steels containing less than 0.80% carbon are called hypo-eutectoid steels, those with between 0.80% and 2.0% carbon are called hypereutectoid steels.

Figure 8.6 shows the cooling of a 0.4% carbon steel, a hypoeutectoid steel, from the austenite phase to room temperature. When the alloy is cooled below temperature T_1, crystals of ferrite start to grow in the austenite. The ferrite tends to grow at the grain boundaries of the austenite crystals. At 723°C the

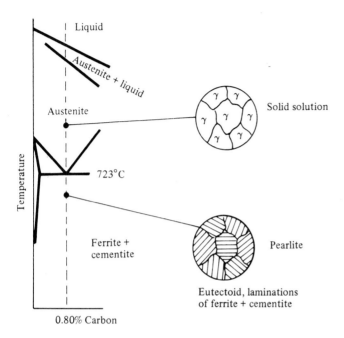

Solid solution

Pearlite

Eutectoid, laminations
of ferrite + cementite

0.80% Carbon

Figure 8.4 Slow cooling of a 0.80% carbon steel

Figure 8.5 Lamellar pearlite × 600 magnification. (From Monks, H. A. and Rochester, D. C., *Technician Structure and Properties of Metals*, Cassell)

remaining austenite changes to the eutectoid structure, i.e. pearlite. The result can be a network of ferrite along the grain boundaries surrounding areas of pearlite (Figure 8.7).

Figure 8.8 shows the cooling of a 1.2% carbon steel, a *hyper*-eutectoid steel,

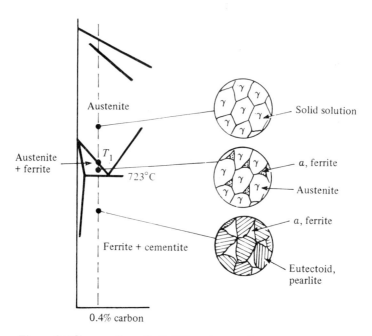

Figure 8.6 Slow cooling of a 0.40% carbon steel

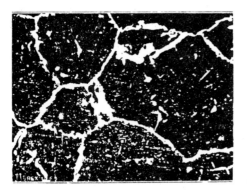

Figure 8.7 A 0.5% carbon steel, slow cooled. Shows network of ferrite. (From Rollason, E. C., *Metallurgy for Engineers*, Edward Arnold)

from the austenite phase to room temperature. When the alloy is cooled below the temperature T_1, cementite starts to grow in the austenite at the grain boundaries of the austenite crystals. At 723°C the remaining austenite changes to the eutectoid structure, i.e. pearlite. The result is a network of cementite along the grain boundaries surrounding areas of pearlite.

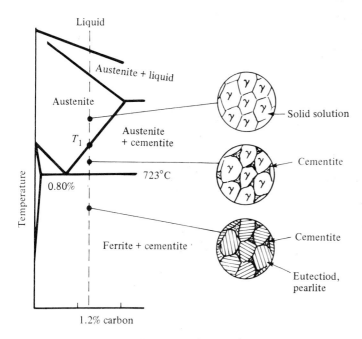

Figure 8.8 Slow cooling of a 1.2% carbon steel

Thus *hypo*-eutectoid carbon steels consist of a *ferrite* network enclosing pearlite and *hyper*-eutectoid carbon steels consist of a *cementite* network enclosing pearlite.

8.3 Critical change points

If water is heated and a graph plotted of temperature with time there will be found to be two discontinuities in the graph where the temperature does not continue to rise at a steady rate despite the heat being supplied at a steady rate. Figure 7.5 showed one of these discontinuities. They occur at 0°C and 100°C when the state of the water changes. When carbon steels are heated, similar discontinuities occur in the temperature–time graph. The temperatures at which there are changes in the rate of temperature rise, for a constant rate of supply of heat, for steels are known as *arrest points* or *critical points*.

Figure 8.9a shows a heating curve for a hypo-eutectoid steel. The lower critical temperature A_1 is the same for all carbon steels and is the temperature 723°C at which the eutectoid change occurs. For the hypo-eutectoid steel this temperature marks the transformation from a steel with a structure of ferrite and cementite to one of ferrite and austenite. The upper critical temperature A_3 depends on the carbon content of the steel concerned and marks the change from a structure of ferrite and austenite to one solely of austenite. For a 0.4% carbon steel this would be temperature T_1 in Figure 8.5.

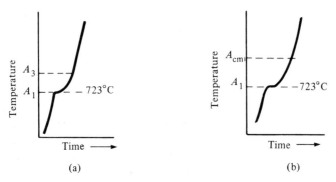

Figure 8.9 Heating curve for a hypo-eutectoid steel, (b) Heating curve for a hyper-eutectoid steel

The heating curve for a hyper-eutectoid steel (Figure 8.9b) has the same lower critical temperature A_1 of 723°C, marking the transformation from a steel with a structure of ferrite and cementite to one of austenite and cementite. The upper critical temperature A_{cm} marks the transformation from a structure of austenite and cementite to one solely austenite. For a 1.2% carbon steel this would be temperature T_1 in Figure 8.8.

Cooling curves give critical points differing slightly from those produced by heating, cooling giving lower values than heating. The heating critical points are generally denoted by the inclusion of the letter c (the abbreviation for the French word for heating—*chauffage*), i.e. Ac_1, Ac_3, Ac_{cm}. The cooling critical points are denoted by the inclusion of the letter r (the abbreviation for the French word for cooling—*refroidissement*), i.e. Ar_1, Ar_3, Ar_{cm}. The letter A

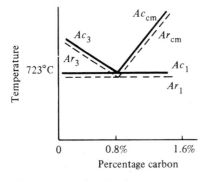

Figure 8.10 Critical points

used with the critical points stands for the term 'arrest'.

Figure 8.10 shows a graph of the critical point temperatures against the percentage carbon in the steel. The graph is restricted to those carbon percentages that result in steels. The graph is, in fact, the iron–carbon diagram given in Figure 8.2. The critical point graph is the thermal equilibrium graph if the heating and cooling graphs were obtained as the result of very slow heating. and cooling rates.

8.4 The effect of carbon content

Ferrite is a comparatively soft and ductile material. Pearlite is a harder and much less ductile material. Thus the relative amounts of these two substances in a carbon steel will have a significant effect on the properties of that steel. Figure 8.11 shows how the percentages of ferrite and pearlite change with percentage carbon and also how the mechanical properties are related to these changes. The data refers only to steels cooled slowly from the austenitic state.

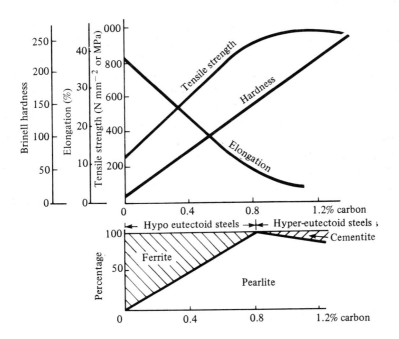

Figure 8.11 The effect of carbon content on the structure and properties of steels

Up to the euctectoid composition carbon steel, i.e. for hypo-eutectoid steels, the decreasing percentage of ferrite and the increasing percentage of pearlite results in an increase in tensile strength and hardness. The ductility decreases, the elongation at fracture being a measure of this. For hyper-eutectoid steels, increasing the amount of carbon decreases the percentage of pearlite and increases the percentage of cementite. This increases the hardness but has little effect on the tensile strength. The ductility also changes little.

Carbon steels are grouped according to their carbon content. *Mild steel* is a group of steels having between 0.10% and 0.25% carbon, *medium-carbon steel* has between 0.20% and 0.50% carbon, *high-carbon steel* has more than 0.5% carbon. Mild steel is a general purpose steel and is used where hardness and tensile strength are not the most important requirements. Typical applications are sections for use as joists in buildings, bodywork for cars and ships, screws, nails, wire. Medium carbon steel is used for agricultural tools, fasteners, dynamo and motor shafts, crankshafts, connecting rods, gears. High carbon steel is used for withstanding wear, where hardness is a more necessary requirement than ductility. It is used for machine tools, saws, hammers, cold chisels, punches, axes, dies, taps, drills, razors. The main use of high carbon steel is thus as a tool steel.

Plain carbon steels can be hardened by rapid cooling from the austenite state. This produces a very hard constituent called *martensite*. The rapid cooling is achieved by quenching in cold water. The resulting steel is hard and brittle, but its properties can be adjusted to a less brittle state by tempering. This involves reheating the steel to a temperature in the range of 200° to 300°C. The temperature used determines the final balance of hardness and brittleness, the lower the temperature the harder and the more brittle the material.

The limitations to the use of plain carbon steels are that high strength steel, up to about 600 N mm^{-2}, can be obtained only by increasing the carbon content to such a level that the material becomes brittle. Another problem is that high hardness is achieved by water quenching. The severity of this rapid rate of cooling is such as often to lead to distortion and cracking of the steel.

8.5 Alloy steels

The term *alloy steel* is used to describe those steels to which one or more alloying elements, in addition to carbon, have been added deliberately in order to modify the properties of the steel. In general the steels have less than 1% carbon and the additions are, in the case of manganese, more than 1.5% or, in the case of silicon, more than 0.5%.

There are a number of ways in which alloying elements can have an effect on the properties of the steel:

(1) They may go into solid solution, giving an increase in strength. In *low-alloy steels*, i.e. those with less than about 3 to 4% of alloying elements other than carbon, the alloying elements enter into solid solution with the ferrite and so give a stronger steel which still has good ductility. Silicon is an example of such an alloying element.

(2) They may form stable, hard carbides or nitrides which may, if in an

appropriate form such as fine particles, increase the strength and hardness. Manganese, chromium and tungsten have this effect.

(3) They could cause the breakdown of cementite and lead to the presence of graphite in the structure. Silicon and nickel have this effect. The result is a decrease in strength and hardness. For this reason silicon and nickel are not used with high carbon steels.

(4) They may lower the temperature at which austenite is formed on heating the steel. Manganese, nickel, copper and cobalt have this effect. The lowering of this temperature means a reduction in the temperature to which the steel has to be heated for hardening by quenching. If a sufficiently high percentage of one of these elements is added to the steel the transformation temperature to austenite may be decreased to such an extent that the austenite is retained at room temperature. With manganese about 11 to 14% produces what is known as an *austenitic steel*. Such steels have relatively good hardness combined with ductility and so are tough.

(5) They may increase the temperature at which austenite is formed on heating the steel. Chromium, molybdenum, tungsten, vanadium, silicon and aluminium have this effect. This raises the temperature to which the steel has to be heated for hardening. If a sufficiently high percentage of one of these elements is added to the steel, the transformation from ferrite to austenite may not take place before the steel reaches its melting point temperature. Such steels are known as *ferritic steels*, 12 to 25% of chromium with a 0.1% carbon steel gives such a type of steel. Such a steel cannot be hardened by quenching.

(6) They could change the critical cooling rate. The *critical cooling rate* is the minimum rate of cooling that has to be used if all the austenite is to be changed into martensite. With the critical cooling rate, or a higher rate of cooling, the steel has maximum hardness. With slower rates of cooling a less hard structure is produced. Most alloying elements reduce the critical cooling rate. The effect of this is to make air or oil quenching possible, rather than water quenching. It also increases the hardenability of steels.

(7) They can influence grain growth. Some elements accelerate grain growth while others decrease grain growth. The faster grain growth leads to large grain structures and consequently to a degree of brittleness. The slower grain growth leads to smaller grain size and so to an improvement in properties. Chromium accelerates grain growth and thus care is needed in heat treatment of chromium steels to avoid excessive grain growth. Nickel and vanadium decrease grain growth.

(8) They can affect the machinability of the steel. Sulphur and lead are elements that are used to improve the chip formation properties of steels.

(9) They can improve the corrosion resistance. Some elements promote the production of adherent oxide layers on the surfaces of the steel and so improve its corrosion resistance. Chromium is particularly useful in this respect. If it is present in a steel in excess of 12% the steel is known as *stainless steel* because of its corrosion resistance. Copper is also used to improve corrosion resistance.

Table 8.1 indicates the main effects of the commonly used alloying elements. As will have been apparent from the preceding discussion an alloying element generally affects the properties of the alloy in more than one way.

Table 8.1   *Effects of alloys on the properties of steels*

Element	Typical amount	Main effects on properties
Aluminium	0.95 to 1.3%	Aids nitriding Restricts grain growth.
Chromium	0.5 to 2% 4 to 18%	Increases hardenability. Improves corrosion resistance.
Copper	0.1 to 0.5%	Improves corrosion resistance.
Lead	Less than 0.2%	Improves machinability
Manganese	0.2 to 0.4% 1 to 2%	Combines with the sulphur in the steel to reduce brittleness. Increases hardenability.
Molybdenum	0.1 to 0.5%	Inhibits grain growth. Improves strength and toughness.
Nickel	0.3 to 65%	Improves strength and toughness. Increases hardenability.
Phosphorus	0.05%	Improves machinability.
Silicon	0.2 to 2%	Increases hardenability. Removes oxygen in steel making.
Sulphur	Less than 0.2%	Improves machinability.
Tungsten		Hardness at high temperatures.
Vanadium	0.1 to 0.3%	Restricts grain growth. Improves strength and toughness.

Structural steels

Mild steel, with tensile strengths up to about 450 N mm^{-2} (MPa), or the high carbon steel, with a tensile strength of the order of 600 N mm^{-2}, has been widely used for structural work, axles to bridges. However, the higher the percentage of carbon in the steel the more brittle the material becomes (see Figure 8.11). There thus is a problem of combining strength with toughness. This can be overcome by the use of alloy steels.

The properties required of a constructional steel are:

(1) High yield strength, this representing the maximum stress to which the material could be exposed in use.
(2) Tough. Not brittle.

(3) Good weldability.
(4) Good corrosion resistance.
(5) Low cost.

Table 8.2 *Properties of typical structural steels*

Composition	Condition	Yield stress ($N\ mm^{-2}$ or MPa)	Tensile strength ($N\ mm^{-2}$ or MPa)	Percentage elongation	Izod value (J)
0.40% C, 0.90% Mn, 1.00% Ni.	Quenched & tempered at 600°C.	490	700	25	90
0.31% C, 0.60% Mn, 3.00% Ni, 1.00% Cr.	Oil quenched & tempered at 600°C.	820	930	23	104
0.40% C, 0.55% Mn, 1.50% Ni, 1.20% Cr, 0.30% Mo.	Oil quenched & tempered at 600°C	990	1080	22	70

These properties can be promoted by the addition of nickel. Unfortunately nickel promotes the breakdown of cementite to give graphite. Thus, in a medium carbon steel, chromium is also added as it promotes the formation of carbides. Such a nickel–chromium steel can show brittleness after tempering, but this effect can be reduced by the addition of a small percentage of molybdenum, about 0.3%. Table 8.2 shows the properties of typical steels.

Tool steels

Tool steels require properties that permit their use for tools that cut and shape metals. Good wear resistance, hardness and general toughness are required.

Plain carbon tool steels obtain their hardness from their high carbon content, between 0.5 and 1.5%. The steels need to be quenched in cold water to obtain maximum hardness. Unfortunately they are rather brittle when very hard, lacking toughness. Where medium hardness with reasonable toughness is required the carbon steel will have about 0.7% carbon. Where hardness is the primary consideration and toughness is not so important, a carbon steel with about 1.2% carbon may be used.

Alloy tool steels are made harder and more wear-resistant by the addition to the steel of elements that promote the production of stable hard carbides. Manganese, chromium, molybdenum, tungsten and vanadium are examples of such elements. A manganese tool steel contains from about 0.7 to 1.0%

carbon and 1.0 to 2.0% manganese. Such a steel is oil-quenched from 780° to 800°C and tempered between about 220° and 245°C. The manganese content may be replaced partially by chromium, such a change improving the toughness. An example of such a steel has 1.00% carbon, 0.45% manganese and 1.40% chromium. The steel is oil-quenched from about 810°C and tempered at 150°C. It is used for ball bearings, cams and instrument pivots.

Shock-resistant tool steels are designed to have toughness under impact conditions, e.g. hand and pneumatic chisels, punches, picks. For such properties, fine grain is necessary and can be achieved by the addition of vanadium.

Tool steels designed for use with hot working processes need to maintain their wear resistance and hardness at the temperature used. Chromium and tungsten, when added to steels, form carbides which are both stable and hard, resulting in steels which retain their properties to high temperatures. An example of such a steel has 0.3% carbon, 3% chromium, 10% tungsten, 0.3% vanadium, 0.3% molybdenum and a maximum of 0.3% manganese. The steel is oil-quenched from 1150°C and tempered at 570°C. It is used for dies in hot forging, drawing and extrusion.

Steels that are able to work effectively at high machining speeds are called high-speed steels. The high speed results in the tool becoming hot: high-speed steels have to retain their properties at temperatures as high as 500°C. Such steels must not become tempered by the high temperatures produced during the machining. The combination of tungsten and chromium is found to give the required properties, the carbides formed by these elements being particularly stable at high temperatures. An example of such a steel has 0.75% carbon, 4.1% chromium, 1.1% vanadium and 18% tungsten. It is quenched in oil or an air blast from 1280°C and double tempered at 560°C (not a true tempering but a precipitation of tungsten–iron carbides and transformation of retained austenite to martensite). At 20°C the steel has a hardness of 875 HV and at 550°C this hardness has only dropped to 555 HV.

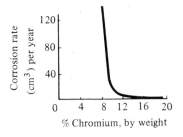

Figure 8.12 The effect of percentage of chromium on corrosion rate

Stainless steels

Stainless steels contain a minimum of 10.5% chromium. The effect of the chromium is to give the steel exceptionally good corrosion resistance, hence the term 'stainless'. Figure 8.12 shows the effect of chromium content on the corrosion rate of a steel. The corrosion rate reaches a very low value when the chromium content is 12%.

There are several types of stainless steels: ferritic, martensitic and austenitic. *Ferritic steels* contain between 12 and 25% chromium and less than 0.1% carbon. This high percentage of chromium with such a small percentage of carbon raises the temperature at which the transformation to austenite occurs to such a level that the steel remains in the ferritic state all the way up to its melting point. Because austenite cannot be formed with these steels, hardening by quenching to give martensite cannot occur. Martensite can only be produced by quenching from the austenite state. Thus such steels cannot be hardened by heat treatment. They can, however, be hardened by work hardening.

Martensitic steels contain between 12 and 18% chromium and between 0.1 and 1.5% carbon. The higher carbon content of these steels enables the austenite transformation to occur before the melting point of the steel is reached. Such steels can thus be quenched to give a martensitic structure.

Austenitic steels contain nickel as well as chromium. The effect of nickel is to lower the temperature at which austenite is produced, the reverse of the effect given by the chromium. A common steel of this type has 0.05% carbon, 18% chromium and 8.5% nickel. The transformation temperature is reduced by this combination to such an extent that the austenite is retained by the steel at room temperature. Such steels cannot be hardened by quenching. They are, however, usually quenched, not to produce martensite, but to minimise the formation of chromium carbide as this causes a reduction in the corrosion resistance of the material.

Table 8.3 gives details of the mechanical properties of examples of the different types of stainless steels.

During welding, when temperatures of the order of 500° to 800°C are realised, these steels may undergo structural changes which are detrimental to the corrosion resistance of the material, the other properties being little affected. The effect is known as *weld decay* (which is in fact a form of intergrain corrosion). One way of overcoming this effect is to '*stabilise*' the steel by adding other elements such as niobium, titanium or columbium.

The following information is taken from the data in the information sheet of a manufacturer of stainless steel (courtesy of Arthur Lee & Sons Ltd).

1 *Austenitic chromium-nickel steels*

This group, typified by the well-known 18% chromium, 8% nickel quality, includes a great many variations, some containing molybdenum, while others are alloyed with columbium or titanium. Some are low in carbon, but each is

Table 8.3 *Mechanical properties of typical stainless steels*

Stainless steel type	Composition	Condition	Yield stress (N mm^{-2} or MPa)	Tensile strength (N mm^{-2} or MPa)	Percentage elongation	Izod (J)	Hardness (HV)
Ferritic	Carbon 0.06% Chromium 13%	Annealed	280	416	20		170
Martensitic	Carbon 0.16% Chromium 12.5%	Oil-quenched from 1000°C	1190	1850	2.5	7	371
		Oil-quenched from 1000°C, tempered at 750°C	370	570	33	134	172
Cutlery steel (a martensitic steel)	Carbon 0.32% Chromium 13%	Oil-quenched from 980°C, tempered at 180°C		1450	8		600
Austenitic	Carbon 0.05% Chromium 18% Nickel 8.5% Manganese 0.8%	Annealed	278	618	50		180
		Cold rolled	803	896	30		

designed for a specific purpose. Their structure is austenitic, a solid solution of chromium, nickel, iron and carbon, possessing the highest degree of corrosion resistance of any of the stainless family. Likewise they have high ductility and while rapid cooling from high temperature puts them in their softest possible condition, they can be equally well cold work-hardened to high values of strength. Inherently they are tough and well adapted for fabrication by deep drawing, roll forming, etc., also readily welded by any known method, or soldered by proper techniques.

2 *Martensitic chromium steels (hardenable)*

This group includes the hardenable types such as the standard cutlery qualities and their variables with higher and lower carbons. These alloys, when cooled from a high temperature, develop greater tensile strength and hardness, and their structure is similar to that of a hardened carbon spring steel. By proper heat-treatment it is possible to develop a wide range of physical properties from 500 N mm^{-2} (MPa) with 25% elongation, to over 1600 N mm^{-2} tensile with 2% to 4% elongation. Corrosion resistance of the alloys in this group is at its optimum in the heat-treated condition. Though not as corrosion-resistant as the austenitic grades, it is adequate for the majority of engineering uses.

3 *Ferritic chromium steels*

This last group is typified by the 16/18% chromium stainless steel; varieties are available containing aluminium and molybdenum which enhance respectively weldability and corrosion resistance. This steel does not harden when cooled rapidly from high temperatures. It is used generally in the fully-softened condition, in which state corrosion resistance achieves its maximum. The alloys in this group work-harden slowly which renders them suitable for fabrication by deep drawing, spinning, roll forming, etc., their structure being somewhat similar to that of a spheroidised carbon steel. Ferritic chromium steels have good oxidation resistance, although their anti-corrosive properties, while not quite equal to the austenitic group, are superior to those of martensitic steels.

In each of these three groups easy machining or freecutting qualities can be provided.

Selection of correct quality

Once it has been decided which group will satisfy the requirements of the application, details can then be studied and the proper modification of the general type chosen. For example, if the application requires forming by cold working, followed by welding, the part to be highly resistant to certain chemicals while exposed at times to temperatures between 450° and 800°C, then from the welding point of view the austenitic group is indicated. Because of the exposure to corrosive conditions after heating to certain temperatures, it is also apparent that a stabilised 18/8 quality would be required, leaving a choice of either 18/8 titanium stabilised, or 18/8 columbium stabilised. If the article has to be finally polished, the user would probably select the columbium stabilised. It is important that all conditions of fabrication and requirements of a new application be known in advance, so that the proper type of stainless steel can be selected.

Applications
Austenitic steels. Window and door frames. Roofing and guttering. Domestic hot water piping. Spoons, forks and knife handles. Kitchen utensils. Washing machines. Hospital equipment. Motor-car hub caps, rim embellishers and bumpers. Wheel spokes. Welding rods and electrodes. Wire ropes. Yacht fittings and masts and all marine applications. Nuts, bolts, screws, rivets, locking wire, split pins. Coiled and flat springs.
Martensitic steels. Flat springs. Coil springs. Scales and rulers. Knives. Spatulas. Kitchen tools and all appliances where high tensile hardness is required with moderate corrosion resistance. Surgical and dental instruments.
Ferritic steels. Mouldings and trim for motor-car bodies, furniture, television sets, gas and electric stoves, refrigerators, etc. Coinage. Spoons and forks. Domestic iron soles. Vehicle silencers. Driving mirror frames. Nuts, bolts, screws, rivets, split pins. Heat-resistant applications such as oil-burner sleeves, or any part where working temperatures do not exceed 800°C.

Maraging steels

Maraging steels are high strength, high alloy steels which can be precipitation-hardened. A typical steel might contain 18% nickel and 7% or 8% cobalt, with other elements such as molybdenum and titanium. The carbon content is kept very low, less than 0.03%. This is because the high nickel content could otherwise lead to the formation of graphite in the structure and this would result in a decrease in strength and hardness.

The steel is heated to about 830°C and then air-cooled. This results in a martensitic structure. The material is now relatively soft and is worked or machined in this condition. The material is then precipitation-hardened by heating it to about 500°C for two or three hours. This causes precipitates of intermetallic compounds to form. The term *maraging* refers to this age-hardening of the martensite. The result is a high strength, hard steel. For example, an 18% nickel–7% cobalt maraged steel might have a tensile strength of about 1900 N mm^{-2} (MPa), a yield stress of 1800 N mm^{-2} (MPa) and a percentage elongation of 11%.

High-temperature steels

Steels for use at high temperatures must provide good resistance to creep and oxidation and suffer no detrimental changes in structure or properties. Carbon steels are limited to about 450°C since at higher temperatures the rate of oxidation can become very high and also there is a marked reduction in stress-bearing capabilities. Low alloy steels have been developed for use at higher temperatures. Such steels include small amounts of elements, such as chromium, vanadium and molybdenum, which combine with the carbon to form carbides. For example, a 1% chromium–$\frac{1}{2}$% molybdenum steel has an oxidation limit of about 550°C and is more resistant to creep than a carbon steel. Such a steel is used for steam pipes.

For temperatures in excess of 550°C the oxidation problem can be reduced by increasing the chromium content. For example, a 12% chromium steel has an oxidation limit of about 575°C and is used for steam turbine rotors and blades.

For higher temperatures, austenitic steels can be used. Such steels do not transform to the ferritic structure on cooling to room temperature, but retain an austenitic structure. This is obtained as a result of adding 18% chromium and 8% nickel. Such a steel has an oxidation limit of about 650°C and superior creep resistance, a typical application being for superheater tubes. Enhanced creep resistance can be obtained by incorporating small amounts of niobium, titanium or molybdenum.

Nickel–chromium alloys (see the next chapter) can be used for higher temperatures. For example, an 80% chromium–20% nickel alloy has an oxidation limit of about 900°C and good creep properties to quite high temperatures.

8.6 Heat treatment of steel

Heat treatment can be defined as the controlled heating and cooling of metals in the solid state for the purpose of altering their properties according to requirements. Heat treatment can be applied to steels to alter their properties by changing grain size and the form of the constituents present.

Annealing is the heat treatment used to make a steel softer and more ductile, remove stresses in the material and reduce the grain size. One form of the annealing process is called *full annealing*. In the case of hypo-eutectoid steels, this involves heating the material to a temperature above the A_3 temperature, holding at that temperature for a period of time and then very slowly cooling it. Typically, the material is heated to about 40°C above the A_3 temperature which has the effect of converting the structure of the steel to austenite. Slow cooling leads to the conversion of the austenite to ferrite and pearlite. The result is a steel in as soft a condition as possible.

A different process has to be used for hyper-eutectoid steels in that heating them to above the A_{cm} temperature turns the entire steel structure into austenite, but slow cooling from that temperature results in the formation of a network of cementite surrounding the pearlite. The cementite is brittle and has the effect of making the steel relatively brittle. To make the steel soft, the original heating is to a temperature only about 40°C above the A_1 temperature.

This converts the structure into austenite plus cementite. Slow cooling from this temperature gives as soft a material as is possible, but not, however, as soft as the full annealing process gives when applied to hypo-eutectoid steels.

Sub-critical annealing, sometimes referred to as *process annealing*, is often used during cold working processes where the material has to be made more ductile for the process to continue. The process involves heating the material to a temperature just below the A_1 temperature, holding it at that temperature for a period of time and then cooling it at a controlled rate, generally just in air rather than cooling in the furnace as with full annealing. This process leads to no change in structure, no austenite being produced, but just a recrystallisation. The process is used for steels having up to about 0.3% carbon. When sub-critical annealing is applied to steel having higher percentages of carbon the effect of the heating is to cause the cementite to assume spherical shapes, hence the process is often referred to as *spheroidizing annealing*. This spheroidizing results in a greater ductility and an improvement in machineability. Figure 8.13 summarises the range of annealing processes.

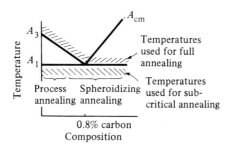

Figure 8.13 Annealing temperatures

Normalising is a heat treatment process similar to full annealing, and is applied to hypo-eutectoid steels. The steel is heated to about 40°C above the A_3 temperature, held at this temperature for a short while, and then cooled freely in air. The effect of the heating is to form an austenite structure. The cooling rate is, however, much faster than that with the annealing process. The result is a finer grained structure, ferrite and pearlite. This finer grain size improves the machinability and gives a slightly harder and stronger material than that given with full annealing.

Both annealing and normalising involve relatively slow cooling of the steel, but what happens if a steel is cooled very quickly, i.e. quenched? Quenching does not allow time for diffusion to occur in the solid solution as in very slow cooling. If a hypo-eutectoid steel is heated to 40°C above the A_3 temperature, all the structure becomes austenitic. Very rapid cooling of this structure does not allow sufficient time for the ferrite structure to be produced, i.e. for the austenite to give up its surplus carbon and so produce ferrite. The result is a new structure called *martensite*. Martensite is a very hard structure, hence the result of such a process is a much harder material. The *hardening* process is thus a sequence of heating to produce an austenitic state and then rapid cooling to produce martensite (Figure 8.14).

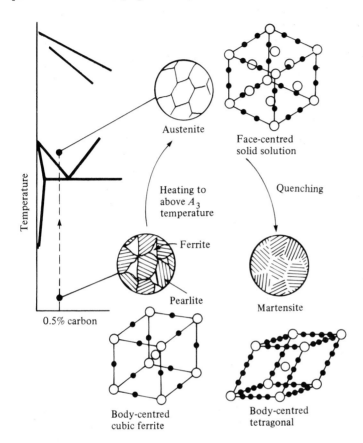

Figure 8.14 Hardening

If a steel is cooled at or above a certain minimum rate, called the *critical cooling rate*, all the austenite is changed into martensite. This gives the maximum hardness. If the cooling is slower than this critical cooling rate a less hard structure is produced.

The rate of cooling depends on the quenching medium used. Water is commonly used and gives a high cooling rate. However, distortion and cracking may be caused by this high cooling rate. Oil gives a slower cooling rate, whereas brine gives a rate even higher than water. The brine may, however, give rise to corrosion problems with the steel. Sodium or potassium hydroxide solution is sometimes used for very high cooling rates. In order to minimise the distortion that can be produced during the quenching process, long items should be quenched vertically, flat sections edgeways. To prevent bubbles of steam adhering to the steel during the quenching, and giving rise to different cooling rates for part of the object, the quenching bath should be agitated. Thick objects offer special problems in that the outer parts of the object cool more rapidly than the inner parts. The result can be an outer layer of martensite and an inner core of pearlite, which gives a variation in mechanical properties between the inner and outer parts. This effect is known as the *mass effect*.

Hypo-eutectoid steels are generally hardened by quenching from about 40°C above the A_3 temperature, though steels with less than about 0.3% carbon cannot be hardened very effectively. Hyper-eutectoid steels are hardened by quenching from about 40°C above the A_1 temperature. This is because to heat the steel above the A_{cm} temperature and then quench it gives rise to a network of cementite and so a brittle structure.

Tempering is the name given to the process in which a steel, hardened as a result of quenching, is reheated to a temperature below the A_1 temperature in order to modify its structure. The result is an increase in ductility at the expense of hardness and strength. The degree of change obtained depends on the temperature to which the steel is reheated, the higher the tempering temperature the lower the hardness but the greater the ductility (Figure 8.15). By combining hardening with tempering, an appropriate balance of mechanical properties can be achieved for a steel.

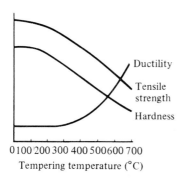

Figure 8.15 The effect of tempering temperature on the properties of a hardened steel

Table 8.4 lists tempering temperatures that give appropriate modification of properties of hardened steel for a range of common tools and components. For example, while a scriber needs hardness and high abrasion resistance, a spring does not need to be so hard but needs more toughness and ductility. Scribers can be relatively much more brittle than springs.

Table 8.4 *Tempering temperatures for various uses of hardened steel*

Steel artefact	Tempering temperature (°C)
Scribers	200
Hacksaw blades	220
Planing and slotting tools	230
Drills, milling cutters	240
Taps, shear blades, dies	250
Punches, reamers	260
Axes, press tools	270
Cold chisels, wood chisels	280
Screwdrivers	290
Saws, springs	300

Internal stresses in a material can be relieved to some extent by heating it to, say, 50° to 100°C below the A_1 temperature. The material is then usually air cooled from that temperature. *Stress relief* by this method is used with welded components before machining, parts requiring machining to accurate dimensions, castings before machining etc.

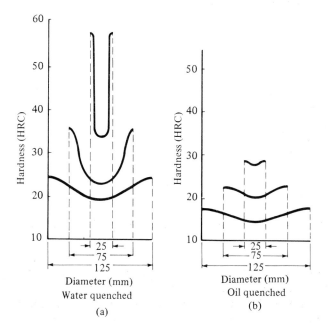

Figure 8.16 Variation of hardness with depth in quenched bars of different diameters, for a 0.48 per cent plain carbon steel

Hardenability

When a block of steel is quenched, the surface can show a rate of cooling different from that of the inner core of the block. Thus the formation of martensite may differ at the surface from the inner core. This means a difference in hardness. Figure 8.16 shows how the hardness varies with depths for a number of bars of different diameters when quenched in water and in oil.

With the water quenching, the hardness in the inner core is quite significantly different from the surface hardness. The larger-diameter bars also show lower surface and core hardness as their increased mass has resulted in a lower overall rate of cooling. With the oil quenching the cooling rates are lower than those of the water-cooled bars and thus the hardness values are lower. The hardness value for a fully martensitic 0.48% plain carbon steel is about 60 HRC. Thus, as the values in Figure 8.16 indicate, the quenching bars are not entirely martensitic.

The term *hardenability* is used as a measure of the depth of hardening introduced into a steel section by quenching (not to be confused with hardness). Hardenability is measured by the response of a steel to a standard test. The *Jominy test* involves heating a standard test piece of the steel to its austenitic state, fixing it in a vertical position and then quenching the lower end by means of a jet of water (Figure 8.17). This method of quenching results in different rates of cooling along the length of the test piece. When the test piece is cool, after the quenching, a flat portion is ground along one side of the test

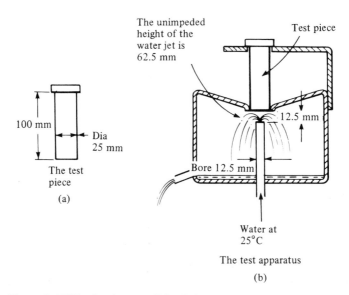

Figure 8.17 The Jominy test (BS 4437)

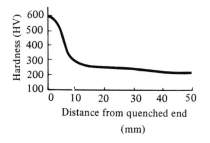

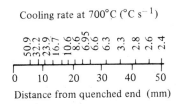

Figure 8.18 Results of the Jominy test for a 0.40 per cent plain carbon steel

Figure 8.19 Cooling rates at different distances from the quenched end of the Jominy test piece

piece, about 0.4 mm deep, and hardness measurements are made along the length of the test piece. Figure 8.18 shows the types of result that are produced. The hardness is greatest at the end on which the cold jet of water played and least at the other end of the bar.

The significant point about the Jominy test results is not that they give the hardness at different distances along the test piece but that they give the hardness at different cooling rates. Each distance along the test piece corresponds to a different rate of cooling (Figure 8.19). The important point, however, is that this applies to both points on the surface and points inside any sample of the steel, provided we know the cooling rates at those points. This applies regardless of the quenching medium used. Figure 8.20, for example,

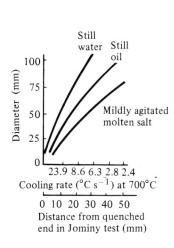

Figure 8.20 Cooling rates at the centres of different diameter bars at 700°C in different media

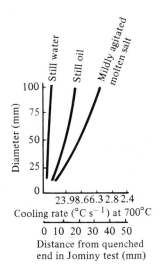

Figure 8.21 Cooling rates at the surfaces of different diameter bars at 700°C in different media

shows how the cooling rate at the centre of circular cross-section bars at 700°C is related to their diameter. As the cooling rates are related to distances along the Jominy test piece, these distances are also given in the figure.

Thus for the Jominy test result given in Figure 8.18, the hardness at 20 mm from the quenched end of the test piece is about 230 HV. This means that, using Figure 8.20, a circular cross-section bar of diameter about 75 mm would have this hardness at its centre when quenched in still water from 700°C. If still oil had been the quenching medium, the hardness at the centre of a 60 mm diameter bar would have been 230 HV.

Figure 8.21 shows, in a similar manner to Figure 8.20, how the cooling rate is related to the diameter of circular cross-section bars for points on the surface. The Jominy test result of 230 HV at 20 mm from the quenched end of the test piece means, using Figure 8.21, that a circular cross-section bar of diameter 50 mm would have this hardness at its surface when quenched from 700°C in mildly agitated molten salt.

Figure 8.22 shows Jominy test results for two different steels. In order to enable the significance of the results to be seen in terms of the hardness at the centre of bars of different diameters, the cooling rates have been transposed into diameter values by means of Figure 8.20. The alloy steel can be said to have better hardenability than the plain carbon steel. The plain carbon steel cannot be used at a diameter greater than about 25 mm with a still water quench if the bar is to be fully hardened. However, the alloy steel is fully hardened for bars with diameters in excess of 100 mm.

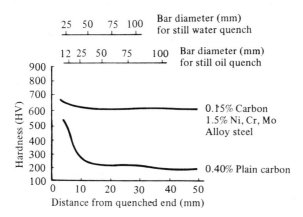

Figure 8.22 Results of Jominy tests

Ruling section

If you look up the mechanical properties of a steel in the data supplied by the manufacturer or other standard tables you will find that different values of the mechanical properties are quoted for different limiting ruling sections. The *limiting ruling section* is the maximum diameter of round bar at the centre of which the specified properties may be obtained. Here is an example of such information:

Steel	Condition	Limiting ruling section (mm)	Tensile strength ($N\ mm^{-2}$ or MPa)	Minimum elongation (%)
070 M 55	Hardened	19	850 to 1000	12
	and	63	770 to 930	14
	tempered	100	700 to 850	14

Different-sized bars of the same steel have different mechanical properties because, during the heat treatment, the cooling rate at the centres of the bars varies according to their size (see Figure 8.16). This results in differences in microstructure and hence in mechanical properties.

Surface hardening

There are many situations where wear or severe stress conditions may indicate the need for a hard surface to a component without the entire component being made hard and possibly brittle. The various methods used for surface hardening can, in the main, be grouped into two categories:

1. Selective heating of the surface layers;
2. Changing the composition of the surface layers.

One selective heating method is called *flame hardening*. This method involves heating the surface of a steel with an oxyacetylene flame and then immediately quenching the surface in cold water (Figure 8.23). The heating transforms the structure of the surface layers to austenite and the quenching changes this austenite to martensite; the result is a hard layer of martensite in

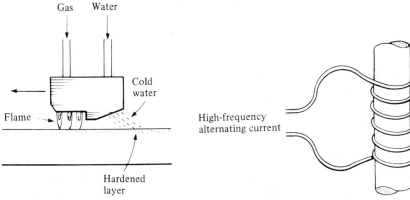

Figure 8.23 A burner with cooling water for flame hardening

Figure 8.24 Induction hardening of a tube or rod

the surface. The depth of hardening depends on the heat supplied per unit surface per unit time. Thus the faster the burner is moved over the surface the less the depth of hardening. This is because less of the steel has its structure transformed into austenite and hence martensite. The temperatures used are typically of the order of 850°C or more, i.e. above the A_3 temperature.

Another method involving selective heating is called *induction hardening*. This method involves placing the steel component within a coil through which a high frequency current is passed (Figure 8.24). This alternating current induces another alternating current to flow in the surface layers of the steel component. This induction is called electromagnetic induction. The induced currents heat the surface layers. The temperatures so produced cause the surface layers to change to austenite. The cooling following the passage of the current transforms the austenite to martensite. The depth of heating produced by this method, and hence the depth of hardening, is related to the frequency of the alternating current used. The higher the frequency the less the hardened depth, but the more rapid the temperature rise. The following data illustrates this point:

Frequency used (kHz)	Depth of hardening (mm)
3	4.0 to 5.0
10	3.9 to 4.0
450	0.5 to 1.1

The form of the induction coil used depends on the shape of the component being hardened, also on the size of the area to be hardened.

The composition of the steel to be surface hardened by selective heating has to be chosen with certain criteria in mind. The inner core of the steel that is to be surface-hardened must have the right mechanical properties. These properties will be unaffected by the surface treatment. This means that the composition of the steel and its heat treatment have to be chosen. The heat treatment has also to be chosen with the selective heating process in mind. The selective heating has to be able to change the microstructure of the steel to austenite in a very short amount of time. Hardened and tempered steels respond well but coarse annealed steels do not. In addition, the composition of the steel must be such that the quenching part of the selective heating process will produce martensite and so harden the steel. This tends to mean carbon contents of 0.4% or more (Figure 8.25).

One surface hardening process that involves changing the composition of the surface layers is *carburising*. This method involves increasing the carbon content of the surface layers, followed by a quenching process to convert the surface layers into martensite. This process is normally carried out on a steel containing less than about 0.2% carbon, the carburising treatment being used to give about 0.7 to 0.8% carbon in the surface layers. This wide difference in carbon content is needed because the quenching treatment following the carburising will affect both the inner core and the surface layers and it is only the surface layers that are to be hardened; the inner core should remain soft and tough. Figure 8.26 shows the results of Jominy tests for a 0.18% carbon (0.46% manganese, 1.68% silicon, 0.03% chromium, 0.21% molybdenum) steel for its core where the percentage of carbon remains unchanged and for the surface where the percentage of carbon has been increased to 0.7% by carburising. Thus with a fast rate of cooling the carburised surface may have a hardness of the order of 60 HRC while the inner core is about 15 HRC.

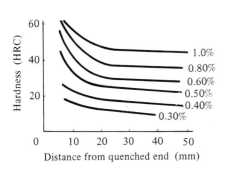

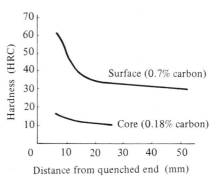

Figure 8.25 Jominy test results showing the effect of percentage carbon on the hardenability of a steel

Figure 8.26 Jominy test results for a carburised steel

There are a number of carburising methods. With *pack carburising* the steel component is heated to above the A_3 temperature while in a sealed metal box which contains charcoal and barium carbonate. The oxygen present in the box reacts with the carbon to produce carbon monoxide. This carbon-rich atmosphere in contact with the hot steel results in carbon diffusing into the surface austenitic layers. Pack carburising tends to be used mainly for large components or where a thick surface layer has to be hardened.

In *gas carburising* the component is heated to above the A_3 temperature in a furnace in an atmosphere of carbon-rich gas. The result is that carbon diffuses into the surface austenitic layers. Gas carburising is the most widely used method of carburising.

Salt bath carburising, or *cyaniding* as it is often known, involves heating the component in a bath of suitable carbon-rich salts. Sodium cyanide is mainly used. The carbon from the molten salt diffuses into the component. In addition there is diffusion of some nitrogen into the component. Both carbon and nitrogen can result in a microstructure which can be hardened. The method tends to produce relatively thin hardened layers with high carbon content. This occurs because the carburising takes place very quickly. One of the problems with this method is the health and safety hazard posed by the poisonous cyanide. Another problem is the removal of salt from the hardened component after the treatment. This can be particularly difficult with threaded parts or blind holes.

Carburising may result in the production of a large grain structure due to the time for which the material is held at temperature in the austenitic state. Though the final product may be hard there may be poor impact properties due to the large grain size. A heat treatment might thus be used to refine the grains. A two stage process is required as the carbon content of the surface is significantly different from that of the inner core of the material.

The first stage involves a heat treatment to refine the grains in the core. The component is heated to just above the A_3 temperature for the carbon content of the core (Figure 8.27). For a core with 0.2% carbon this is about 870°C. The component is then quenched in oil. The result is a fine grain core, but the surface layers are rather coarse martensite. The martensite is refined by heating to above the A_1 temperature for the carbon content of the surface layers; if this is 0.9% carbon then the temperature would be about 760°C. The component is then water-quenched. This second-stage treatment has little effect on the core but refines the martensite in the outer layers. The treatment may then be followed by a low temperature tempering, e.g. about 150°C, to relieve internal stresses produced by the treatment.

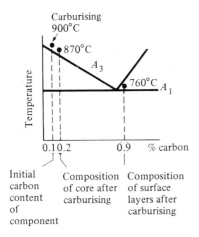

Figure 8.27 Heat treatment for a carburised steel

Nitriding involves changing the surface composition of a steel by diffusing nitrogen into it: hard compounds, nitrides, are produced. The process is used with those alloy steels that contain elements that form stable nitrides, e.g. aluminium, chromium, molybdenum, tungsten, vanadium. Prior to the nitriding treatment the steel is hardened and tempered to the properties required of the core. The tempering temperature does, however, need to be in the region of 560° to 750°C. The reason for this is that the nitriding process requires a temperature up to about 530°C, and this must not be greater than the tempering temperature, as the nitriding process would temper the steel and so change the properties of the core.

Unlike carburising, nitriding is carried out at temperatures below the stable austenitic state. The process consists of heating a component in an atmosphere of ammonia gas and hydrogen, the temperature being of the order of 500° to 530°C. The time taken for the nitrogen to react with the elements in the surface of the steel is often as much as 100 hours. The depth to which the nitrides are formed in the steel depends on the temperature and the time allowed for the reaction. Even with such long times, the depth of hardening is unlikely to exceed about 0.7 mm. After the treatment the component is allowed to cool slowly in the ammonia–hydrogen atmosphere. With most nitriding conditions a thin white layer of iron nitrides is formed on the surface of the component. This layer adversely affects the mechanical properties of the steel, being brittle and generally containing cracks. It is therefore removed by mechanical means or chemical solutions.

Because with nitriding no quenching treatment is involved, cracking and distortion are less likely than with other surface hardening treatments. Very high surface hardnesses can be obtained with special alloys. The hardness is retained at temperatures up to about 500°C, whereas that produced by carburising tends to decrease and the surface tends to become softer at temperatures of the order of 200°C. The capital cost of the plant is, however, higher than that associated with pack carburising.

Carbonitriding is the name given to the surface hardness process in which both carbon and nitrogen are allowed to diffuse into a steel when it is in the austenitic–ferritic condition. The component is heated in an atmosphere containing carbon and ammonia and the temperatures used are about 800° to 850°C. The nitrogen inhibits the diffusion of carbon into the steel and, with the temperatures and times used being smaller than with carburising, this leads to relatively shallow hardening. Though this process can be used with any steel that is suitable for carburising it is used generally only for mild steels or low alloy steels.

Ferritic nitrocarburising involves the diffusion of both carbon and nitrogen into the surface of a steel. The treatment involves a temperature below the A_1 temperature when the steel is in a ferritic condition. A very thin layer of a compound of iron, nitrogen and carbon is produced at the surfaces of the steel. This gives excellent wear and anti-scuffing properties. The process is mainly used on mild steel in the rolled or normalised condition.

Table 8.5 compares the main surface hardening treatments. In comparing the processes the term *case depth* is used, which is defined graphically (Figure 8.28) in terms of a graph of carbon or nitrogen content against depth under the surface of the steel. A straight line is drawn on the graph such that it passes through the surface carbon content value and so that the area between the line and the core carbon content C_c is the same as that between the actual carbon content graph curve and the core carbon content C_c. The point at which this line meets the core carbon content line gives the case depth.

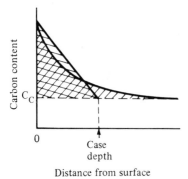

Figure 8.28 Defining case depth

Table 8.5   *Comparison of surface hardening treatments*

Process	Temperature (°C)	Case depth (mm)	Case hardness (HRC)	Main use
Pack carburising	810–1100	0.25–3	45–65	Low carbon and carburising alloy steels. Large case depths, large components
Gas carburising	810–980	0.07–3	45–65	Low carbon and carburising alloy steels. Large numbers of components.
Cyaniding	760–870	0.02–0.7	50–60	Low carbon and light alloy steels. Thin case.
Nitriding	500–530	0.07–0.7	50–70	Alloy steels. Lowest distortion.
Carbo-nitriding	700–900	0.02–0.7	50–60	Low carbon and low alloy steels.
Flame hardening	850–1000	Up to 0.8	55–65	0.4 to 0.7% carbon steels, selective heating.
Induction hardening	850–1000	0.5–5	55–65	0.4 to 0.7% carbon steels, selective heating.

Heat treatment equipment

The industrial heat treatment of metal requires:

(1) Furnaces to heat components to the required measured temperature;
(2) Quenching equipment.

An important consideration before the choice of furnace is made is whether the atmosphere in which the components are heated is to be oxidising, i.e. generally air, or some inert gas to prevent oxidisation. A *muffle furnace* (Figure 8.29) is one in which the components are contained in a chamber, the so-called muffle chamber, and the heat is supplied externally to the chamber. This enables the atmosphere in the chamber to be controlled. A non-muffle furnace has no control over the atmosphere surrounding the components.

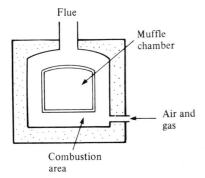

Figure 8.29 The basic form of a gas-fired muffle furnace

Heating an iron alloy in an oxidising atmosphere can result in the carbon in the outer layers of the material combining with the oxygen to give carbon dioxide, which escapes with the flue gases and so leaves the outer layers of the alloy with less carbon than the inner layers. This effect is known as *decarburisation*. Scaling and tarnishing can also occur when an alloy is heated in an oxidising atmosphere. Heating in a non-oxidising atmosphere can eliminate costly cleaning processes.

The *salt-bath furnace* consists essentially of a pot containing a molten salt, which may be heated directly by an electric current being passed through it or by external heating of the pot. Salt-bath furnaces give an even and rapid heating of components immersed in the salt. By suitable choice of the salt, carburisation can be promoted and so surface hardening occur. No oxidisation takes place for the components immersed in the salt as air is excluded.

Quenching may involve the immersion of the hot components in water, brine, oil or sodium or potassium hydroxide solution. The quenching tanks are designed so that the quenching liquid remains, as nearly as possible, at a constant temperature. In the case of brine or oil baths, these may stand in large tanks through which water flows, the water carrying the heat away and so keeping the quenching liquids at a near constant temperature during the quenching.

There are a number of hazards during heat treatment processes:

(1) Hot metals are being handled, so care should be taken in approaching any piece of metal during the heat treatment process.

(2) Quenching baths and gas-fired furnaces give off dangerous fumes. The heat treatment area needs to be well ventilated and fume extraction systems should be used.

(3) Fire hazards are present with furnaces and also oil quenching tanks. The immersion of a piece of hot metal in oil leads to some of the oil being vaporised and the possibility of a fire. If this happens, the air supply to the tank should be cut off by covering the tank with a lid.

(4) Explosions are possible with gas-fired furnaces if the furnace becomes charged with a gas–air mixture before the lighter is operated. The lighter should be operated before the gas is introduced into the combustion chamber.

(5) Some of the materials used in heat treatment are poisonous, e.g. sodium cyanide is used for salt-bath furnaces.

8.7 Coding systems for steels

In Great Britain the standard codes for the specification of steels are specified by the *British Standards Institution*. The following codes are those specified to the system given in BS 970 for wrought steels, published between 1970 and 1972. The system uses six symbols.

(1) The first three digits designate the type of steel.

000 to 199	Carbon and carbon-manganese types, the numbers being 100 times the manganese content.
200 to 240	Free cutting steels, the second and third numbers having an approximate relationship to 100 times the mean sulphur content.
250	Silicon-manganese spring steels.
300 to 499	Stainless and heat resistant valve steels.
500 to 999	Alloy steels.

(2) The fourth symbol is a letter.

A   The steel is supplied to a chemical composition determined by chemical analysis.
H   The steel is supplied to hardenability specification.
M   The steel is supplied to mechanical property specification.
S   The material is stainless.

(3) The fifth and sixth digits correspond to 100 times the mean percentage carbon content of the steel.

The following examples illustrate the use of the coding.

070M20   A carbon/carbon-manganese type of steel, supplied to mechanical property specification and having 0.20% carbon with 0.70% manganese.

150M19   A carbon/carbon-manganese type of steel supplied to mechanical property specification and having 0.19% carbon with 1.50% manganese.

220M07   A free-cutting steel supplied to mechanical property specification and having 0.07% carbon content. The sulphur content would be about 0.20%.

040A04   A carbon/carbon-manganese type of steel supplied to a chemical composition specification and having 0.04% carbon with 0.40% manganese.

Alloy steels have the first three digits in the code between 500 and 999. These are a few examples from this range of steels.

503M40   A nickel steel having 1.0% nickel and 0.40% carbon, being supplied to a mechanical property specification. The number 503 indicates that the steel is a nickel steel.

526M60   A chromium steel with 0.75% chromium and 0.60% carbon, to a mechanical property specification.

530M40   Also a chromium steel, having 1% chromium with 0.40% carbon to a mechanical property specification.

640M40   A nickel–chromium steel with 1.25% nickel, 0.5% chromium and 0.40% carbon, to a mechanical property specification.

653M31   Also a nickel chromium steel having 3% nickel, 1% chromium and 0.31% carbon, to a mechanical property specification.

A wide range of steels is available. As the specification of a steel is usually according to its tensile strength when in the hardened and tempered condition, a steel can be specified by means of a code letter, the letter indicating the tensile stength range in which the steel falls when in the hardened and tempered condition. The letter is said to refer to the *condition* of the steel. Table 8.6 gives condition codes and examples of steels which, by composition and size of section, fall into the various ranges.

Table 8.7 indicates the type of information available for steels when specified to BS 970. Thus, if one of the carbon or carbon–manganese steels in Table 8.7 had to be selected to have a hardness of about 200 HB when used in the hardened and tempered condition with a limiting ruling section of 63 mm, the choice would be between 080M40, 150M19 and 212M36. If there was also the condition that the Izod impact value should be not less than 40 J, the choice would be narrowed to 150M19.

Table 8.6 *Specificity of steels by condition codes*

Condition code	Tensile strength range $(N\,mm^{-2}\,or\,MPa)$
P	550 to 700
Q	629 to 770
R	700 to 850
S	770 to 930
T	850 to 1000
U	930 to 1080
V	1000 to 1150
W	1080 to 1240
X	1150 to 1300
Y	1240 to 1400
Z	1540 minimum

Examples

Condition	Grade of steel	Limiting ruling section (mm)
P	070M20	19 (0.7% Mn)
P	070M26	29
Q	070M26	13
Q	080M30	29 (0.8% Mn)
R	080M36	13
R	070M55	100 (0.7% Mn)
S	070M55	63
T	070M55	19
R	605M40	150 (1½% Mn, 0.27% Mo)
S	605M40	100
T	605M40	63
U	605M40	29
T	830M31	250 (3% Ni, 1% Cr, 0.2% Mo)
U	830M31	150
V	830M31	100
W	830M31	63
U	826M40	250 (1½% Ni, 2½% Cr, 2½% Mo)
V	826M40	250
W	826M40	250
X	826M40	150
Y	826M40	150
Z	826M40	100

Similarly, if there were a requirement for an alloy steel from Table 8.7 to meet a specification of yield stress 585 mm^{-2} (MPa) and a limiting ruling section of 100 mm, then possible choices would be 605M36 and 640M40. A steel to meet the specification of tensile strength in the range 850–1000 N mm^{-2} (MPa) and a limiting ruling section of 250 mm, from the same list, would be 817M40. With this size of ruling section the Izod impact value would be 41 J.

Table 8.7 *Mechanical properties of hardened and tempered steels related to tensile strength range*

Carbon and carbon–manganese steels

Condition	P					Q					R				
R_m	550–700					620–770					700–850				
HB	152–207					179–229					201–255				
Grade	LRS	R_e	A	I	$R_{p0.2}$	LRS	R_e	A	I	$R_{p0.2}$	LRS	R_e	A	I	$R_{p0.2}$
070M20	19	355	20	40	340										
070M26	29	355	20	40	325	13	415	16	33	400					
080M40						63	385	16	33	355	19	465	16	33	450
150M19	150	340	18	54	310	63	430	16	54	400	29	510	16	40	480
212M36	100	340	20	33	310	63	400	18	33	370	13	495	16	54	480

Alloy steels

Condition	R					S					T				
R_m	700–850					770–930					850–1000				
HB	201–255					223–227					248–302				
Grade	LRS	R_e	A	I	$R_{p0.2}$	LRS	R_e	A	I	$R_{p0.2}$	LRS	R_e	A	I	$R_{p0.2}$
593M40	63	525	17	47	510	22	585	15	27	570					
530M60	100	525	17	54	510	63	585	15	54	570	29	680	13	54	665
605M36	250	495	15	33	480	100	585	15	54	570	63	680	13	54	665
	150	525	17	54	510										
640M40	150	525	17	54	510	100	585	15	54	570	63	680	13	54	665
653M31						150	585	15	40	570	100	680	13	54	665
709M40	250	495	15	33	480	250	555	13	27	540					
						150	585	15	54	570	100	680	13	54	665
817M40											250	650	13	41	635
											150	680	13	54	665

R_m denotes tensile strength (N mm^{-2} or MPa)
HB denotes Brinell hardness number
LRS denotes limiting ruling section (mm)
R_e denotes yield stress (N mm^{-2} or MPa)
A denotes elongation, minimum on 5.65 S_0 gauge length (%)
I denotes minimum Izod impact value (J)
$R_{p0.2}$ denotes minimum 0.2% proof stress (N mm^{-2} or MPa)

Other countries have other code systems for specifying steels. A common one is that developed by the *American Society of Automotive Engineers*, the code being referred to as *SAE* numbering. A four-symbol code is used, the first two numbers indicating the type of steel and the third and fourth numbers indicating 100 times the percentage of carbon content.

1000 series   Carbon steels; 10XX is plain carbon with a maximum manganese content of 1.00%, 11XX is resulphurised, 12XX resulphurised and rephosphorised, 13XX has 1.75% manganese, etc.

2000 series	Nickel steels; 23XX are nickel steels having 3.50% nickel, 25XX has 5.00% nickel.
3000 series	Nickel chromium steels; 32XX has 1.75% nickel and 1.07% chromium, 34XX has 3.00% nickel and 0.77% chromium.
4000 series	Chromium molybdenum steels; 40 XX has 0.20 or 0.25% molybdenum, 41XX has 0.50%, 0.80% or 0.95% chromium with 0.12%, 0.20% 0.25% or 0.30% molybdenum.
5000 series	Chromium steels containing up to 1% chromium.
6000 series	Chromium vanadium steels.
7000 series	Steels containing up to 6% tungsten.
8000 series	Steels containing nickel, chromium and molybdenum.
9000 series	Silicon manganese steels.

There are some specially specified steels which have modified SAE numbers.

| XXBXX | The 'B' denotes boron intensified steels, the letter being incorporated in the middle of the normal specifications. |
| XXLXX | The 'L' denotes leaded steels, the letter being incorporated in the middle of the normal specification. |

These examples illustrate the use of the coding.

| 1330 | A carbon steel having 0.30% carbon and 1.75% manganese. |
| 4012 | A steel having 0.12% carbon, 0.20% molybdenum and 0.90% manganese. |

Tool steels are specified by BS 4659, the coding used being based on that of the American Iron and Steel Institute (AISI), the only difference being that in Britain the American codes are prefixed by a B:

BS 4659 code	AISI code	
BW	W	A water-hardening tool steel.
BO	O	Oil-hardening tool steel for cold work.
BA	A	Medium alloy hardening for cold work.
BD	D	High-carbon and high-chromium content for cold work.
BH	H	Chromium or tungsten base for hot work.
BT	T	Tungsten base, high-speed steel.
BM	M	Molybdenum base, high-speed steel.
BL	L	Low alloy tool steel for special applications.

The tool steel code letters are followed by a number to denote a particular steel composition. Thus BW5 is a water-hardened tool steel having 1.10% carbon and 0.50% chromium. BO7 is an oil-hardened tool steel having 1.20% carbon, 1.75% tungsten and 0.75% chromium.

8.8 Cast irons

The term 'cast iron' arises from the method by which the material is produced. Pig iron is remelted in a furnace and the properties of the iron are modified by additions of other materials such as steel scrap. The resulting iron alloy is cast. Cast irons have more than about 2% carbon and often significant amounts of silicon as well as smaller amounts of other materials. Cast irons are used widely because of their ease of melting and hence use where components are to be produced by casting. They are also relatively cheap.

The structure, and hence physical properties, of cast irons is affected by:

(a) the carbon content,
(b) the rate of cooling of the iron from the liquid state to room temperature,
(c) the presence of other elements.

In carbon steels the carbon exists in the form of cementite, a compound formed between iron and carbon. In cast iron the carbon occurs as graphite or cementite. Graphite is a soft, grey substance used in the 'lead' in pencils. Cementite, however, is very hard. The higher the graphite content of an iron, the more grey the appearance of the iron. The various cast irons can be classified by the presence or otherwise of the carbon as graphite and the type of structure of the iron. Figure 8.30 shows the general classification. The formation of graphite in a cast iron is affected by the rate of cooling in that a high rate of cooling tends to hinder the formation of graphite and so give a structure with more cementite. The presence of other elements can also have a significant effect on the production of graphite.

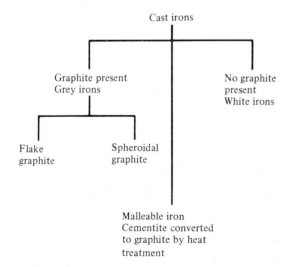

Figure 8.30 A general classification of cast irons

Grey and white cast irons

Figure 8.2 (p. 209) includes that part of the iron–carbon diagram of relevance to cast irons, i.e. iron alloys containing more than 2.0% carbon. At a temperature of about 1147°C a eutectic occurs for alloys with 4.3% carbon.

If we consider the cooling of an iron from the liquid state we have the following sequence of events. With very slow cooling, when solidification starts to occur the result is the formation of austenite in the liquid. At 1147°C the liquid solidifies to give austenite plus graphite. As the temperature is further lowered, the graphite increases as a result of precipitation from the austenite. At 723°C the remaining austenite changes to ferrite plus graphite. The result at room temperature is a structure of ferrite and graphite. This type of cast iron is known as a *grey iron* because of the grey appearance of its freshly fractured surface. This grey iron is referred to as a ferritic form of grey iron.

A faster cooling rate can lead to a structure in which the austenite changes to a mixture of ferrite and pearlite at 723°C, or, with an even faster cooling rate, entirely pearlite. The results are structures involving graphite with either ferrite and pearlite or just pearlite. The greater the amount of pearlite in these grey irons, the harder the material.

When the iron solidifies from the liquid state the graphite usually forms as flakes. Hence the structure at room temperature has graphite flakes in ferrite, ferrite and pearlite, or pearlite, depending on the rate of cooling (Figure 8.31).

Figure 8.31 A grey cast iron. The structure consists of black graphite flakes in a matrix of pearlite. (Courtesy of BCIRA)

With even faster cooling a different type of structure is formed. The solidification at 1147°C gives austenite and cementite. As the temperature is further lowered the cementite grows due to precipitation from the austenite. At 723°C the remaining austenite changes to pearlite. The result at room temperature is a structure of cementite and pearlite (Figure 8.32). This type of cast iron is known as a *white iron* because of the white appearance of its freshly fractured surface. White iron, because of its high cementite content, is hard and brittle. This makes it difficult to machine and hence of limited use. The main use is where a wear-resistant surface is required.

A casting will often have sections with differing thicknesses. The rate of cooling of a part of a casting will depend on the thickness of the section concerned, the thinner the section the more rapidly it will cool. This means that there is likely to be a variation in the properties throughout the casting, the thinner section possibly giving white iron while the thicker part gives a grey iron. There is thus a need to consider carefully the way in which a casting cools if the properties of the resultant casting are to be controlled.

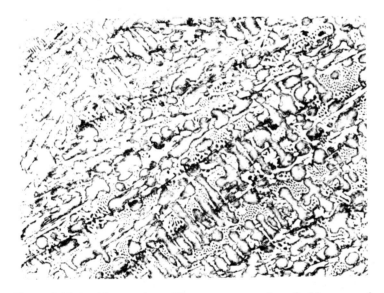

Figure 8.32 A white cast iron. The structure consists of white cementite and dark pearlite. (Courtesy of BCIRA)

Some elements when included with the carbon in cast iron promote the formation of graphite, others promote the production of the carbide. Silicon and nickel promote the formation of graphite by causing the cementite to become unstable and precipitate carbon. Thus the effect of including silicon with the carbon is to increase the change of a grey iron being produced, rather than a white iron. Figure 8.33 illustrates this. With, say, a 3% carbon iron and normal cooling we can have the following situations:

(a) With no silicon a white iron is produced.
(b) With 1% silicon a pearlitic grey iron is produced.
(c) With 2% silicon a grey iron with both ferrite and pearlite is produced.
(d) With 3% silicon a ferritic grey iron is produced.

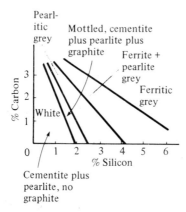

Figure 8.33 The effect of silicon on the structure of cast iron during normal cooling.

The type of iron produced depends on the rate of cooling as well as the silicon content. The silicon content can, however, be chosen for a particular casting, which will cool at some particular rate, so that the required type of cast iron is produced.

Sulphur often exists in cast iron and has the effect of stabilising cementite, i.e. the opposite effect to that of silicon. The presence of sulphur thus favours the production of white iron rather than grey iron. The sulphur thus increases the hardness of the cast iron and also increases the brittleness.

The addition of small amounts of manganese to a sulphur-containing iron enables the sulphur to combine with the manganese to form manganese sulphide. This removal of the free sulphur has the effect of increasing the chance of grey iron being produced. With higher amounts of manganese, a reaction occurs between the manganese and the carbon with the production of manganese carbide which considerably hardens the cast iron.

The addition of phosphorus to the iron has little effect on the amounts of cementite or graphite but does increase the fluidity of the iron in casting. It does this by reacting with the iron to produce iron phosphide which has a low melting point.

The following is a typical composition for a grey iron:

Carbon 3.2 to 3.5% (most as graphite)
Silicon 1.3 to 2.3%
Sulphur 0.10%
Manganese 0.5 to 0.7%
Phosphorus 0.15 to 1.0%

A typical white iron could have the following composition:

Carbon 3.3% (all in cementite)
Silicon 0.5%
Sulphur 0.15%
Manganese 0.5%
Phosphorus 0.5%

Malleable cast irons

Malleable cast irons are produced by the heat treatment of white cast irons. Three forms of malleable iron occur: *whiteheart*, *blackheart* and *pearlitic*. Malleable irons have better ductility than grey cast irons and this, combined with their high tensile strength, makes them a useful material.

In the blackheart process, white iron castings are heated in a non-oxidising atmosphere to 900°C and soaked at that temperature for two days or more. This causes the cementite to break down. The result is spherical aggregates of graphite in austenite. The casting is then cooled very slowly, resulting in the austenite changing into ferrite and more graphite (Figure 8.34). Figure 8.35 shows the form of the product.

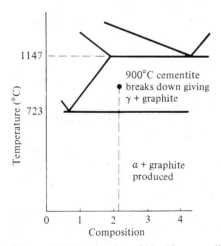

Figure 8.34 Production of blackheart malleable iron

The whiteheart process also involves heating white iron castings to about 900°C and soaking at that temperature. But in this process the castings are packed in canisters with haematite iron ore. This gives an oxidising atmosphere. Where the casting is thin, the carbon is oxidised to form a gas and so leaves the casting. In the thicker sections of the casting only the carbon in the surface layers leaves. The result, after very slow cooling, is a ferritic

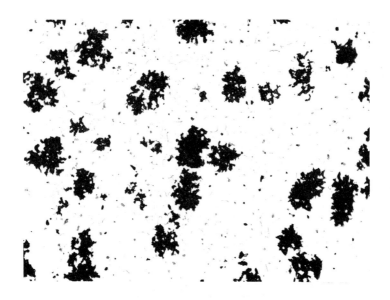

Figure 8.35 Blackheart malleable iron. The structure consists of 'rosettes' of graphite in a matrix of ferrite. (Courtesy of BCIRA)

structure in the thin sections of the casting and, in the thick sections, a ferrite outer layer with a ferrite plus pearlite inner core.

Pearlite maleable iron is produced by heating a white iron casting in a non-oxidising atmosphere to 900°C and then soaking at that temperature. This causes the cementite to break down and give spherical aggregates of graphite in austenite, as with blackheart iron. However, if a more rapid cooling is used a pearlitic structure is produced. This pearlitic malleable iron has a higher tensile strength than blackheart iron.

Pearlitic malleable irons can be produced by other methods. One method involves adding 1% manganese to the iron. The manganese inhibits the production of graphite. The higher strength pearlitic malleable irons are produced by quenching the iron from 900°C and then tempering.

Nodular cast irons

Nodular iron, or *spheroidal-graphite (SG) iron* or *ductile iron* as it is sometimes called, has the graphite in the iron in the form of nodules of spheres. Magnesium or cerium is added to the iron before casting. The effect of these materials is to prevent the formation of graphite flakes during the slow cooling of the iron; the graphite forms nodules instead. At room temperature the structure of the cast iron is mainly pearlitic with nodules of graphite (Figure 8.36). The resulting material is more ductile than a grey iron.

A heat treatment process can be applied to a pearlitic nodular iron to give a microstructure of graphite nodules in ferrite. The treatment is to heat to

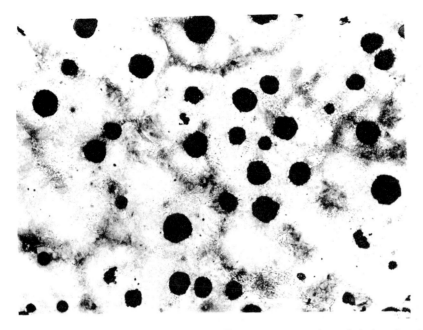

Figure 8.36 Spheroidal graphite iron. The structure consists of dark spheroidal graphite in a matrix of pearlite. (Courtesy of BCIRA)

900°C, soak at that temperature and then slowly cool. This ferritic form is more ductile, but has less tensile strength, than the pearlitic form.

Properties and uses of cast irons

Table 8.8 gives typical properties of cast irons. With the exception of the white

Table 8.8 *Typical properties of cast irons*

Type of material	Condition	Tensile strength ($N\ mm^{-2}\ or\ MPa$)	Elongation (%)	Hardness (HB)
Grey iron	As cast	150–400	0.5–0.7	130–300
White iron	As cast	230–460	0	400
Blackheart	Annealed	290–340	6–12	125–140
Whiteheart	Annealed	270–410	3–10	120–180
Pearlitic malleable	Normalised	440–570	3– 7	140–240
Nodular ferritic	As cast	370–500	7–17	115–215
Nodular pearlitic	As cast	600–800	2– 3	215–305

iron, all the other cast irons listed give good to very good machining. Similarly for casting, the white iron gives only fair castings while the other cast irons give good to very good castings. Fair welds can be achieved with the cast irons, apart from the white iron for which welding is poor. White iron has a very high abrasion resistance.

The following are typical uses of the various cast irons:

Grey iron	Water pipes, motor cylinders and pistons, machine castings, crankcases, machine tool beds, manhole covers
White iron	Wear-resistant parts, such as grinding mill parts, crusher equipment
Blackheart	Wheel hubs, pedals, levers, general hardware, brake shoes
Whiteheart	Wheel hubs, bicycle and motor-cycle frame fittings
Pearlitic malleable	Camshafts, gears, couplings, axle housings
Nodular ferritic	Heavy duty piping
Nodular pearlitic	Crankshafts

In selecting a cast iron for a particular application the following considerations are taken into account:

Grey irons	Very good machinability and stability. Able to damp out vibrations, i.e. considered to have a good damping capacity. Excellent wear resistance due to the graphite giving a self-lubricating effect for metal-to-metal contacts. Relatively poor tensile strength and ductility, but good strength in compression. Not good for shock loads
White iron	Excellent abrasion resistance. Very hard. Virtually unmachinable so has to be cast to the required shape and dimensions
Malleable iron	Good machinability and stability. Higher tensile strength and ductility than grey iron. Better for shock loads then grey iron
Nodular iron	High tensile strengths with reasonable ductility. Good machinability and wear characteristics, but not as good as grey iron. Better for shock loads than grey iron (about the same as malleable iron)

Problems

1 Sketch and label the steel section of the iron–carbon system, using the terms austenite, ferrite and cementite.
2 How do the structures of *hypo*-eutectoid and *hyper*-eutectoid steels differ at room temperature as a result of being slowly cooled from the austenite state?

3 Describe the form of the microstructure of a slowly cooled steel having the eutectoid structure.

4 What would be the expected structure of a 1.1% carbon steel if it were cooled slowly from the austenitic state?

5 Explain the terms points A_1, A_3 and A_{cm}.

6 Explain how the percentage of carbon present in a carbon steel affects the mechanical properties of the steel.

7 Carbon steel is used for the following items. Which type of carbon steel would be most appropriate for each item?
   (a) Railway track rails.
   (b) Ball bearings.
   (c) Hammers.
   (d) Reinforcement bars for concrete work.
   (e) Knives.

8 In what ways do alloying elements, other than carbon, have an effect on the properties of steel?

9 In what main ways do the following elements affect the properties of steel when they are alloyed with it?
   (a) Manganese, (b) chromium, (c) molybdenum, (d) sulphur, (e) silicon.

10 What elements added to a steel can improve the corrosion properties of that steel?

11 What elements are usually added to a steel to increase its machinability?

12 State two elements that, when added to steel, restrict grain growth.

13 What is meant by austenitic steels and ferritic steels?

14 A commonly used structural steel has the following composition: 0.40% carbon, 0.55% manganese, 1.50% nickel, 1.20% chromium, 0.30% molybdenum. What are the effects of the nickel, chromium and molybdenum on the properties of the steel?

15 What are the properties required of (a) tool steels in general, (b) tool steels for use in hot working processes, (c) shock-resistant tool steels?

16 What are stainless steels?

17 What is meant by 'free-cutting' steels?

18 How can (a) ferritic steels, (b) martensitic steels be hardened?

19 What will be the form of the microstructure of a 0.5% carbon steel after the following treatment: Heat and soak at 805°C and then a very slow cool to room temperature?

20 How would the answer to Problem 19 have differed if the steel, instead of being slowly cooled from 805°C to room temperature, had been quenched?

21 Explain how a 0.4% carbon steel would be hardened. Give details of the temperatures involved.

22 Describe the differences between full annealing and normalising.

23 State what is meant by the critical cooling rate.

24 How does increasing the temperature at which a carbon steel is tempered change the final properties of the steel?

25 How would the mechanical properties of 0.6% carbon steels differ if the following heat treatments were applied?
   (a) Heat and soak at 800°C and then quench in cold water.
   (b) Heat and soak at 800°C and then quench in oil.
   (c) Heat and soak at 800°C and then slowly cool in the furnace.
26 A carbon steel with 1.1% carbon is to be given a full annealing treatment. What temperature and cooling rate are necessary for such a treatment?
27 Why are cylindrical objects quenched vertically?
28 A cold chisel is tempered at a temperature of 280°C while a scriber is tempered at 200°C. How does the hardness of the steel differ for the two items as a result of the differing tempering temperatures? Why are the components required to have different hardnesses? What would happen if there was an error and the tempering temperature for the cold chisel was as high as 380°C?
29 Explain, by reference to an iron–carbon thermal equilibrium diagram, the procedures used for the case hardening of a low carbon steel.
30 Describe the functions of the muffle and salt-bath furnaces?
31 State the form of heat treatments needed to effect the following changes:
   (a) a 0.2% carbon steel to be made as soft as possible;
   (b) a 1.0% carbon steel to be made as soft as possible:
   (c) a 0.4% carbon steel to be made as hard as possible;
   (d) a 0.2% carbon steel to be case hardened.
32 Explain the term 'hardenability'.
33 Describe the Jominy test.
34 Explain how Jominy test results can be used to determine the hardness distribution that would occur across a section of a steel component when it is quenched.
35 State the type of steel and the percentages of the various constituents for the steels specified by the following BS 970 codes.
   (a) 040A12, (b) 070M26, (c) 150M19, (d) 210M15, (e) 503A37, (f) 653M31.
36 A steel is specified as having the condition 'R' when at a certain ruling section. What is meant by the condition 'R' and why is the ruling section specified?
37 Tool steel is specified as being 'BD' according to BS 4659. What is the significance of the letters BD?
38 A chromium-base tool steel is required for hot work. What code letters should be looked for in tool steels quoted to BS 4659 specifications?
39 Explain the effects on the microstructure and properties of cast iron of (a) cooling rate, (b) carbon content, (c) the addition of silicon, manganese, sulphur and phosphorus.
40 Describe the conditions under which (a) grey iron, (b) white iron are produced.
41 In what way does the section thickness of a casting affect the structure of the cast iron?

42 How do the mechanical properties of malleable irons compare with those of grey irons?

43 Describe the way in which the structures of nodular irons and malleable irons differ from those of grey irons?

44 Which types of cast iron have high ductilities?

45 Which types of cast irons have high tensile strengths?

46 Describe the forms of the microstructure of (a) blackheart and (b) whiteheart cast irons.

47 Why is it important to know whether a casting will require any machining before deciding on the material to be used?

48 How does the presence of graphite as flakes or nodules affect the properties of the cast iron?

49 Which type of cast iron would be most suitable for a situation where there was a high amount of wear anticipated?

50 Which types of cast iron would you suggest for the following applications? Justify your answers.
   (a) Sewage pipe.
   (b) Crankshaft in an internal combustion engine.
   (c) Brake discs.
   (d) Manhole cover.

9

Non-ferrous alloys

9.1 The range of alloys

The term *ferrous alloys* is used for those alloys having iron as the base element, e.g. cast iron and steel. The term *non-ferrous alloys* is used for those alloys which do not have iron as the base element, e.g. alloys of aluminium. The following are some of the non-ferrous alloys in common use in engineering:

Aluminium alloys   Aluminium alloys have a low density, good electrical and thermal conductivity, high corrosion resistance. Typical uses are metal boxes, cooking utensils, aircraft bodywork and parts.

Copper alloys   Copper alloys have good electrical and thermal conductivity, high corrosion resistance. Typical uses are pump and valve parts, coins, instrument parts, springs, screws. The names brass and bronze are given to some forms of copper alloys.

Magnesium alloys   Magnesium alloys have a low density, good electrical and thermal conductivity. Typical uses are castings and forgings in the aircraft industry.

Nickel alloys   Nickel alloys have good electrical and thermal conductivity, high corrosion resistance, can be used at high temperatures. Typical uses are pipes and containers in the chemical industry where high resistance to corrosive atmospheres is required, food processing equipment, gas turbine parts. The names Monel, Inconel and Nimonic are given to some forms of nickel alloys.

Titanium alloys   Titanium alloys have a low density, high strength, high corrosion resistance, can be used at high temperatures. Typical uses are in aircraft for compressor discs, blades and casings, in chemical plant where high resistance to corrosive atmospheres is required.

Zinc alloys	Zinc alloys have good electrical and thermal conductivity, high corrosion resistance, low melting points. Typical uses are as car door handles, toys, car carburettor bodies—components that in general are produced by die casting.

Non-ferrous alloys have, in general, these advantages over ferrous alloys:

(a) good resistance to corrosion without special processes having to be carried out;
(b) most non-ferrous alloys have a much lower density and hence lighter weight components can be produced;
(c) casting is often easier because of the lower melting points;
(d) cold working processes are often easier because of the greater ductility;
(e) higher thermal and electrical conductivities;
(f) more decorative colours.

Ferrous alloys have these advantages over non-ferrous alloys:

(a) generally greater strengths;
(b) generally greater stiffness, i.e. larger values of Young's modulus;
(c) better for welding.

9.2 Aluminium

Pure aluminium has a density of 2.7×10^3 kg m^{-3}, compared with that of 7.9×10^3 kg m^{-3} for iron. Thus for the same size component the aluminium version will be about one third of the mass of an iron version. Pure aluminium is a weak, very ductile, material. It has an electrical conductivity about two thirds that of copper but weight for weight is a better conductor. It has a high thermal conductivity. Aluminium has a great affinity for oxygen and any fresh metal in air rapidly oxidises to give a thin layer of the oxide on the metal surface. This surface layer is not penetrated by oxygen and so protects the metal from further attack. The good corrosion resistance of aluminium is due to this thin oxide layer on its surface.

Aluminium of high purity (99.5% pure aluminium, or greater) is too weak a material to be used in any other capacity than a lining for vessels. It is used in this way to give a high corrosion resistant surface.

Aluminium of commercial purity, 99.0 to 99.5% aluminium, is widely used as a foil for sealing milk bottles, thermal insulation, and kitchen foil for cooking. The presence of a relatively small percentage of impurities in aluminium considerably increases the tensile strength and hardness of the material.

The mechanical properties of aluminium depend not only on the purity of the aluminium but also upon the amount of work to which it has been subject. The effect of working the material is to fragment the grains. This results in an increase in tensile strength and hardness and a decrease in ductility. By controlling the amount of working different degrees of strength and hardness can be produced. These are said to be different *tempers*. The properties of aluminium may thus, for example, be referred to as that for the annealed condition, the half-hard temper and the fully hardened temper.

Table 9.1 shows typical properties of aluminium.

Table 9.1   *Typical properties of aluminium*

Composition (%)	Condition	Tensile strength ($N\ mm^{-2}$ or MPa)	Elongation (%)	Hardness (HB)
99.99	Annealed	45	60	15
	Half hard	82	24	22
	Full hard	105	12	30
99.8	Annealed	66	50	19
	Half hard	99	17	31
	Full hard	134	11	38
99.5	Annealed	78	47	21
	Half hard	110	13	33
	Full hard	140	10	40
99	Annealed	87	43	22
	Half hard	120	12	35
	Full hard	150	10	42

Aluminium alloys

Aluminium alloys can be divided into two groups, wrought alloys and cast alloys. Each of these can be divided into two further groups:

(a) those alloys which are not heat treatable,
(b) those alloys which are heat treated.

The term *wrought material* is used for a material that is suitable for shaping by working processes, e.g. forging, extrusion, rolling. The term *cast material* is used for a material that is suitable for shaping by a casting process.

The non-heat treatable wrought alloys of aluminium do not significantly respond to heat treatment but have their properties controlled by the extent of the working to which they are subject. A range of tempers is thus produced. Common alloys in this category are aluminium with manganese or magnesium. A common aluminium–manganese alloy has 1.25% manganese, the effect of this manganese being to increase the tensile strength of the aluminium. The alloy still has a high ductility and good corrosion properties. This leads to uses such as kitchen utensils, tubing and corrugated sheet for building. Aluminium–magnesium alloys have up to 7% magnesium. The greater the percentage of magnesium, the greater the tensile strength (Figure 9.1). The alloy still has good ductility. It has excellent corrosion resistance and thus finds considerable use in marine environments, e.g. constructional materials for boats and ships.

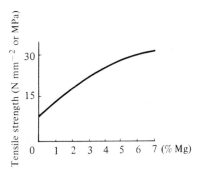

Figure 9.1 The effect on the tensile strength of magnesium content in annealed aluminium–magnesium alloys

The heat treatable wrought alloys can have their properties changed by heat treatment. Copper, magnesium, zinc and silicon are common additions to aluminium to give such alloys. Figure 9.2 shows the thermal equilibrium diagram for aluminium–copper alloys. When such an alloy, say 3% copper–97% aluminium, is slowly cooled, the structure at about 540°C is a solid solution of the α phase. When the temperature falls below the solvus temperature a copper–aluminium compound is precipitated. The result at room temperature is α solid solution with this copper–aluminium compound

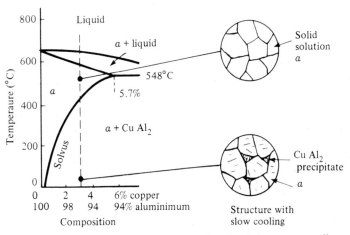

Figure 9.2 Thermal equilibrium diagram for aluminium–copper alloys

precipitate (CuAl$_2$). The precipitate is rather coarse, but this structure of the alloy can be changed by heating to about 500°C, soaking at that temperature, and then quenching, to give a supersaturated solid solution, just α phase with no precipitate. This treatment, known as *solution treatment*, results in an unstable situation. With time a fine precipitate will be produced. Heating to, say, 165°C for about ten hours hastens the production of this fine precipitate (Figure 9.3). The microstructure with this fine precipitate is both stronger and harder. The treatment is referred to as *precipitation hardening* (see Chapter 7 for more information).

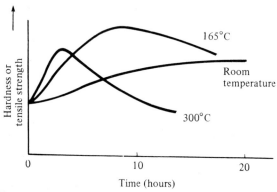

Figure 9.3 The effects of time and temperature on hardness and strength for an aluminium–copper alloy

A common group of heat treatable wrought alloys is based on aluminium with copper. Thus one form has 4.0% copper, 0.8% magnesium, 0.5% silicon and 0.7% manganese. This alloy is known as Duralumin. The heat treatment process used is solution treatment at 480°C, quenching and then precipitation hardening either at room temperature for about 4 days or at 165°C for 10 hours. This alloy is widely used in aircraft bodywork. The presence of the copper does, however, reduce the corrosion resistance and thus the alloy is often clad with a thin layer of high-purity aluminium to improve the corrosion resistance (Figure 9.4).

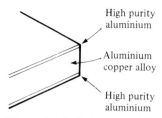

Figure 9.4 A clad duraluminium sheet

The precipitation hardening of an aluminium–copper alloy is due to the precipitate of the aluminium–copper compound. The age hardening of aluminium–copper–magnesium–silicon alloys is due to the precipitates of both an aluminium–copper compound $CuAl_2$ and an aluminium–copper–magnesium compound $CuAl_2Mg$. Other heat treatable wrought alloys are based on aluminium with magnesium and silicon. The age hardening with this alloy is due to the precipitate of a magnesium–silicon compound, Mg_2Si. A typical alloy has the composition 0.7% magnesium, 1.0% silicon and 0.6% manganese. This alloy is not as strong as the duralumin but has greater ductility. It is used for ladders, scaffold tubes, container bodies, structural members for road and rail vehicles. The heat treatment is solution treatment at 510°C with precipitation hardening by quenching followed by precipitation hardening of 10 hours at about 165°C. Another group of alloys is based on aluminium–zinc–magnesium–copper, e.g. 5.5% zinc, 2.8% magnesium, 0.45% copper, 0.5% manganese. These alloys have the highest strength of the aluminium alloys and are used for structural applications in aircraft and spacecraft.

An alloy for use in the casting process must flow readily to all parts of the mould and on solidifying it should not shrink too much and any shrinking should not result in fractures. In choosing an alloy for casting, the type of casting process being used needs to be taken into account. In sand casting the mould is made of sand bonded with clay or a resin. The cooling rate with such a method is relatively slow. A material for use by this method must give a material of suitable strength after a slow cooling process. With die casting the mould is made of metal and the hot metal is injected into the die under pressure. This results in a fast cooling. A material for use by this method must

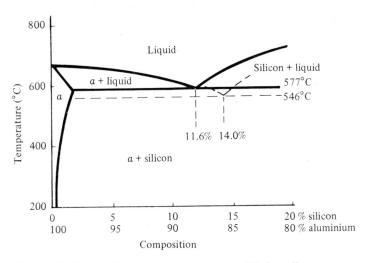

Figure 9.5 The aluminium–silicon thermal equilibrium diagram

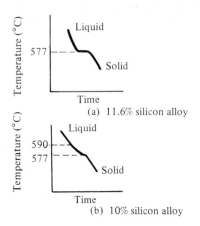

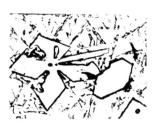

Figure 9.6 Cooling curves, (a) 11.6% silicon alloy, (b) 10% silicon alloy

Figure 9.7 A 13% cast silicon alloy showing silicon with eutectic. (From Rollason E. C., *Metallurgy for Engineers*, Edward Arnold)

develop suitable strength after fast cooling.

A family of aluminium alloys that can be used in the 'as cast' condition, i.e. no heat treatment is used, has aluminium with between 9 and 13% silicon. These alloys can be used for both sand and die casting. Figure 9.5 shows the thermal equilibrium diagram for aluminium–silicon alloys. The addition of silicon to aluminium increases its fluidity, between about 9 to 13% giving a suitable fluidity for casting purposes. The eutectic for aluminium–silicon alloys has a composition of 11.6% silicon. An alloy of this composition changes from the liquid to the solid state without any change in temperature, and alloys close to this composition solidify over a small temperature range (Figure 9.6). This makes them particularly suitable for die casting where a quick change from liquid to solid is required in order that a rapid ejection from the die can permit high output rates.

For the eutectic composition the microstructure shows a rather coarse eutectic structure of α phase and silicon. For an alloy having more silicon than the 11.6% eutectic value, the microstructure consists of silicon crystals in eutectic structure (Figure 9.7). The coarse eutectic structure, together with the presence of the embrittling silicon crystals, result in rather poor mechanical properties for the casting. The structure can, however, be made finer and the silicon crystal formation prevented by a process known as *modification*. This involves adding about 0.005 to 0.15% metallic sodium to the liquid alloy before casting. This produces a considerable refinement of the eutectic structure and also causes the eutectic composition to change to about 14.0% silicon. This displacement of the eutectic point is indicated on Figure 9.5 by the dashed line. Thus for a silicon content below 14%, the structure, as modified, has α phase crystals in a finer eutectic structure (Figure 9.8). The result is an increase in both tensile strength and ductility and so a much better casting material.

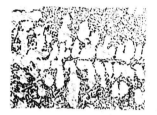

Figure 9.8 The same material as in Figure 9.7, but modified by sodium, showing aluminium with a fine eutectic (From Rollason, E. C., *Metallurgy for Engineers*, Edward Arnold)

The aluminium–silicon alloy is widely used for both sand and die casting, being used for many castings in cars, e.g. sumps, gear boxes and radiators. It is also used for pump parts, motor housings and a wide variety of thin walled and complex castings.

Other cast alloys that are not heat treated are aluminium–silicon–copper alloys, e.g. 5.0% silicon and 3.0% copper, and aluminium–magnesium–manganese alloys, e.g. 4.5% magnesium and 0.5% manganese. The silicon-copper alloys can be both sand and die cast, the magnesium–manganese alloys are however only suitable for sand casting. They have excellent corrosion resistance and are often used in marine environments.

The addition of copper, magnesium and other elements to aluminium alloys, either singly or in some suitable combination, can enable the alloy to be heat treated. Thus an alloy having 5.5% silicon and 0.6% magnesium can be subjected to solution treatment followed by precipitation hardening to give a high-strength casting material. Another heat treatable casting alloy has 4.0% copper, 2.0% nickel and 1.5% magnesium.

Table 9.2 shows typical properties of aluminium alloys.

Table 9.2 *Typical properties of aluminium alloys*

Composition	Condition	Tensile strength ($N\ mm^{-2}$ or MPa)	Elongation (%)	Hardness (HB)
Wrought, non-heat treated alloys				
1.25% Mn	Annealed	110	30	30
	Hard	180	3	50
2.25% Mg	Annealed	180	22	45
	$\frac{3}{4}$ hard	250	4	70
5.0% Mg	Annealed	300	16	65
	$\frac{1}{4}$ hard	340	8	80

Wrought, heated treated alloys

4.0% Cu, 0.8% Mg, 0.5% Si, 0.7% Mn	Annealed	180	20	45
	Solution treated, precipitation hardened	430	20	100
4.3% Cu, 0.6% Mg, 0.8% Si, 0.75% Mn	Annealed	190	12	45
	Solution treated, precipitation hardened	450	10	125
0.7% Mg, 1.0% Si, 0.6% Mn	Annealed	120	15	47
	Solution treated, precipitation hardened	300	12	100
5.5% Zn, 2.8% Mg, 0.45% Cu, 0.5% Mn	Solution treated, precipitation hardened	500	6	170

Cast, non-heat treated alloys

12% Si	Sand cast	160	5	55
	Die cast	185	7	60
5% Si, 3% Cu	Sand cast	150	2	70
	Die cast	170	3	80
4.5% Mg, 0.5% Mn	Sand cast	140	3	60

Cast, heat treated alloys

5.5% Si, 0.6% Mg	Sand cast, solution treated, precipitation hardened	235	2	85
4.0% Cu, 2% Ni, 1.5% Mg	Sand cast, solution treated, precipitation hardened	275	1	110

The effects of the various alloying elements used with aluminium can be summarised as:

Copper	Increases strength. Precipitation heat treatment possible. Improves machineability
Manganese	Improves ductility. Improves, in combination with iron, the castability
Magnesium	Improves strength. Precipitation heat treatment possible with more than about 6%. Improves the corrosion resistance
Silicon	Improves castability, giving an excellent casting alloy. Improves corrosion resistance
Zinc	Lowers castability. Improves strength when combined with other alloying elements

9.3 Copper

Copper has a density of 8.93×10^3 kg m^{-3}. It has very high electrical and thermal conductivity and can be manipulated readily by either hot or cold working. Pure copper is very ductile and relatively weak. The tensile strength and hardness can be increased by working; this does, however, decrease the ductility. Copper has good corrosion resistance. This is because there is a surface reaction between copper and the oxygen in the air which results in the formation of a thin protective oxide layer.

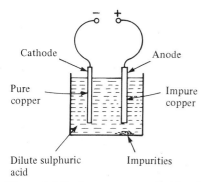

Figure 9.9 Basic arrangement for the electrolytic refining of copper

Very pure copper can be produced by an electrolytic refining process. An impure slab of copper is used as the anode while a pure thin sheet of copper is used as the cathode. The two electrodes are suspended in a warm solution of dilute sulphuric acid (Figure 9.9). The passage of an electric current through the arrangement causes copper to leave the anode and become deposited on the cathode. The result is a thicker, pure copper cathode, while the anode effectively disappears, the impurities have fallen to the bottom of the container. The copper produced by this process is often called *cathode copper* and has a purity greater than 99.99%. It is used mainly as the raw material for the production of alloys, though there is some use as a casting material.

Electrolytic tough pitch high-conductivity copper is produced from cathode copper which has been melted and cast into billets, and other suitable shapes, for working. It contains a small amount of oxygen, present in the form of cuprous oxide, which has little effect on the electrical conductivity of the copper. This type of copper should not be heated in an atmosphere where it can combine with hydrogen because the hydrogen can diffuse into the metal and combine with the cuprous oxide to generate steam. This steam can exert sufficient pressure to cause cracking of the copper.

Fire refined tough pitch high-conductivity copper is produced from impure copper. In the fire refining process, the impure copper is melted in an oxidising atmosphere. The impurities react with the oxygen to give a slag which is removed. The remaining oxygen is partially removed by poles of green hardwood being thrust into the liquid metal, the resulting combustion removes oxygen from the metal. The resulting copper has an electrical conductivity almost as good as the electrolytic tough pitch high-conductivity copper.

Oxygen-free high-conductivity copper can be produced if, when cathode copper is melted and cast into billets, there is no oxygen present in the atmosphere. Such copper can be used in atmospheres where hydrogen is present.

Another method of producing oxygen-free copper is to add phosphorus during the refining. The effect of small amounts of phosphorus in the copper is a very marked decrease in the electrical conductivity, of the order of 20%. Such copper is known as *phosphorus deoxidised copper* and it can give good welds, unlike the other forms of copper.

The addition of about 0.5% arsenic to copper increases its tensile strength, especially at temperatures of about 400°C. It also improves its corrosion resistance but greatly reduces the electrical and thermal conductivities. This type of copper is known as *arsenical copper*.

Electrolytic tough pitch high-conductivity copper finds use in high-grade electrical applications, e.g. wiring and busbars. Fire-refined tough pitch high-conductivity copper is used for standard electrical applications. Tough pitch copper is also used for heat exchangers and chemical plant. Oxygen-free high-conductivity copper is used for high-conductivity applications where hydrogen may be present, electronic components and as the anodes in the electrolytic refining of copper. Phosphorus deoxidised copper is used in chemical plant where good weldability is necessary and for plumbing and general pipework. Arsenical copper is used for general engineering work, being useful to temperatures of the order of 400°C.

Table 9.3 shows typical properties of the various forms of copper.

Table 9.3   *Typical properties of the various types of copper*

Composition	Condition	Tensile strength $(N\ mm^{-2}\ or\ MPa)$	Elongation (%)	Hardness (HB)
Electrolytic tough-pitch high conductivity copper				
99.90 min	Annealed	220	50	45
0.05 oxygen	Hard	400	4	115
Fire refined tough-pitch high conductivity copper				
99.85 min	Annealed	220	50	45
0.05 oxygen	Hard	400	4	115
Oxygen-free high-conductivity copper				
99.95 min	Annealed	220	60	45
	Hard	400	6	115

Phosphorus deoxidised copper

99.85 min 0.013–0.05 P	Annealed	220	60	45
	Hard	400	4	115

Arsenical copper

99.20 min 0.05 oxygen, 0.3–0.5 As	Annealed	220	50	45
	Hard	400	4	115

Copper alloys

The most common elements with which copper is alloyed are zinc, tin, aluminium and nickel. The copper–zinc alloys are referred to as brasses, the copper–tin alloys as tin bronzes, the copper–aluminium alloys as aluminium bronzes and the copper–nickel alloys as cupronickels, though where zinc is also present they are called nickel silvers. A less common copper alloy involves copper and beryllium.

The copper–nickel thermal equilibrium diagram is rather simple as the two metals are completely soluble in each other in both the liquid and solid states. The copper–zinc, copper–tin and copper–aluminium thermal equilibrium diagrams are however rather complex. In all cases, the α phase solid solutions have the same types of microstructure and are ductile and suitable for cold working. When the amount of zinc, tin or aluminium exceeds that required to saturate the α solid solution, a β phase is produced. The microstructures of these β phases are similar and alloys containing this phase are stronger and less ductile. They cannot be readily cold worked and are hot worked or cast.

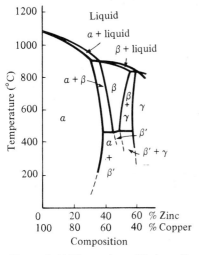

Figure 9.10 Thermal equilibrium diagram for copper–zinc alloys

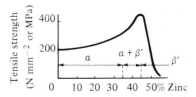
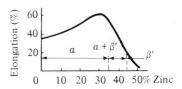

Figure 9.11 Strength and ductility for copper–zinc alloys

Further additions lead to yet further phases which are hard and brittle.

The *brasses* are copper–zinc alloys containing up to about 43% zinc. Figure 9.10 shows the relevant part of the thermal equilibrium diagram. Brasses with between 0 and 35% zinc solidify as α solid solutions, usually cored (Figure 9.12). These brasses have high ductility and can readily be cold worked. *Gilding brass*, 15% zinc, is used for jewellery because it has a colour resembling that of gold and can so easily be worked. *Cartridge brass*, 30% zinc and frequently referred to as *70/30 brass*, is used where high ductility is required with relatively high strength. It is called cartridge brass because of its use in the production of cartridge and shell cases. The brasses in the 0 to 30% zinc range all have their tensile strength and hardness increased by working, but the ductilities decrease.

Brasses with between 35 and 46% zinc solidify as a mixture of two phases (Figure 9.13). Between about 900°C and 453°C the two phases are α and β. At 453°C this β phase transforms to a low temperature modification referred to as β' phase. Thus at room temperature the two phases present are α and β'. The presence of the β' phase produces a drop in ductility but an increase in tensile strength to the maximum value for a brass (Figure 9.11). These brasses are

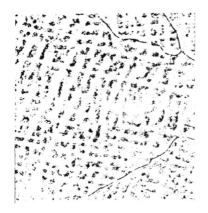
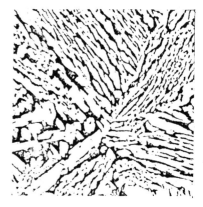

Figure 9.12 A 70% copper–30% zinc alloy. The structure is heavily cored

Figure 9.13 A 60% copper–40% zinc alloy. This consists of light α phase in a matrix of dark β'

known as *alpha-beta or duplex brasses*. They are not cold worked but have good properties for hot forming processes, e.g. extrusion. This is because the β phase is more ductile than the β' phase and hence the combination of α plus β gives a very ductile material. The hot working should take place at temperatures in excess of 453°C. The name *Muntz metal* is given to a brass with 60% copper–40% zinc.

The addition of lead to Muntz metal improves considerably the machining properties, without significantly changing the strength and ductility. *Leaded Muntz metal* has 60% copper, 0.3 to 0.8% lead and the remainder zinc.

Copper–zinc alloys containing just the β' phase have little industrial application. The presence of γ phase in a brass results in a considerable drop in strength and ductility, a weak brittle product being obtained.

Copper–tin alloys are known as *tin bronzes*. Figure 9.14 shows the thermal equilibrium diagram for such alloys. The dashed lines on the diagram indicate the phase that can occur with extremely slow cooling. The structure that normally occurs with up to about 10% tin is predominantly α solid solution. Higher percentage tin alloys will invariably include a significant amount of the δ phase. This is a brittle intermetallic compound, the α phase being ductile.

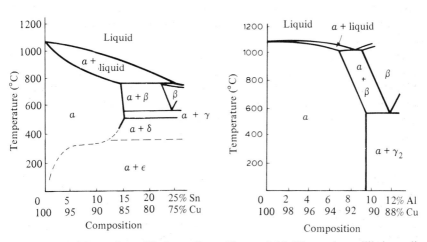

Figure 9.14 Thermal equilibrium diagram for tin bronzes

Figure 9.15 Thermal equilibrium diagram for aluminium bronzes

Bronzes that contain up to about 8% tin are α bronzes and can be cold worked. In making bronze, oxygen can react with the metals and lead to a weak alloy. Phosphorus is normally added to the liquid metals to act as a deoxidiser. Some of the phosphorus remains in the final alloy. This type of alloy is known as *phosphor bronze*. These alloys are used for springs, bellows, electrical contacts, clips, instrument components. A typical phosphor bronze might have about 95% copper, 5% tin and 0.02 to 0.40% phosphorus.

The above discussion refers to wrought phosphor bronzes. Cast phosphor bronzes contain between 5 and 13% tin with as much as 0.5% phosphorus. A typical cast phosphor bronze used for the production of bearings and high grade gears has about 90% copper, 10% tin and a maximum of 0.5% phosphorus. This material is particularly useful for bearing surfaces; it has a low coefficient of friction and can withstand heavy loads. The hardness of the material occurs by virtue of the presence of both δ phase and a copper–phosphorus compound.

Casting bronzes that contain zinc are called *gunmetals*. This reduces the cost of the alloy and also makes unnecessary the use of phosphorus for deoxidation as this function is performed by the zinc. *Admiralty gunmetal* contains 88% copper, 10% tin and 2% zinc. This alloy finds general use for marine components, hence the word 'Admiralty'.

Copper–aluminium alloys are known as *aluminium bronzes*. Figure 9.15 shows the thermal equilibrium diagram for such alloys. Up to about 9% aluminium gives *alpha bronzes*, such alloys containing just the α phase. Alloys with up to about 7% aluminium can be cold worked readily. *Duplex alloys* with about 10% aluminium are used for casting. Aluminium bronzes have high strength, and good resistance to corrosion and wear. These corrosion and wear properties arise because of the thin film of aluminium oxide formed on the surfaces. Typical applications of such materials are high-strength and highly corrosion-resistant items in marine and chemical environments, e.g. pump casings, gears, valve parts.

Alloys of copper and nickel are known as *cupronickels*, though if zinc is also present they are referred to as *nickel silvers*. Copper and nickel are soluble in each other in both the liquid and solid states, they thus form a solid solution whatever the proportions of the two elements. They are thus α phase over the entire range and suitable for both hot and cold working over the entire range. The alloys have high strength and ductility, and good corrosion resistance. The 'silver' coinage in use in Britain is a 75% copper–25% nickel alloy. The addition of 1 to 2% iron to the alloys increases their corrosion resistance. A 66% copper–30% nickel–2% manganese–2% iron alloy is particularly resistant to corrosion, and erosion, and is used for components immersed in moving sea water.

Nickel silvers have a silvery appearance and find use for items such as knives, forks and spoons. The alloys can be cold worked and usually contain about 20% nickel, 60% copper and 20% zinc.

Copper alloyed with small percentages of beryllium can be precipitation heat treated to give alloys with very high tensile strengths, such alloys being known as *beryllium bronzes*, or *beryllium copper*. The alloys are used for high-conductivity, high-strength electrical components, springs, clips and fastenings. For more details the reader is referred to E. G. West, *The Selection and Use of Copper-rich Alloys* (Oxford University Press).

Table 9.4 shows typical properties of copper alloys.

Table 9.4 *Typical properties of copper alloys*

Composition (%)	Condition	Tensile strength (N mm^{-2} or MPa)	Elongation (%)	Hardness (HB)
Brasses				
90 copper, 10 zinc	Annealed	280	48	65
80 copper, 20 zinc	Annealed	320	50	67
70 copper, 30 zinc	Annealed	330	70	65
	Hard	690	5	185
60 copper, 40 zinc	Annealed	380	40	75
Tin bronzes				
95 copper 5 tin, 0.02–0.40 phosphorus	Annealed	340	55	80
	Hard	700	6	200
91 copper, 8–9 tin, 0.02–0.40 phosphorus	Annealed	420	65	90
	Hard	850	4	250
Gunmetal				
88 copper, 10 tin, 2 zinc	Sand cast	300	20	80
Aluminium bronzes				
95 copper, 5 aluminium	Annealed	370	65	90
	Hard	650	15	190
88 copper, 9.5 aluminium, 2.5 iron	Sand cast	545	30	110
Cupronickels				
87.5 copper, 10 nickel, 1.5 iron, 1 manganese	Annealed	320	40	155
75 copper, 25 nickel, 0.5 manganese	Annealed	360	40	90
	Hard	600	5	170
Nickel silver				
64 copper, 21 zinc, 15 nickel	Annealed	400	50	100
	Hard	600	10	180
Beryllium bronzes				
98 copper, 1.7 beryllium, 0.2 to 0.6 cobalt and nickel	Solution treated, precipitation hardened	1200	3	370

There are a considerable number of copper alloys. The following indicates the type of selection that could be made.

Electrical conductors	Electrolytic tough-pitch, high-conductivity copper
Tubing and heat exchangers	Phosphorus deoxidised copper is generally used. Muntz metal, cupronickel or naval brass (62% copper–37% zinc–1% tin) is used if the water velocities are high
Pressure vessels	Phosphorus deoxidised copper, copper-clad steel or aluminium bronze
Bearings	Phosphor bronze. Other bronzes and brasses with some lead content are used in some circumstances
Gears	Phosphor bronze. For light duty gunmetals, aluminium bronze or die cast brasses may be used
Valves	Aluminium bronze
Springs	Phosphor bronze, nickel silver, basis brass are used for low cost springs. Beryllium bronze is the best material

9.4 Magnesium

Magnesium has a density of 1.7×10^3 kg m^{-3} and thus a very low density compared with other metals. It has an electrical conductivity of about 60% of that of copper, as well as a high thermal conductivity. It has a low tensile strength, needing to be alloyed with other metals to improve its strength. Under ordinary atmospheric conditions magnesium has good corrosion resistance, which is provided by an oxide layer that develops on the surface of the magnesium in air. However, this oxide layer is not completely impervious, particularly in air that contains salts, and thus the corrosion resistance can be low under adverse conditions. Magnesium is only used generally in its alloy form, the pure metal finding little application.

Magnesium alloys

Because of the low density of magnesium, the magnesium-base alloys have low densities. Thus magnesium alloys are used in applications where lightness is the primary consideration, e.g. in aircraft and spacecraft. Aluminium alloys have higher densities than magnesium alloys but can have greater strength. The strength-to-weight ratio for magnesium alloys is, however, greater than that of aluminium alloys. Magnesium alloys also have the advantage of good machinability and weld readily.

Magnesium–aluminium–zinc alloys and magnesium–zinc–zirconium are the main two groups of alloys in general use. Small amounts of other elements are also present in these alloys. The composition of an alloy depends on whether it is to be used for casting or working, i.e. as a wrought alloy. The cast alloys can often be heat treated to improve their properties.

A general-purpose wrought alloy has about 93% magnesium–6% aluminium–1% zinc–0.3% manganese. This alloy can be forged, extruded and welded, and has excellent machinability. A high-strength wrought alloy has 96.4% magnesium–3% zinc–0.6% zirconium. A general-purpose casting alloy has about 91% magnesium–8% aluminium–0.5% zinc–0.3% manganese. A high-strength casting alloy has 94.8% magnesium–4.5% zinc–0.7% zirconium. Both these casting alloys can be heat treated.

Table 9.5 shows typical properties of magnesium alloys.

Table 9.5   *Properties of typical wrought and cast magnesium alloys*

Composition (%)	Condition	Tensile strength ($N\ mm^{-2}$ or MPa)	Elongation (%)	Hardness (HB)
Wrought alloys				
93 magnesium, 6	Forged	290	8	65
aluminium, 1 zinc,	Extruded	310	8	70
0.3 manganese				
96.4 magnesium,				
3 zinc, 0.6				
zirconium				
Cast alloys				
91 magnesium, 8	As cast	140	2	55
aluminium, 0.5	Heat treated	200	6	75
zinc, 0.6 zirconium				
94.8 magnesium,	Heat treated	230	5	70
4.5 zinc, 0.7				
zirconium				

9.5 Nickel

Nickel has a density of 8.88×10^3 kg m^{-3} and a melting point of 1455°C. It possesses excellent corrosion resistance, hence it is used often as a cladding on a steel base. This combination allows the corrosion resistance of the nickel to be realised without the high cost involved in using entirely nickel. Nickel has good tensile strength and maintains it at quite elevated temperatures. Nickel can be both cold and hot worked, has good machining properties and can be joined by welding, brazing and soldering.

Nickel is used in the food processing industry in chemical plant, and in the petroleum industry, because of its corrosion resistance and strength. It is also used in the production of chromium-plated mild steel, the nickel forming an intermediate layer between the steel and the chromium. The nickel is electroplated on to the steel.

Nickel alloys

Nickel is used as the base metal for a number of alloys with excellent corrosion resistance and strength at high temperatures. One group of alloys is based on nickel combined with copper. A common nickel–copper alloy is known as *Monel* which has 68% nickel, 30% copper and 2% iron. It is highly resistant to sea water, alkalis, many acids and superheated steam. It has also high strength, hence its use for steam turbine blades, food processing equipment and chemical engineering plant components.

Another common nickel alloy is known as *Inconel* which contains 78% nickel, 15% chromium and 7% iron. The alloy has a high strength and excellent resistance to corrosion at both normal and high temperatures. It is used in chemical plant, aero-engines, as sheaths of electrical cooker elements, steam turbine parts and heat treatment equipment.

The *Nimonic* series of alloys are basically nickel–chromium alloys, essentially about 80% nickel and 20% chromium. They have high strength at high temperatures and are used in gas turbines for discs and blades.

Table 9.6 shows typical properties of nickel alloys.

Table 9.6   *Typical properties of nickel alloys*

Composition (%)	Condition	Tensile strength ($N mm^{-2}$ or MPa)	Elongation (%)	Hardness (HB)
68 nickel, 30	Annealed	500	40	110
copper, 2 iron	Cold worked	840	8	240
78 nickel, 15	Annealed	700	35	170
chromium, 7 iron	Cold worked	1050	15	290

9.6 Titanium

Titanium has a relatively low density, 4.5×10^3 kg m^{-3}, just over half that of steel. It has a relatively low tensile strength when pure but alloying gives a considerable increase in strength. Because of the low density of titanium its alloys have a high strength-to-weight ratio. Also, it has excellent corrosion resistance. However, titanium is an expensive metal, its high cost reflecting the difficulties experienced in the extraction and forming of the material; the ores are quite plentiful.

Titanium alloys

The main alloying elements used with titanium are aluminium, copper, manganese, molybdenum, tin, vanadium and zirconium. A ductile heat-treatable alloy has 97.5% titanium–2.5% copper and it can be welded and formed. A higher-strength alloy, which can also be welded and formed, has 92.5% titanium–5% aluminium–2.5% tin. A very high strength alloy has 82.5% titanium–11% tin–4% molybdenum–2.25% aluminium–0.25% silicon. This alloy can be heat-treated and forged.

The titanium alloys all show excellent corrosion resistance, have good strength-to-weight ratios, can have high strengths, and have good properties at high temperatures. They are used for compressor blades, engine forgings, components in chemical plant, and other duties where their properties make them one of the few possible choices despite their high cost.

Table 9.7 shows typical properties of titanium alloys.

Table 9.7 *Typical properties of titanium alloys*

Composition (%)	Condition	Tensile strength ($N\ mm^{-2}\ or\ MPa$)	Elongation (%)	Hardness (HB)
97.5 titanium, 2.5 copper	Heat treated	740	15	360
92.5 titanium, 5 aluminium, 2.5 tin	Annealed	880	16	360
82.5 titanium, 11 tin, 4 molybdenum, 2.25 aluminium, 0.25 silicon	Heat treated	1300	15	380

9.7 Zinc

Zinc has a density of 7.1×10^3 kg m^{-3}. Pure zinc has a melting point of only 419°C and is a relatively weak metal. It has good corrosion resistance, due to the formation of an impervious oxide layer on the surface. Zinc is frequently used as a coating on steel in order to protect that material against corrosion, the product being known as galvanised steel.

Zinc alloys

The main use of zinc alloys is for die-casting. They are excellent for this purpose by virtue of their low melting points and the lack of corrosion of dies used with them. The two alloys in common use for this purpose are known as alloy A and alloy B. *Alloy A*, the more widely used of the two, has the composition of 3.9 to 4.3% (max) aluminium, 0.03% (max) copper, 0.03 to

0.06% (max) magnesium, the remainder being zinc. *Alloy B* has the composition 3.9 to 4.3% (max) aluminium, 0.75 to 1.25% (max) copper, 0.03 to 0.06% (max) magnesium, with the remainder being zinc. Alloy A is the more ductile, alloy B has the greater strength.

The zinc used in the alloys has to be extremely pure so that little, if any, other impurities are introduced into the alloys, typically the required purity is 99.99%. The reason for this requirement is that the presence of very small amounts of cadmium, lead or tin renders the alloy susceptible to inter-crystalline corrosion. The products of this corrosion cause a casting to swell and may lead to failure in service.

After casting, the alloys undergo a shrinkage which takes about a month to complete; after that there is a slight expansion. A casting can be *stabilised* by annealing at 100°C for about 6 hours.

Zinc alloys can be machined and, to a limited extent, worked. Soldering and welding are not generally feasible.

Zinc alloy die-castings are widely used in domestic appliances, for toys, car parts such as door handles and fuel pump bodies, optical instrument cases.

The properties that might be obtained with zinc die-casting alloys are shown below.

Composition (%)	Condition	Tensile strength (N mm^{-2} or MPa)	Elongation (%)	Hardness (HB)
Alloy A	As cast	285	10	83
Alloy B	As cast	330	7	92

9.8 Comparison of non-ferrous alloys

The range of strengths that are obtained with the non-ferrous alloys considered in this chapter, are as follows:

Alloy	Tensile strength (N mm^{-2} or MPa)
Aluminium	100 to 550
Copper	200 to 1300
Magnesium	150 to 350
Nickel	400 to 1300
Titanium	400 to 1600
Zinc	200 to 350

Thus, if the main consideration in the choice of an alloy is that it must have high strength, then the choice, if limited to the alloys listed above, would be titanium, with the second choice nickel or copper alloys.

In some applications it is not just the strength of a material that is important but the strength/weight ratio. This is particularly the case in aircraft or spacecraft where not only is strength required but also a low mass. The densities of the alloys and the tensile strength/density ratios are shown below.

Alloy	Density $(10^3 \, kg \, m^{-3})$	Tensile strength/density $(N \, mm^{-2} \div 10^3 \, kg \, m^{-3})$
Aluminium	2.7	37 to 200
Copper	8	25 to 160
Magnesium	1.8	110 to 190
Nickel	8.9	47 to 146
Titanium	4.5	89 to 356
Zinc	6.7	30 to 52

Titanium gives the best possible strength/weight ratio. Magnesium, with its relatively low strength, has however a fairly high strength/weight ratio by virtue of its very low density. Aluminium, magnesium and titanium alloys are widely used in aircraft.

Another consideration that may affect the choice of an alloy is its corrosion resistance. In general the most resistant are titanium alloys, with the least resistant being magnesium. The rough order of descending corrosion resistance is: titanium, copper, nickel, zinc, aluminium and magnesium.

A vital factor in considering the choice of a material is the cost. The relative costs of the alloys, the costs all being relative to aluminium, are listed below. The costs are given in terms of the cost per unit mass.

Alloy	Relative cost per unit mass
Aluminium	1
Copper	1 to 2
Magnesium	3
Nickel	5
Titanium	25
Zinc	0.5

Titanium is the most expensive of the alloys, zinc the least expensive. However, in practice the use of a high-strength material may mean that a thinner section can be used and so less mass of alloy is required.

The process to be used to shape the material is another constraint on the selection of the material. With ductile materials, the drag forces on a cutting tool are high and the swarf tends to build up. These lead to poor machinability. Annealed aluminium has a high ductility and thus has poor machinability. A half-hard aluminium alloy can, however, have a good machinability, the ductility being much less. Materials can have their machinability improved by introducing other materials, such as lead, into the alloy concerned. The introduced material is in the form of particles which help to break up the swarf into small chips. Very hard materials may present machining problems in that the cutting tool needs to be harder than the material being machined.

Problems

1 What is the effect on the strength and ductility of aluminium of (a) the purity of the aluminium, (b) the temper?
2 Describe the effect on the strength of aluminium of the percentage of magnesium alloyed with it.
3 Describe the solution treatment and precipitation hardening processes for aluminium–copper alloys.
4 Describe the features of aluminium–silicon alloys which makes them suitable for use with die-casting.
5 What is the effect of heat treatment on the properties of an aluminium alloy, such as a 4.3% copper–0.6% magnesium–0.8% silicon–0.75% manganese?
6 Ladders are often made from an aluminium alloy. What are the properties required of that material in this particular use?
7 Explain with the aid of thermal equilibrium diagram how the addition of a small amount of sodium to the melt of an aluminium–silicon alloy changes the properties of alloys with, say, 12% silicon.
8 What are the differences between the following forms of copper: electrolytic tough-pitch, high-conductivity copper; fire-refined, tough-pitch, high-conductivity copper; oxygen-free, high-conductivity copper; phosphorus-deoxidised copper; arsenical copper.
9 Which form of copper should be used in an atmosphere containing hydrogen?
10 What is the effect of cold work on the properties of copper?
11 What is the effect of the percentage of zinc in a copper–zinc alloy on its strength and ductility?
12 What brass composition would be most suitable for applications requiring (a) maximum tensile strength, (b) maximum ductility, (c) the best combined tensile strength and ductility?
13 In general, what are the differences in (a) composition, (b) properties of α phase and duplex alloys of copper?
14 The 'silver' coinage used in Britain is made from a cupro-nickel alloy. What are the properties required of this material for such a use?
15 What are the constituent elements in (a) brasses, (b) phosphor bronzes and (c) cupro-nickels?
16 The name Muntz metal is given to a 60% copper–40% zinc alloy. Some forms of this alloy also include a small percentage of lead. What is the reason for the lead?

17 What percentage of aluminium would be likely to be present in an aluminium bronze that is to be cold worked?

18 What is meant by the term strength-to-weight ratio? Magnesium alloys have a high strength-to-weight ratio; of what significance is this in the uses to which the alloys of magnesium are put?

19 How does the corrosion resistance of magnesium alloys compare with other non-ferrous alloys?

20 What are the general characteristics of the nickel–copper alloys known as Monel?

21 Though titanium alloys are expensive compared with other non-ferrous alloys, they are used in modern aircraft such as *Concorde*. What advantages do such alloys possess which outweighs their cost?

22 What problems can arise when impurities are present in zinc die-casting alloys?

23 What is the purpose of 'stabilising annealing' for zinc die-casting alloys?

24 Describe the useful features of zinc die-casting alloys which makes them so widely used.

25 This question is based on the aluminium–silicon equilibrium diagram given in Figure 9.5.
(a) At what temperature will the solidification of a 85% aluminium–15% silicon start? At what temperature will it be completely solid?
(b) The above alloy is modified by the addition of a small amount of sodium. At what temperature will solidification start? At what temperature will it be completely solid?

26 This question is based on the copper–zinc thermal equilibrium diagram given in Figure 9.10.
(a) What is the temperature at which a 85% copper–15% zinc alloy begins to solidify?
(b) The pouring temperature used for casting is about 200°C above the liquidus temperature. What is the pouring temperature for the above alloy?

27 This question is based on the copper–zinc thermal equilibrium diagram given in Figure 9.10.
(a) Which of the following brasses would you expect to be just α solid solution? (i) 10% zinc–90% copper, (ii) 20% zinc–80% copper, (iii) 40% zinc–60% copper.
(b) How would you expect the properties of the above brasses to differ? How are the differences related to the phases present in the alloys?

28 This question is based on the copper–nickel thermal equilibrium diagram given in Figure 7.8.
(a) How does the copper–nickel thermal equilibrium diagram differ from that of copper–zinc?
(b) Over what range of compositions will copper–nickel alloys be α solid solution?

29 The following group of questions is concerned with justifying the choice of a particular alloy for a specific application.

(a) Why are zinc alloys used for die-casting?

(b) Why are magnesium alloys used in aircraft?

(c) Why are milk bottle caps made of an aluminium alloy?

(d) Why are titanium alloys extensively used in high-speed aircraft?

(e) Why are domestic water pipes made of copper?

(f) Why is cartridge brass used for the production of cartridge cases?

(g) Why are the ribs of hang gliders made of an aluminium alloy?

(h) Why are nickel alloys used for gas-turbine blades?

(i) Why is brass used for cylinder lock keys?

(j) Why are kitchen pans made of aluminium or copper alloys?

(k) Why is copper used for gaskets?

(l) Why are electrical cables made of copper?

(m) Why are cupro-nickels used for tubes in desalination plants?

30 Compare the properties of ferrous and non-ferrous alloys, commenting on the relative ease of processing.

10

Non-metals

10.1 Polymer structure

The plastic washing-up bowl, the plastic measuring rule, the plastic cup –
these are examples of the use of polymeric materials. The molecules in these
plastics are very large. A molecule of oxygen consists of just two oxygen atoms
joined together. A molecule in a plastic may have thousands of atoms all joined
together in a long chain. The backbones of these long molecules are chains of
carbon atoms. Carbon atoms are able to bond together strongly to produce long
chains of carbon atoms to which other atoms can become attached.

The term *polymer* is used to indicate that a compound consists of many
repeating structural units. The prefix 'poly' means many. Each structural unit
in the compound is called a *monomer*. Thus the plastic polyethylene is a
polymer which has as its monomer the substance ethylene. For many plastics
the monomer can be determined by deleting the prefix 'poly' from the name of
the polymer. Figure 10.1 shows the basic form of a polymer.

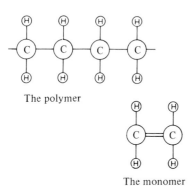

The polymer

The monomer

Figure 10.1 The polymer, polyethylene

If you apply heat to a plastic washing-up bowl the material softens. Removal
of the heat causes the material to harden again. Such a material is said to be

thermoplastic. The term implies that the material becomes 'plastic' when heat is applied.

If you applied heat to a plastic cup you might well find that the material did not soften but charred and decomposed. Such a material is said to be a *thermosetting plastic*.

Another type of polymer is the elastomers. Rubber is an elastomer. An *elastomer* is a polymer which by its structure allows considerable extensions which are reversible.

The thermoplastic, thermosetting and elastomer materials can be distinguished by their behaviour when forces are applied to them to cause stretching. Thermoplastic materials are generally flexible and relatively soft; if heated they become softer and more flexible. Thermosetting materials are rigid and hard with little change with an increase in temperature. Elastomers can be stretched to many times their initial length and still spring back to their original length when released. These different types of behaviours of polymers can be explained in terms of differences in the ways the long molecular chains are arranged inside the material.

Figure 10.2 shows some of the forms the molecular chains can take. These forms can be described as linear, branched and cross-linked. The linear chains have no side branches or chains or cross-links with other chains. Linear chains can move readily past each other. If, however, the chain has branches there is a reduction in the ease with which chains can move past each other. This shows itself in the material being more rigid and having a higher strength. If there are cross-links a much more rigid material is produced in that the chains cannot slide past each other at all.

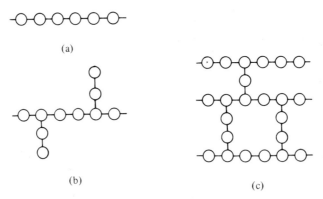

(a)

(b)

(c)

Figure 10.2 (a) Linear polymer chain, (b) Branched polymer chain, (c) Cross-linked polymer

Polyethylene, a thermoplastic material commonly known as polythene, has linear molecular chains (in the high density version). Polyethylene is easily stretched and is not rigid. Because the chains are independent of each other they can easily flow past each other and so the material has a relatively low melting point, no energy being needed to break bonds between chains. The absence of bonds between chains also means that, as none are broken when the material is heated, the removal of heat allows the material to revert to its initial harder state.

Some thermoplastic materials have molecules with side branches; the effect of these is to give a harder and more rigid material. Polypropylene is such a material, being harder and more rigid than polyethylene. Another consequence of a material having branched chains is that, as they do not pack so readily together in the material as linear chains, the material will generally have a lower density than the linear chain material.

Thermosetting materials are cross-linked polymers and are rigid. As energy is needed to break bonds before flow can occur, thermosetting materials have higher melting points than thermoplastic materials having linear or branched chains. Also the effect of heat is not reversible; when heat causes bonds to break, an irreversible change to the structure of the material is produced. Bakelite is an example of a thermosetting material. It can withstand temperatures up to 200°C, but most thermoplastics are not used above 100°C.

Elastomers have linear molecular chains. In the material these chains are all tangled up and there is no order in the packing of the molecular chains in the material. These tangled chains give a relatively open structure with the large amount of empty space between the tangled chains. When forces are applied to the material the chains are able to move very easily within the voids. It is this which accounts for the very high extensions possible with elastomers.

Copolymers

The term *homopolymer* is used to describe those polymers that are made up of just one monomer: for instance, polyethylene is made up of only the monomer

(a) Alternating

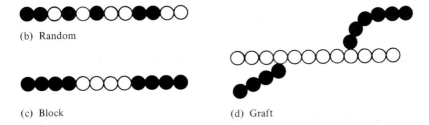

(b) Random

(c) Block

(d) Graft

Figure 10.3 Structures of copolymers made up of two monomers

ethylene. Other types of polymers, *copolymers*, can be produced by combining two or more monomers in a single polymer chain. Figure 10.3 shows four possible types of structure of copolymers based on two monomers.

Crystallinity in polymers

A crystal can be considered to be an orderly packing-together of atoms. The molecular chains of a polymer may be completely tangled up in a solid with no order whatsoever (Figure 10.4). Such a material is said to be *amorphous*. There is, however, the possibility that the polymer molecules can be arranged in an orderly manner within a solid. Thus Figure 10.5 shows linear polymer molecules folded to give regions of order. The orderly parts of such polymeric materials are said to be *crystalline*. Because long molecules can easily become tangled up with each other, polymer materials are often only partially crystalline, i.e. parts of the material have orderly arrangements of molecules while other parts are disorderly.

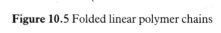

Figure 10.4 A linear amorphous polymer. Individual atoms are not shown, the chains just being represented by lines

Figure 10.5 Folded linear polymer chains

Not all polymers can give rise to crystallinity. It is most likely to occur with simple linear chain molecules. Branched polymer chains are not easy to pack together in a regular manner, the branches get in the way. If the branches are completely regularly spaced along the chain then some crystallinity is possible; irregularly spaced branches make crystallinity improbable. Cross-linked polymers cannot be rearranged due to the links between chains and so crystallinity is not possible.

There are many polymers based on the form of the polyethylene molecule. Despite being linear molecules, they do not always give rise to crystalline structures. PVC is essentially just the polyethylene molecule with some of the hydrogen atoms replaced with chlorine atoms (Figure 10.6). The molecule does not, however, give rise to crystalline structures. This is because the chlorine atoms are rather bulky and are not regularly spaced along the molecular chain. It is this lack of regularity which makes packing of the PVC molecular chains too difficult. Polypropylene has a molecule rather like that of polythene but with some of the hydrogen atoms replaced with CH_3 groups.

These are, however, regularly spaced along the molecular chain and thus orderly packing is possible and so some degree of crystallinity.

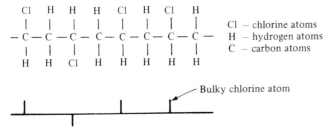

Figure 10.6 The basic form of a PVC molecule. The chlorine atoms are generally irregularly arranged on the different sides of the chain, so rendering orderly packing of the chains difficult

Table 10.1 shows the form of the molecular chains and the degree of crystallinity possible for some common polymers.

Table 10.1 *Chain structure of some common polymers*

Polymer	Form of chain	Possible crystallinity
Polyethylene	Linear	95%
	Branched	60%
Polypropylene	Regularly spaced side groups on linear chain	60%
Polyvinyl chloride	Irregularly spaced bulky chlorine atoms (Fig. 10.6)	0
Polystyrene	Irregularly spaced bulky side groups	0

When an amorphous polymer is heated, it shows no definite melting temperature but progressively becomes less rigid. The molecular arrangement in an amorphous material is all disorderly, just like that which occurs in a liquid. It is for this reason, i.e. no structural change occurring, that no sharp melting point occurs. For crystalline polymers there is an abrupt change at a particular temperature. Thus if the density of the polymer were being measured as a function of temperature, an abrupt change in density would be seen at a particular temperature. At this temperature the crystallinity of the polymer disappears, the structure changing from a relatively orderly one below the temperature to a disorderly one above it. The temperature at which the crystallinity disappears is defined as being the *melting point* of the polymer. The following are the melting points of some common polymers.

Polymer	Crystallinity	Melting point (°C)
Polyethylene	95%	138
Polyethylene	60%	115
Polypropylene	60%	176
Polyvinyl chloride	0%	212★
Polystyrene	0%	—

★This is not a clear melting point but a noticeable softening region of temperature.

The degree of crystallinity of a polymer affects its mechanical properties. The more crystalline a polymer, the higher its tensile modulus. Thus the linear-chain form of polyethylene in its crystalline form has the molecules closely packed together. Greater forces of attraction can exist between the chains when they are closely packed. The result is a stiffer material, a material with a higher tensile modulus. The branched form of polyethylene has a lower crystallinity and thus a lower tensile modulus. This is because the lower degree of crystallinity means that the molecules are not so closely packed together, an orderly structure being easier to pack closely than a disorderly one (you can get more clothes in a drawer if you pack them in an orderly manner than if you just throw them in). The farther apart the molecular chains, the lower the forces of attraction between them and so the less stiff, i.e. more flexible, the material and hence the lower the tensile modulus. The following are the tensile modulus and tensile strength values for polyethylene with different degrees of crystallinity:

Polymer	Crystallinity	Tensile modulus $(kN\ mm^{-2}\ or\ GPa)$	Tensile strength $(N\ mm^{-2}\ or\ MPa)$
Polyethylene	95%	21 to 38	0.4 to 1.3
Polyethylene	60%	7 to 16	0.1 to 0.3

The more cross-links there are with a polymer, the stiffer the polymer is, i.e. the higher its tensile modulus. A highly cross-linked polymer may be a rather hard, brittle substance.

Glass transition temperature

PVC, without any additives and at room temperature, is a rather rigid material. It is often used in place of glass. But if it is heated to a temperature of about 87°C a change occurs, the PVC becomes flexible and rubbery. The PVC below this temperature gives only a moderate elongation before breaking, above this temperature it stretches a considerable amount and behaves rather like a strip of rubber.

Polythene is a flexible material at room temperature and will give considerable extensions before breaking. If, however, the polythene is cooled to below about −120°C it becomes a rigid material.

The temperature at which a polymer changes from a rigid to a flexible material is called the *glass transition temperature*. The material is considered to be changing from a glass-like material to a rubber-like material. The following are the glass transition temperatures for some common thermoplastics.

Material	Glass transition temperature (°C)
Polyethylene	−120
Polypropylene	−10
Polyvinyl chloride	87
Polystyrene	100

At glass transition temperature the tensile modulus shows an abrupt change from a high value for the glass-like material to a low value for the rubber-like material. Figure 10.7 shows the form of the stress–strain graph for a polymer both below and above this transition temperature. Below the transition temperature the material has the type of stress–strain graph characteristic of a relatively brittle substance; above the transition temperature the graph is more like that of a rubber-like material in that the polymer may be stretched to many times its original length.

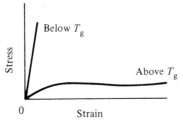

Figure 10.7 Stress–strain graphs for a polymer below and above its glass transition temperature T_g

Above the glass transition temperature the polymer chains are reasonably free to move and thus the application of stress causes molecular chains to uncoil and/or slide over one another, hence producing large extensions. When the temperature is decreased the density of the polymer increases and the chains become more closely packed. At the glass transition temperature the packing has become close enough to hinder the movement of the polymer chains to such an extent that the application of stress results in little extension.

The value of the glass transition temperature varies from one polymer to another. It is lowest for those polymers that have linear molecules, e.g. polythene, which are flexible and easily packed together in the solid. Polymers having branched chains have higher glass transition temperatures, e.g. polyvinyl chloride, and are thus not easily packed, if at all, in an orderly manner.

The effects of temperature and time on the mechanical properties

For polymers the variation of the tensile modulus with temperature is of considerable significance in that, for instance, it determines whether the plastic spoon used to stir the coffee is stiff and useful for the job concerned or far from stiff and not particularly useful. The tensile modulus of polymers decreases with an increase in temperature. The overall way in which the modulus changes with temperature is the same for all non-crystalline polymers, the differences between polymers being just the actual temperatures at which the change occurs. Figure 10.8 shows the general type of variation that occurs with temperature—note that the modulus axis of the graph is on a log scale.

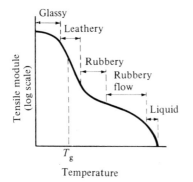

Figure 10.8 Variation of tensile modulus with temperature for an amorphous polymer

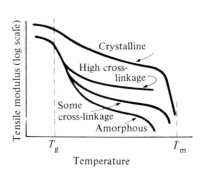

Figure 10.9 Variation of tensile modulus with temperature for polymers

In the glassy range of the graph the tensile modulus is at its maximum value and is less dependent on temperature that at other temperatures. In this region the polymer is stiff and not so easily stretched. This is because the polymer chains are too closely packed to slide past each other and the only way the polymer can extend is by stretching the chains themselves.

The region between the glassy region and the rubbery region is referred to as the leathery region. This represents the transition of the polymer from a tightly packed structure where no chain movement was possible to a state where the polymer chains can uncoil from their tangled state, but spring back to their tangled state when the applied stress is removed.

At yet higher temperatures the chains begin to slip past each other when stress is applied. This is the rubbery flow region. At still higher temperatures the polymer becomes liquid and complete movement of polymer chains past each other occurs. When chains move past each other the deformation is irrecoverable.

A plastic coffee spoon at room temperature is fairly rigid and if stressed may break in a brittle manner. The spoon is in the glassy region. If the spoon is in boiling water it behaves completely differently, very little stress has to be applied to cause a permanent irrecoverable deformation due to chain slipping occurring (the rubbery flow region).

The above discussion has considered what happens when stress is applied to polymers, but no mention was made of whether the stress was applied quickly or slowly. Time is needed for uncoiling or movement of polymer chains. A polymer that, at some particular temperature, may be rubbery with a slow application of stress, may be quite glass-like with a faster application of stress. The effects of increasing the rate of application of stress is to make the polymer more brittle and have a higher tensile modulus.

The effect of crystallinity on the behaviour of a polymer when subject to stress is to make it more stiff, i.e. have a higher tensile modulus. The polymer also becomes more brittle. Figure 10.9 shows how the tensile modulus varies with temperature for a crystalline polymer.

The graph also shows how cross-links in a polymer affect its properties. With an increasing amount of cross-linkage, the rubbery and flow ranges disappear. With a high amount of cross-linkage the material is hard and fairly brittle.

Temperature and polymer use

Amorphous polymers tend to be used below their glass transition temperature T_g. They are, however, formed and shaped at temperatures above the glass transition temperature when they are in a soft condition. Crystalline polymers are used up to their melting temperature T_m. They can be hot-formed and shaped at temperatures above T_m, being cold-formed and shaped at temperatures between T_g and T_m.

Polythene is a crystalline polymer, with a melting point of 138°C and a glass transition temperature of -120°C for the form that gives 95% crystallinity. The maximum service temperature of polythene items is about 125°C, i.e. just below the melting point. The form of polythene with 60% crystallinity has a melting point of 115°C and a maximum service temperature of about 95°C.

Polystyrene is an amorphous polymer with a glass transition temperature of 100°C. It has a maximum service temperature of about 80°C, i.e. just below the glass transition temperature.

Orientation

Figure 10.10 shows a typical stress–strain graph for a crystalline polymer, e.g. polythene. When the stress reaches point 'A' the material shows a sudden large reduction in cross-sectional area at some point (Figure 10.11). After this initial necking a considerable increase in strain takes place at essentially a constant stress, as the necked area gradually spreads along the entire length of the material. When the entire piece of material has reached the necked stage an increase in stress is needed to increase the strain further.

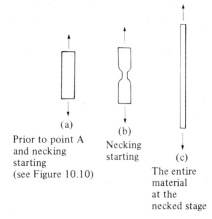

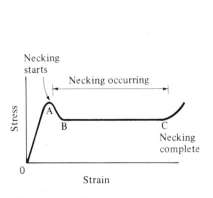

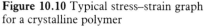

Figure 10.10 Typical stress–strain graph for a crystalline polymer

Figure 10.11 Necking with a polymer. (a) Prior to point 'A' and necking starting, (b) necking starting, (c) the entire material at the necked stage

The above sequence of events can be explained by considering the orientation of the polymer chains. Prior to necking starting, the polymer chains are folded to give regions of order in the material (as in Figure 10.4). When necking starts the polymer chains unfold to give a material with the chains lying along the direction of the forces stretching the material. As the necking spreads along the material so more of the chains unfold and line up. Eventually when the entire material is at the necked stage all the chains have lined up. Material that has reached this state shows a different behaviour and is said to be *cold drawn*. It is completely crystalline, i.e. all the chains are packed in a very orderly manner.

The above sequence of events only tends to occur if the material is stretched slowly and sufficient time elapses for the molecular chains to slide past each other. If a high strain rate is used, the material is likely to break without becoming completely orientated. You can try pulling a strip of polythene for yourself and see the changes. (A strip cut from a polythene bag can be used and you can pull it with your hands.)

With crystalline or semi-crystalline polymers, orientation is produced at temperatures both below and above the glass transition temperature. With amorphous polymers orientation is produced at temperatures above the glass transition temperature. Below that temperature an amorphous polymer is too brittle and breaks. The effect of the chains becoming orientated is to give a harder, stronger material. The effect can be considered to be similar to work hardening with metals.

In order to improve the strength of polymer fibres, e.g. polyester fibres, they are put through a drawing operation to orientate the polymer chains.

Stretching a polymer film causes orientation of the polymer chains in the direction of the stretching forces. The result is an increase in strength and stiffness in the stretching direction. Such stretching is referred to as uniaxial stretching and the effect as *uniaxial orientation*. The material is, however, weak and has a tendency to split if forces are applied in directions other than those of the stretching forces. For the polyester fibres this does not matter as the forces will be applied along the length of the fibre; with film this could be a serious defect. The problem can be overcome by using a biaxial stretching process in which the film is stretched in two directions at right angles to each other. The film has then *biaxial orientation*.

Rolling through compression rollers is similar in effect to the drawing operation and results in an uniaxial orientated product. Extrusion has a similar effect.

Orientation can be obtained by both cold and hot working processes. Hot working involves working at temperatures just below the glass transition temperature. Under such conditions orientation can be produced without any internal stresses being developed. Cold working, at lower temperature than those of hot working, requires more energy to produce orientation but also results in internal stresses being produced.

If orientated polymers are heated to above their glass transition temperature they lose their orientation. On cooling they are no longer orientated and are in the same state as they were before the orientation process occurred. This effect is made use of with shrinkable films. The polymer film is stretched, and so made longer even when the stretching force is removed. If it is then wrapped around some package and heated, the film contracts back to its initial, prestretched state. The result is a plastic film tightly fitting the package. The film is said to show *elastic memory*.

Plasticisation

The term *plastic* is used to describe materials based on polymers. Such materials invariably contain other substances which are added to the polymers to give required properties. *Stabilisers* are substances added in order to protect the plastic against ultraviolet radiation. *Fillers* may be added to change the strength properties, e.g. the addition of glass fibres, or to reduce friction, e.g. graphite. An important group of additives are called *plasticisers*. Their primary purpose is to enable the molecular polymer chains to slide more easily past each other.

Internal plasticisation involves modifying the polymer chain by the introduction into the chain of bulky side groups. An example of this is the plasticisation of polyvinyl chloride by the inclusion of some 15% of vinyl acetate in the polymer chains. These bulky side groups have the effect of forcing the polymer chains farther apart, thus reducing the attractive forces between the chains and so permitting easier flow of chains past each other.

A more common method of plasticisation is called *external plasticisation*. It involves a plasticiser being added to the polymer, after the chains have been produced. This plasticiser may be a liquid which disperses throughout the plastic, filling up the spaces between the chains. The effect of the liquid is the same as adding a lubricant between two metal surfaces, the polymer chains slide more easily past each other. The effect of the plasticiser is to weaken the attractive forces that exist between the polymer chains. The plasticiser decreases the crystallinity of polymers as it tends to hinder the formation of orderly arrays of polymer chains. The plasticiser also reduces the glass transition temperature. The effect on the mechanical properties is to reduce the tensile strength and increase the flexibility. The effects of plasticiser on the mechanical properties of PVC can be summarised as:

	Tensile strength $(N\ mm^{-2}\ or\ MPa)$	Elongation (%)
No plasticiser	52 to 58	2 to 40
Low plasticiser	28 to 42	200 to 250
High plasticiser	14 to 21	350 to 450

The PVC with no plasticiser is a rigid material, with low plasticiser content the PVC is pliable, with very high plasticiser content the PVC is soft and rubbery.

10.2 Thermoplastics

Thermoplastics are classed as *polyolefins* if they are based on the ethylene monomer, as *vinyls* if a different atom or group of atoms replaces one or more of the hydrogen atoms in the ethylene, and as *non-ethnic* where a non-ethylene monomer is involved. The main properties and uses of thermoplastics commonly used in engineering are outlined below and in Table 10.2. For fuller information about thermoplastics and their use the reader is referred to *The Selection and Use of Thermoplastics* by P. C. Powell (Oxford University Press).

Polyethylene (polythene), a polyolefin, is referred to as *high density* when the linear chain polymer is involved and as *low density* when the branched form is used. The high density polyethylene has the greater crystallinity. Low and high density polymers may be blended to give plastics with properties intermediate between those quoted above. Figure 10.12 shows the stress–strain graph for low density polythene.

Low density polyethylene is used mainly in the form of films and sheeting, e.g. polythene bags, 'squeeze' bottles, ball-point pen tubing, wire and cable insulation. High density polythene is used for piping, toys, filaments for fabrics, household ware. Both forms have excellent chemical resistance, low moisture absorption and high electrical resistance. The additives commonly used with polyethylene are carbon black as a stabiliser, pigments to give

Table 10.2 Properties of typical thermoplastics used in engineering

	Crystallinity (%)	Density (10^3 kg m^{-3})	T_g (°C)	T_m (°C)	Tensile strength (N mm^{-2} or MPa)	Tensile modulus (kN mm^{-2} or GPa)	Elongation (%)
Polythene							
High density	95	0.95	−120	138	22–38	0.4–1.3	50–800
Low density	60	0.92	−120	115	8–16	0.1–0.3	100–600
Polypropylene	60	0.90	−10	176	30–40	1.1–1.6	50–600
PVC							
With no plasticiser	0	1.4	87	—	52–58	2.4–4.1	2–40
Low plasticiser content	0	1.3	—	—	28–42	—	200–250
Polystyrene							
No additives	0	1.1	100	—	35–60	2.5–4.1	2–40
Toughened	0	1.1	—	—	17–24	1.8–3.1	8–50
ABS	0	1.1	71–112	—	17–58	1.4–3.1	10–140
Polycarbonate	0	1.2	150	—	55–65	2.1–2.4	60–100
Acrylic	0	1.18	0	—	50–70	2.7–3.5	5–8
Polyamides	Can be varied from low to high percentage						
Nylon 6		1.13	50	225	75	1.1–3.1	60–320
Nylon 6.6		1.1	57	265	80	2.8–3.3	60–300
Nylon 6.10		1.1	50	228	60	1.9–2.1	85–230
Nylon 11		1.1	—	—	50	0.6–1.5	70–300
Polyethylene terephthalate	60	1.3	69	267	50–70	2.1–4.4	60–100
Polyacetals							
Polyformaldehyde	High	1.41	−73	180	50–70	3.6	15–75
Polyoxymethylene	High	1.41	−76	180	70	3.6	15–75
PTFE	90	2.2	−120	327	14–35	0.4	200–600

coloured forms, glass fibres to give increased strength and butyl rubber to prevent inservice cracking.

Polypropylene, a polyolefin, is used mainly in its crystalline form, being a linear polymer with side-groups regularly arranged along the chain. The presence of these side-groups gives a more rigid and stronger polymer (Figure 10.13) than polyethylene in its linear form.

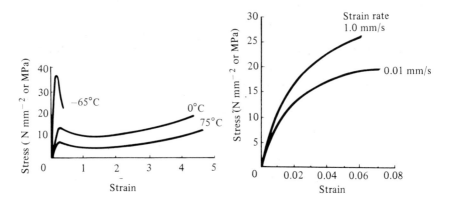

Figure 10.12 Stress–strain graph for low density polythene

Figure 10.13 Stress–strain graph for polypropylene

Polypropylene is used for crates, containers, fans, car fascia panels, tops of washing machines, cabinets for radios and television sets, toys, chair shells.

Polyvinyl chloride (PVC), a vinyl, is a linear chain polymer with bulky side-groups (see Figure 10.6) which prevent crystalline regions occurring. The material is hard and rigid. It is often used with a plasticiser to give a more flexible plastic.

The rigid form of PVC, i.e. the unplasticised form, is used for piping for waste and soil drainage systems, rainwater pipes, lighting fittings, curtain rails. Plasticised PVC is used for the fabric of 'plastic' raincoats, bottles, shoe soles, garden hose piping, gaskets, inflatable toys.

Polystyrene, a vinyl, is a linear chain polymer with bulky side-groups which prevent crystalline regions occurring. Polystyrene with no additives is a brittle, transparent material. A toughened form of polystyrene is produced by blending polystyrene with rubber particles.

Polystyrene is an excellent electrical insulator and is widely used in electrical equipment. Other applications are packaging of cosmetics, toys, boxes. A widely used form of polystyrene is as expanded polystyrene, which is a rigid foam used for insulation and packaging. Toughened polystyrene is less brittle than ordinary polystyrene and finds use as cups in vending machines, casing for cameras, projectors, radios, television sets, vacuum cleaners.

Acrylonitrile-butadiene-styrene terpolymer (ABS) is produced by forming polymer chains with three different polymer materials, polystyrene, acrylonitrile and butadiene. It gives an amorphous material which is tough, stiff and abrasion-resistant.

ABS is widely used as the casing for telephones, vacuum cleaners, hair driers, radios, television sets, typewriters, luggage, boat shells, food containers.

Polycarbonate, a non-ethnic, is an amorphous thermoplastic having a linear chain with bulky side groups. It is tough, stiff and strong. It is used for applications where the plastics are required to be resistant to impact abuse and (for plastics) relatively high temperatures. Typical applications are transparent street lamp covers, infant-feeding bottles, machine housings, safety helmets, housings for car lights, and tableware such as cups and saucers.

Acrylics are completely transparent thermoplastics, mostly based on methyl methacrylate which is a vinyl, having linear chains with bulky side groups and so giving an amorphous structure. They give a stiff, strong material with outstanding weather resistance.

Because of its transparency acrylic is used for windscreens, light fittings, canopies, lenses for car lights, signs and nameplates. Opaque acrylic sheet is used for the production of domestic baths, shower cabinets, basins and lavatory cisterns.

Polyamides, or nylons as they are better known, are linear non-ethnic polymers and give crystalline structures. There are a number of common polyamides: nylon 6, nylon 6.6, nylon 6.10 and nylon 11. The numbers refer to the numbers of carbon atoms in each of the reacting substances used to give the polymer. The full stops separating the two numbers are sometimes omitted, e.g. nylon 66 is nylon 6.6. The two most used nylons are nylon 6 and nylon 6.6. Nylon 6.6 has a higher melting point than nylon 6 and is also stronger and stiffer. Nylon 11 has a lower melting point than either nylon 6 or 6.6; it is also more flexible. Nylon 6.10 has properties intermediate between nylon 11 and nylon 6.6.

In general, nylon materials are strong, tough and have relatively high melting points. But they do tend to absorb moisture. The effect of this is to reduce their tensile strength. Nylon 6.6 can absorb quite large amounts of moisture; nylon 11, however, absorbs considerably less.

Nylons often contain additives, e.g. a stabiliser or flame retardant substance. Glass spheres or glass fibres are added to give improved strength and rigidity. Molybdenum disulphide is an additive to nylon 6 to give a material with very low frictional properties.

Nylon is used for the manufacture of fibres for clothing, gears, bearings, bushes, housings for domestic and power tools, electric plugs and sockets.

Polyesters are available in a thermoplastic form, usually *polyethylene terephthalate*, a non-ethnic polymer. This is a linear chain polymer with side groups. It gives crystalline structures and is below its glass transition point at normal temperatures. If it is rapidly quenched from a melt to below its glass transition point an amorphous structure is produced, the molecular chains not having sufficient time to become packed in an orderly arrangement.

The polyester has properties similar to nylon. It is used widely in fibre form for the production of clothes. Other uses are for electrical plugs and sockets, push-button switches, wire insulation, recording tapes, insulating tapes, gaskets.

Polyacetals, polyformaldehyde or polyoxymethylene, are linear non-ethnic polymers giving rise to crystalline structures. They are stiff, strong polymers and maintain their properties at relatively high temperatures. They are, however, adversely affected by ultraviolet light and thus have to be used with a stabiliser.

Typical applications are pipe fittings, parts for water pumps and washing machines, car instrument housings, bearings, gears, hinges and window catches, seat belt buckles.

Polytetrafluoroethylene (PTFE) is a linear polymer like polyethylene, the only difference being that instead of hydrogen atoms there are fluorene atoms. It has a very high crystallinity as manufactured, about 90%, though this degree of crystallinity can be reduced during processing to about 50% if quench cooled, or 75% with slow cooling. Tough and flexible, PTFE can be used over a wide range of temperature, 250°C down to almost absolute zero, and still retain the very important property of not being attacked by any reagent or solvent. It also has a very low coefficient of friction. No known material can be used to bond it satisfactorily to other materials.

PTFE is a relatively expensive material and is not processed as easily as other thermoplastics. It tends to be used where its special properties, i.e. resistance to chemical attack and very low coefficient of friction, are needed. Journal bearings with a PTFE surface can be used without lubrication because of the low coefficient of friction; they can even be used at temperatures up to about 250°C. Piping carrying corrosive chemicals at temperatures up to 250°C are made of PTFE. Other applications for PTFE are gaskets, diaphragms, valves, O-rings, bellows, couplings, dry and self-lubricating bearings, coatings for frying pans and other cooking utensils (known as 'non-stick'), coverings for rollers handling stickly materials, linings for hoppers and chutes, and electrical insulating tape.

10.3 Thermosetting polymers

Thermoplastic polymers soften if heated, and can be reshaped, the new shape being retained when the plastic cools. The process can be repeated. *Thermosetting polymers* cannot be softened and reshaped by heating. They are plastic in the initial stage of manufacture but once they have set they cannot be resoftened. The atoms in a thermosetting material form a three-dimensional structure of cross-linked chains (see Figure 10.2c). The bonds linking the chains are strong and not easily broken. Thus the chains cannot slide over one

Table 10.3 *Properties of typical thermosets used in engineering*

	Density (10^3 kg m^{-3})	Tensile strength (N mm^{-2} or MPa)	Tensile modulus (kN mm^{-2} or GPa)	Elongation (%)	Max. service temp (°C)
Phenol formaldehyde					
Unfilled	1.25–1.30	35–55	5.2–7.0	1–1.5	120
Wood flour filler	1.32–1.45	40–55	5.5–8.0	0.5–1	150
Asbestos filler	1.60–1.85	30–55	0.1–11.5	0.1–0.2	180
Urea formaldehyde					
Cellulose filler	1.5–1.6	50–80	7.0–13.5	0.5–1	80
Melamine formaldehyde					
Cellulose filler	1.5–1.6	55–85	7.0–10.5	0.5–1	95
Epoxy resin					
Plain weave glass fabric (60–65% fabric)	1.8	200–420	21–25		200

another but are essentially fixed in the position they occupied when the polymer was solidifying during its formation.

Thermosetting polymers are stronger and stiffer than thermoplastics and generally they can be used at higher temperatures than thermoplastics. As they cannot be shaped after the initial reaction in which the polymer chains are produced, the processes by which thermosetting polymers can be shaped are limited to those where the product is formed from the raw polymer materials. No further processing is possible (other than possibly some machining) and this limits the processes available to essentially just *moulding*. A number of different moulding methods are used but essentially all involve the combining together of the chemicals in a mould so that the cross-linked chains are produced while the material is in the mould. The result is a thermosetting polymer shaped to the form dicated by the mould.

The main properties and uses of common thermosets are outlined below and in Table 10.3.

Phenolics give highly cross-linked polymers. *Phenol formaldehyde* was the first synthetic plastic and is known as *Bakelite*. The polymer is opaque and initially light in colour. It does, however, darken with time and so is always mixed with dark pigments to give dark-coloured materials. It is supplied in the form of a moulding powder, including the resin, fillers and other additives such as pigments. When this moulding powder is heated in a mould the cross-linked polymer chain structure is produced. The fillers account for some 50 to 80% of the total weight of the moulding powder. Wood flour, a very fine soft wood sawdust, when used as a filler increases the impact strength of the plastic, asbestos fibres improve the heat properties, and mica the electrical resistance.

Phenol formaldehyde mouldings are used for electrical plugs and sockets, switches, door knobs and handles, camera bodies and ashtrays. Composite materials involving the phenolic resin being used with paper or an open-weave fabric, e.g. a glass fibre fabric, are used for gears, bearings, and electrical insulation parts.

Amino-formaldehyde materials, generally *urea formaldehyde* and *melamine formaldehyde*, give highly cross-linked polymers. Both are used as moulding powders, like the phenolics. Cellulose and wood flour are widely used as fillers. Hard, rigid, high-strength materials are produced.

Both materials are used for table ware, e.g. cups and saucers, knobs, handles, light fittings and toys. Composites with open-weave fabrics are used as building panels and electrical equipment.

Epoxide materials are generally used in conjunction with glass, or other, fibres to give hard and strong composites. Epoxy resins are excellent adhesives giving very high adhesive strengths.

The unfilled epoxide has a tensile strength of 35 to 80 N mm^{-2}, considerably less than that of the composite. The composite is used for boat hulls and table tops.

Polyesters can be produced as either thermosets or thermoplastics. The thermoset form is mainly used with glass, or other, fibres to form hard and strong composites. Such composites are used for boat hulls, architectural panels, car bodies, panels in aircraft, and stackable chairs. They have a maximum service temperature of the order of 200°C.

10.4 Elastomers

Elastomers are polymers which show very large strains when subject to stress and which will return to their original dimensions when the stress is removed. Elastomers are essentially amorphous polymers with a glass transition temperature below their service temperature. The polymer structure is that of linear chain molecules with some cross-linking between chains (Figure 10.14). Without the cross-linking, the molecular chains might slide past each other and give permanent deformations. The presence of these cross-links ensures that the material is elastic and returns to its original dimensions when the stress is removed. However, if there are too many cross-links the material becomes inflexible.

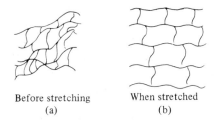

Before stretching   When stretched
(a)   (b)

Figure 10.14 Stretching an elastomer which has cross-links

The main properties and uses of common elastomers are outlined below and in Table 10.4.

Natural rubber is, in its crude form, just the sap from a particular tree. The addition of sulphur to the rubber produces cross-links, the amount of cross-linkage being determined by the amount of sulphur added. The process of producing cross-links is called *vulcanisation*. Anti-oxidants and plasticisers are also added to the rubber.

Butadiene styrene rubbers, commonly called *SBR* or *GR-S* or *Buna S* rubbers, are synthetic rubbers. They are cheaper than natural rubber SBR is used in the manufacture of tyres, hosepipes, conveyor belts and cable insulation.

Butyl rubbers, known as *isobutylene isoprene* or *GR-I*, have the important property of extreme impermeability to gases. They are thus widely used for the inner linings of tubeless tyres, steam hoses and diaphragms.

Nitrile rubbers, known as *butadiene acrylonitrile* or *Buna N*, are extremely

Table 10.4 *Properties of common elastomers*

	Tensile strength ($N\ mm^{-2}$ or MPa)	Elongation (%)	T_g (°C)	Service temp. range (°C)	Resistance to oils and greases	Resistance to water
Natural rubber	20	800	−73	−50 to + 80	Poor	Good
SBR	24	600	−58	−50 to + 80	Poor	Good
Butyl	20	900	−53	−50 to +100	Poor	Good
Nitrile	28	700	−10	−50 to +100	Excellent	Good
Neoprene	25	1000	−48	−50 to +100	Good	Fair
Polyurethane	40	650		−55 to +125	Poor	Good

resistant to organic liquids and used for such applications as hoses, gaskets, seals, tank linings, rollers, valves.

Neoprene, known as *polychloroprene*, has good resistance to oils and a variety of other chemicals, also having good weathering characteristics. It is used for oil and petrol hoses, gaskets, seals, diaphragms and chemical tank linings.

Polyurethane elastomers are widely used for both flexible and rigid foams. As flexible foams, they have properties similar to those of natural rubber but are less flammable and have better resistance to ageing. Hence they are widely used as cushioning or for packaging. Rigid foams are used in structural and insulation panels. Polyurethanes have higher tensile strengths than other rubbers.

10.5 Thermal, electrical, optical and chemical properties of polymers

Solid polymers have very low thermal conductivities when compared with metals, and cellular polymers even lower. For this reason they can be used for thermal insulation. Thus polyurethane foam has a thermal conductivity of about 0.4 W m^{-1} K^{-1}, compared with values of hundreds for metals.

Polymers have higher thermal expansivities than metals, in general about 2 to 10 times greater. Polypropylene has a linear thermal expansivity of about 9 $\times 10^{-5}$ K^{-1} while copper has a value of 1.7×10^{-5} K^{-1}.

Solid polymers have specific heat capacities about twice those of metals. For example, low density polythene has a specific heat capacity of 2.2 kJ kg^{-1} K^{-1}, copper 0.38 kJ kg^{-1} K^{-1}.

Polymers are electrical insulators, having resistivities which are about 10^{20} times greater than those of metals. If a piece of polymer is rubbed with a piece of cloth it will become electrically charged, the high resistivity of the polymer allowing large electrostatic charges to accumulate. Such charges can give rise to problems. For example, in the production of polymer sheets, charges on the sheets can result in their sticking together or in dust sticking to them.

Polymer transparency ranges from highly transparent to completely opaque to visible light. Polypropylene, for example, is translucent and a 1 mm thickness allows about 10% of the incident light to be transmitted through it. Acetal plastics are virtually opaque. PVC in a 1 mm thickness will allow over 90% of incident light to be transmitted through it.

In general, polymers are resistant to weak acids, weak alkalis and salt solutions but not necessarily resistant to organic solvents, oils and fuels. The effect of solvents on polymers is usually that the polymer dissolves. The degree of resistance depends on the polymer concerned. Thus, for example, polythene is almost completely insoluble in all organic solvents at room temperature, but above 70°C it dissolves in many of them. Acrylics, however, have a poor resistance to organic solvents at room temperature.

Polymers are generally affected by exposure to the atmosphere and to sunlight. The effect can show as a slow ageing process with the material becoming more brittle. This is due to bonds being formed between neighbouring molecular chains. Such effects may be reduced by the inclusion of suitable additives with the polymer in the plastic or rubber.

Some polymers when stressed and in contact with certain environments can develop brittle cracking. This effect is known as *environmental stress cracking*. Polythene, for example, can show this effect in the presence of detergents.

Elastomers can be particularly affected by atmospheric ozone. If the elastomer is under stress in such conditions, cracking can occur; this effect is known as ozone cracking. Stabilisers are usually included with the polymer in the rubber to inhibit such cracking.

10.6 Ceramics

The term *ceramics* covers a wide range of materials, e.g. brick, concrete, stone, clay, glasses and refractory materials. Ceramics are formed from combinations of one or more metals with a non-metallic element such as oxygen, nitrogen or carbon. Ceramics are usually hard and brittle, good electrical and thermal insulators, and have good resistance to chemical attack. They tend to have a low thermal shock resistance, because of their low thermal conductivity and a low thermal expansivity: think of the effect of pouring a very hot liquid into a drinking glass.

Ceramics are generally crystalline, though amorphous states are possible. Thus if silica in the molten state is cooled very slowly it crystallises at the freezing point. However, if the molten silica is cooled more rapidly it is unable to get all its atoms into the orderly arrangements required of a crystal (see Section 6.1) and the resulting solid is a disorderly arrangement which is called a glass (Figure 10.15). The temperature at which the molten silica turns into a glass (the glass transition temperature, T_g) depends on the rate of cooling from the molten material.

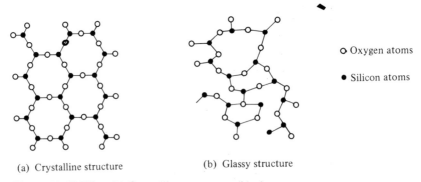

(a) Crystalline structure   (b) Glassy structure

○ Oxygen atoms

● Silicon atoms

Figure 10.15 Silica. (a) Crystalline structure, (b) glassy structure

Table 10.5 *Properties of typical engineering ceramics*

	Density (10^3 kg m^{-3})	Coefficient of expansion (10^{-6} K^{-1})	Melting point (°C)	Thermal conductivity (W m^{-1} K^{-1})	Tensile strength (MN m^{-2} or MPa)	Compressive strength (MN m^{-2} or MPa)	Tensile modulus (GN m^{-2} or GPa)	Resistivity (Ω m)
Alumina	3.9	8	2040	12–30	400	3800	380	Greater than 10^{17}
Silicon carbide	3.1	4.5	Decomposes at 2300	40–100	200–800	1400	200–400	
Silicon nitride	3.2	2.9	Sublimes at 1900	4–16	200–900		150–300	

Some common engineering ceramics with their properties and typical applications are described below and in Table 10.5.

Alumina is an oxide of aluminium and is widely used for electrical insulators. It is used for sparking plug insulators where the material withstands rapid fluctuations of temperature and pressure, high voltages and also maintains gas-tight joints with the metal conductor and base. Its high melting point means that it can be used as a refractory material for the lining of high temperature furnaces. It has high compressive strength and resistance to wear and so can be used for tool tips and grinding tools (corundum used for emery paper and grinding wheels is a mixture of alumina and iron oxide).

Silicon nitride is a compound of silicon and nitrogen. Nitrides, in general, are fairly brittle and oxidise readily. Silicon nitride has high thermal conductivity with a low thermal expansion and so has good thermal shock resistance. It also has high strength and so finds uses in heat exchangers, furnace components, crucibles and high temperature bearings.

Silicon carbide is a compound of silicon with carbon. Carbides in general have very high melting points, however most of them cannot be used unprotected at high temperatures because they oxidise; the exception is silicon carbide. It has a high thermal conductivity combined with a low thermal expansion and so is resistant to thermal shock. It has very good abrasion resistance and chemical inertness. It is used for ball and roll bearings, combustion tubes, rocket nozzles and high temperature furnaces.

Glasses

The basic ingredient of most glasses is sand, i.e. silica—silicon dioxide. Ordinary window glass is made from a mixture of sand, limestone (calcium carbonate) and soda ash (sodium carbonate). Heat-resistant glasses such as Pyrex are made by replacing the soda ash by boric oxide. Many of the mechanical properties of glasses are almost independent of their chemical composition. They thus tend to have a tensile modulus of about 70 GN m^{-2} (GPa). The tensile strength in practice is markedly affected by microscopic defects and surface scratches and for design purposes a value of about 50 MN m^{-2} (MPa) is generally used. They have low ductility, being brittle. They have low thermal expansivity and low thermal conductivity. They are electrical insulators, with resistivities of the order of 10^{14} Ω m or higher. They are resistant to many acids, solvents and other chemicals. The maximum service temperature tends to be about 500°C to 1300°C, depending on the composition of the glass.

	Density (10^3 kg m^{-3})	Coefficient of expansion (10^{-6} K^{-1})	Max. service temperature (°C)	Tensile modulus (GN m^{-2} or GPa)
Soda–lime–silica glass	2.5	9.2	460	70
Pyrex	2.2	3.2	490	67

10.7 Composites

The term *composite* is used for a material composed of two different materials bonded together with one serving as the matrix surrounding fibres or particles of the other. A common example of a composite is reinforced concrete. This has steel rods embedded in the concrete (Figure 10.16). The composite enables loads to be carried that could not be carried by the concrete alone. Concrete itself is a composite, without the presence of steel reinforcement. It is made by mixing cement, sand, aggregate and water. Stone chips or gravel are often used as the aggregate. The resulting concrete consists of the aggregate in a matrix (Figure 10.17).

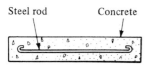

Figure 10.16 Reinforced concrete, steel rods in a matrix

Figure 10.17 Concrete aggregate in a matrix

There are many examples of composite materials encountered in everyday components. Many plastics are glass fibre or glass particle reinforced. Vehicle tyres are rubber reinforced with woven cords. Wood is a natural composite material with tubes of cellulose bonded by a natural plastic called lignin (Figure 10.18). Cermets, widely used for cutting tool tips, are composites involving ceramic particles in a metal matrix.

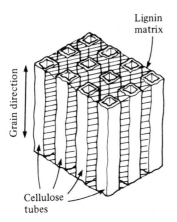

Figure 10.18 Wood, cellulose fibres in a lignin matrix

Continuous fibres in a matrix

Consider a composite rod made up of continuous fibres, all parallel to the rod axis, in a matrix (Figure 10.19). These could be glass fibres in a plastic, or steel reinforcement rods in concrete. Each element in the composite has a share of the applied force, thus

Total force = force on fibres + force on matrix

But the stress on the fibres is equal to force on them divided by their cross-sectional area. Similarly the stress on the matrix is equal to the force on the matrix divided by its area. Hence

Total force = (stress on fibres × area of fibres) + (stress on matrix × area of matrix)

Dividing both sides of the equation by the total area of the composite gives

$$\frac{\text{total force}}{\text{total area}} = \left(\text{stress on fibres} \times \frac{\text{area of fibres}}{\text{total area}}\right) + \left(\text{stress on matrix} \times \frac{\text{area of matrix}}{\text{total area}}\right)$$

The fraction of the cross-sectional area that is fibre is given by the area of the fibres divided by the total area, similarly the fraction of the cross-section that is matrix is the area of the matrix divided by the total area. The total area divided by the total force is the stress applied to the composite. Thus

stress on composite = (stress on fibres × area fraction fibres) + (stress on matrix × area fraction matrix)

If the fibres are firmly bonded to the matrix then the elongation or contraction of the fibres and matrix must be the same as and equal to that of the composite as a whole. Thus

strain on composite = strain on fibres = strain on matrix

Dividing both sides of the stress equation by the strain gives an equation in terms of the tensile moduli (stress/strain = tensile modulus). Thus

modulus of composite = (modulus of fibres × area fraction fibres) + (modulus of matrix × area fraction matrix)

Suppose that we have glass fibres with a tensile modulus of 76 kN mm^{-2} in a matrix of polyester having a tensile modulus of 3 kN mm^{-2}. Then, if the fibres occupy 60% of the cross-sectional area, the tensile modulus of the composite will be given by

modulus of composite = 76 × 0.6 + 3 × 0.4
= 46.8 kN mm^{-2} or GPa

The composite has a tensile modulus considerably greater than that of the polyester.

Not only has the composite a higher tensile modulus but also a higher strength than that of the matrix material. Thus for the 60% glass fibres in polyester the tensile strength may be about 800 N mm^{-2} for the composite when the matrix has a tensile strength of only about 50 N mm^{-2}.

For glass fibre-polyester composites with long fibres all in the same direction and all parallel to the axis of the composite along which the force is applied, Figure 10.20 shows how the tensile modulus depends on the percentage of the area occupied by the fibres and Figure 10.21 how the tensile strength is affected by this factor. Such composites usually have between 40% and 80% of the cross-sectional area as glass fibre. The following is the data for the glass fibres and polyester when separate.

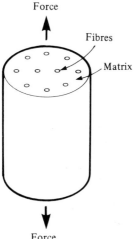

Figure 10.19 Continuous fibres in a matrix

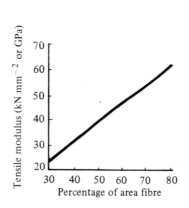

Figure 10.20 The effect of the percentage of cross-sectional area of glass fibre on the tensile modulus of a glass fibre-polyester composite

	Tensile modulus *(kN mm^{-2} or GPa)*	*Tensile strength* *(N mm^{-2} or MPa)*
Polyester	2 to 4	20 to 70
E-glass fibres	76	1200 to 1800

Two further examples of composites with continuous fibres follow.

(a) A column of reinforced concrete has steel reinforcing rods running through the entire length of the column and parallel to the column axis. The concrete has a modulus of elasticity of 20 kN mm^{-2} (GPa) and the steel 210 kN mm^{-2} (GPa), and the steel rods occupy 10% of the cross-sectional area of the column. Thus

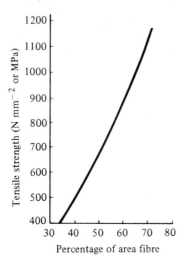

Figure 10.21 The effect of the percentage of cross-sectional area of glass fibre on the tensile strength of a glass fibre-polyester composite

modulus of elasticity of the composite $= 210 \times 0.1 + 20 \times 0.9$
$$= 39 \text{ kN mm}^{-2} \text{ or GPa}$$

(b) Carbon fibres with a tensile modulus of 400 kN mm^{-2} (GPa) are used to reinforce aluminium with a tensile modulus of 70 kN mm^{-2} (GPa). The fibres are long and parallel to the axis along which the load is applied, and occupy 50% of the cross-sectional area of the composite. Thus

tensile modulus of the composite $= 400 \times 0.5 + 70 \times 0.5$
$$= 235 \text{ kN mm}^{-2} \text{ or GPa}$$

Reinforced plastics

Reinforced plastics consist of a stiff, strong material combined with the plastic. Glass fibres are probably the most common additive. The fibres may be long lengths, running through the length of the composite, or discontinuous short lengths randomly orientated within the composite. Another form of composite uses glass fibre mats or cloth in the plastic. The effect of the additives is to increase both the tensile strength and the tensile modulus of the plastic, the amount of change depending on both the form the additive takes and the amount of it. The continuous fibres give the highest tensile modulus and tensile strength composite but with a high directionality of properties. The strength along the direction of the fibres could be perhaps 800 N mm^{-2} (MPa) while that at right-angles to the fibre direction may be as low as 30 N mm^{-2} (MPa), i.e. just about the strength of the plastic alone. Randomly orientated short fibres do not lead to this directionality of properties but do not give such high strength and tensile modulus. The composites with glass fibre mats or cloth tend to give tensile strength and modulus values intermediate between those of the continuous and short length fibres. The following are examples of the strength and modulus values obtained with reinforced polyester.

Material	Percentage weight of glass	Tensile modulus $(kN\ mm^{-2}\ or\ GPa)$	Tensile strength $(N\ mm^{-2}\ or\ MPa)$
Polyester without reinforcement	0	2 to 4	20 to 70
With short fibres	10 to 45	5 to 14	40 to 180
With plain weave cloth	45 to 65	10 to 20	250 to 350
With long fibres	50 to 80	20 to 50	400 to 1200

Dispersion-strengthened metals

The strength of a metal can be increased by small particles dispersed throughout it. Thus solution treatment followed by precipitation hardening for an aluminium–copper alloy can lead to a fine dispersion of an aluminium–copper compound throughout the alloy. The result is a material of higher tensile strength.

Aluminium alloy	Condition	Tensile strength $(N\ mm^{-2}\ or\ MPa)$
4.0% Cu, 0.8% Mg, 0.5% Si, 0.7% Mn.	Annealed	190
	Solution treated Precipitation hardened	430

Another way of introducing a dispersion of small particles throughout a metal involves *sintering*. This process consists of compacting powdered metal in a die and then heating it to a temperature high enough to knit together the particles in the powder. If this is done with aluminium the result is a fine dispersion of aluminium oxide (about 10%) throughout an aluminium matrix. The aluminium oxide occurs because aluminium in the presence of oxygen is coated with aluminium oxide. When the aluminium powder is compacted, much of the surface oxide film becomes separated from the aluminium and becomes a fine powder dispersed throughout the metal. The aluminium oxide powder, a ceramic, dispersed throughout the aluminium matrix gives a stronger material than that which would have been given by the aluminium alone. At room temperature the tensile strength of the sintered aluminium powder is about 400 N mm^{-2} (MPa), compared with that of about 90 N mm^{-2} (MPa) for commercial annealed aluminium. The sintered aluminium has an advantage over the precipitation hardened aluminium alloy in that it retains its strength better at high temperatures (Figure 10.22). This is because, at the higher temperatures, the precipitate particles tend to coalesce or go into solution in the metal.

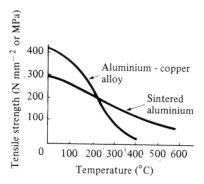

Figure 10.22 The effect of temperature on the tensile strength of an aluminium–copper alloy and sintered aluminium

Other dispersion-strengthened metals, involving ceramics, have been developed. Thorium oxide particles (about 7%) in nickel give a composite which has good strength properties at high temperatures.

Cermets

Cermets are composites consisting of ceramics in a matrix of metal. The ceramics used have high strengths, high values of tensile modulus, high hardness, but are by themselves brittle substances. By comparison, the metals are weaker and less stiff, but they are ductile. By combining the ceramic with the metal, up to 80% ceramic, a composite can be produced which is strong, hard and tough. The composite has the ceramic in particle form in the metal matrix.

Typical ceramics used in cermets are carbides (tungsten, titanium, silicon, molybdenum), borides (chromium, titanium, molybdenum) and oxides (aluminium, chromium, magnesium), the metals being cobalt, chromium, nickel, iron or tungsten. A typical cermet used for cutting tool bits consists of tungsten carbide in a matrix of cobalt.

Laminated materials

Plywood is an example of laminated material. It is made by gluing together thin sheets of wood with their grain directions at right angles to each other (Figure 10.23). The grain directions are the directions of the cellulose fibres in the wood and thus the resulting structure, the plywood, has fibres in mutually perpendicular directions. Thus whereas the thin sheet had properties that were directional, the resulting laminate has no such directionality.

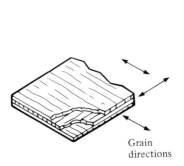

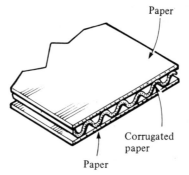

Figure 10.23 Plywood

Figure 10.24 Corrugated cardboard

The term *laminated wood* is generally used to describe the product obtained by sticking together thin sheets of wood but with the grain of each layer parallel to the grain of the others. Large wooden arches and beams in modern buildings are likely to be laminated rather than made of a solid piece of wood. By carefully choosing the wood used to build up the beam, a better quality beam can be produced than could be made of wood in its natural state.

Metals can be laminated by cladding or plating them with other metals to give improved corrosion resistance. Examples include the cladding of aluminium–copper alloy with aluminium, the galvanising of steel with a layer of zinc, and the plating of steel food containers with tin.

Corrugated cardboard is another form of laminated structure (Figure 10.24), consisting of paper corrugations sandwiched between layers of paper. The resulting structure is much stiffer, in the direction parallel to the corrugations, than the paper alone. A similar type of material is produced with metals, a metal *honeycomb structure* sandwiched between thin sheets of metal. Such a structure has good stiffness and is very light. Aluminium is often used for both the honeycomb and the sheets.

Problems

1 How is the structure of an amorphous polymer different from that of a crystalline polymer?
2 How does the form of the polymer molecular chain determine the degree of crystallinity possible with the polymer?
3 Explain how the crystallinity of a polymer affects its properties.
4 What is the glass transition temperature?
5 PVC has a glass transition temperature of 87°C. How would its properties below this temperature differ from those above it?

6 A polypropylene article was designed for use at room temperature. What difference might be expected in its behaviour if used at about $-15°C$. The glass transition temperature for polypropylene is $-10°C$.

7 When a piece of polythene is pulled it starts necking at one point. Further pulling results in no further reduction in the cross-section of the material at the necked region but a spread of the necked region along the entire length of the material. Why doesn't the material just break at the initial necked section instead of the necking continuing?

8 Why are the properties of a cold drawn polymer different from those of the undrawn polymer?

9 Why are polyester fibres cold drawn before use?

10 Polypropylene is a crystalline polymer with a glass transition temperature of $-10°C$ and a melting point of $176°C$. What would be its normal maximum service temperature? At what temperatures would the polymer be hot formed?

11 Explain what is meant by internal and external plasticisation.

12 What is the effect of a plasticiser on the mechanical properties of a polymer?

13 How do the properties of high and low density polythene differ?

14 How do the properties of polystyrene with no additives and toughened polystyrene differ?

15 To what temperature would high density polythene have to be heated to be hot formed?

16 Which is stiffer at room temperature, PVC with no plasticiser or polypropylene?

17 The casing of a telephone is made from ABS. How would the casing behave if someone left a burning match or cigarette against it?

18 What are the special properties of PTFE which render it useful despite its high price and processing problems?

19 How do the mechanical properties of thermosets differ, in general, from those of thermoplastics?

20 What is Bakelite and what are its mechanical properties?

21 Describe how cups made of melamine formaldehyde with a cellulose filler might be expected to behave in service.

22 What is meant by vulcanisation?

23 Would Buna S or Buna N be the better rubber to use for a pipe carrying oil?

24 Explain how rubbers can be stretched to several times their length and return to the same initial length when released.

25 Figure 10.25 shows the effect on the Charpy impact strength for nylon 6 of percentage of water absorbed. As the percentage of water absorbed increases, is the material becoming more or less brittle?

26 Figure 10.26 shows the stress–strain graphs for two forms of ABS plastic, one containing 20% by weight glass fibre and the other without such fibres. Estimate from the graph (a) the tensile strength and (b) the tensile modulus for both forms.

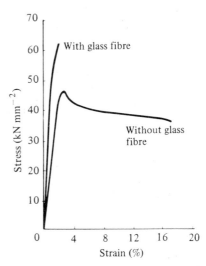

Figure 10.26 Stress–strain graphs for ABS plastics Novodur PHGV (reinforced) and PH-AT (un-reinforced). (Courtesy of Bayer UK Ltd)

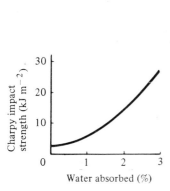

Figure 10.25 The effect of water absorption on the impact strength of nylon 6

27 Explain the significance of the following information from a manufacturer's data sheet for neoprene rubber. (Courtesy of Du Pont Ltd.)

Recovery from deformation

The recovery of a material after being held under load is important in the design of many rubber products. Neoprene exhibits a relatively low degree of permanent deformation from compression (compression set). For example, specimens of a standard compound, compressed to 75% of their original thickness and then released, behaved as follows:

Held 22 hours at 70°C–15% compression set.

Held 70 hours at 100°C–35% compression set.

Permanent deformation due to elongation (permanent set) will be approximately 5% for most neoprene products.

28 Figure 10.27 shows how the tensile modulus of a polycarbonate plastic Makrolon 2800 changes with temperature. How is the behaviour of the material under stress changing as the temperature increases?

29 Which polymers would be suitable for the following applications? (a) An ashtray. (b) A garden hose pipe. (c) A steam hose. (d) Insulation for electric wires. (e) The transparent top of an electrical meter. (f) A plastic raincoat. (g) A toothbrush. (h) A camera body.

30 Describe the basic properties of ceramics.

31 Explain the effect of the rate of cooling from the molten state on the resulting structure of silica.

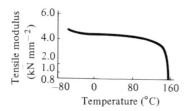

Figure 10.27 The effect of temperature on the tensile modulus of Makrolon 2800. (Courtesy of Bayer UK Ltd)

32 Explain the reasons for the use of ceramic materials as (a) sparking plug insulators, (b) tool tips, (c) rocket nozzles, (d) furnace components.

33 Describe the basic properties of glasses.

34 Explain the term 'composite'.

35 Give examples of composites involving (a) plastics and (b) metals.

36 Describe how the mechanical properties of a fibre composite depend on the form of the fibres and their orientation.

37 Calculate the tensile modulus of a composite consisting of 45% by volume of long glass fibres, tensile modulus 76 kN mm^{-2} (GPa), in a polyester matrix, tensile modulus 4 kN mm^{-2} (GPa). In what direction does your answer give the modulus?

38 In place of the glass fibres referred to in Problem 37, carbon fibres are used. What would be the tensile modulus of the composite if the carbon fibres had a tensile modulus of 400 kN mm^{-2} (GPa)?

39 What is a cermet?

40 Explain how plywood gets its stiffness.

41 Explain how corrugated cardboard gets its stiffness.

42 Explain what is meant by the term 'dispersion strengthened metals' and give an example of one.

PART THREE

Production Planning

This part of the book consists of four chapters on the various aspects of planned production: planning and control of the production process, quality control, the interaction between workers and production, and costing.

Chapter 11   Production planning and control

This gives an overview of the key aspects of production planning and control, i.e. types of production operation, plant layout, maintenance planning, capacity planning, aggregate planning, scheduling, inventory management, network analysis, progressing and control.

Chapter 12   Quality control

Quality, the cost of quality control, acceptance sampling and process control form the subject of this chapter.

Chapter 13   Work design

The basic concepts involved in the design of the work situation and task are discussed: work study, method study, ergonomics, motion economy, job design and safety legislation.

Chapter 14   Costing

Cost accounting is discussed in terms of its basic cost elements: direct labour costs, direct materials costs, overheads, depreciation, costing of jobs, batches and processes, and marginal costing. Break-even analysis is explained and the chapter concludes with a look at the factors involved in the selection of a manufacturing process.

11

Production planning and control

11.1 Introduction

This chapter gives an overview of the key aspects of production planning and control. For more detailed information the reader is referred to more specialist texts, e.g. *Essentials of Production and Operations Management* by R. Wild (Holt, Rinehart and Winston).

In a general sense *production planning and control* can be considered to have a number of facets:

1 *Process design*. This involves the selection of processes and the layout of the facilities.
2 *Capacity planning*. This aims to match the level of production operations to the level of demand.
3 *Inventory management*. This is the management of stocks of raw materials, part-finished goods and finished goods.
4 *Scheduling of operations*. This is the planning of activities so that the available capacity is efficiently and effectively used.

Types of production operation

Production operations can be classified as:

1 *Job or unique product production*. This is the production of single unique items, generally to a customer's order and specification.
2 *Small batch production*. This generally differs little from unique product production, a small batch rather than a single product being produced.
3 *Large batch production*. This involves the production of large numbers of standard products and is often known as *mass production, line production* or *flowline production*. The products are very often made to stock against future customers' orders.
4 *Process production*. This term is used to describe the virtually continuous processes used where essentially chemical processes are involved.

The different types of operation have different production implications. Thus an organisation engaging in unique product production, to customers' orders, will require a highly flexible production operation in order to cope with the diversity of products. It will probably also require skilled workers in that only they are likely to be able to cope with constantly changing requirements. An organisation engaged in mass production will, however, often be producing for stock rather than directly for customers' orders. It will probably have a rigid production system requiring high capital investment in order to produce the standardised products most economically.

Choice of process

Generally the choice of type of production operation will be determined by the product and the industry concerned. The term 'choice' is often likely to be inappropriate in that the type of production operation is likely to be obvious. Thus, for instance, in the case of domestic appliances mass production is almost invariably required because the large quantities to be produced and the economics of such large number production dictate that type of operation. However, in the production of large items of machinery or civil engineering the products are generally one-off, against a customer order rather than in anticipation of customer orders, and thus job or unique product production is required.

Products where there is a reasonably constant high sales volume, where the product design is standard and does not change often and the potential life of the product is long, are likely to require large batch production methods in order to be economic.

11.2 Plant layout

The term *plant layout* is used to describe the arrangement of the various facilities required in the production process. The prime aim is to arrange the facilities to provide efficient operation. This can be considered to involve designing the layout to achieve:

1 Maximum utilisation of the space, facilities and labour.
2 The minimum manufacturing time.
3 The minimum amount of movement of materials and work-in-progress.
4 Safe operation.
5 A socially acceptable layout for the workforce.

Essentially there are three main systems of production plant layout:

Layout by process
Layout by product
Group layout

Each of the systems has its own characteristics (Figure 11.1) which make it appropriate in some situations. The main factors involved in determining which system might be the most appropriate are the type of production operation involved, e.g. unique product or perhaps mass production, and the range of products that have to be produced.

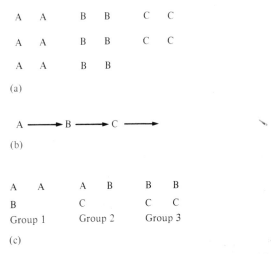

Figure 11.1 Plant layout. (a) Process layout: A is one type of process, e.g. milling, B is another, e.g. drilling, and C another, e.g. grinding. Each product takes its own operational sequence through the processes. (b) Product layout: all products follow the same sequence of processes. (c) Group layout: a sequence of processes occurring at each group. Products take their own operational sequence between groups but not generally within groups

With a *process layout* all the plant concerned with a particular process is grouped together. Thus all the milling machines would be together, all the grinding machines together in separate groups, etc. Where unique production is concerned this could be the best layout. It permits great flexibility in scheduling in that the schedule does not have to dictate which machine the workpiece should go to, but just which type of machine. It also gives flexibility in that complex jobs requiring processing through many processes can be dealt with alongside those requiring just a few processes. It is a versatile layout system. It does have the disadvantage of requiring a lot of travel time between the relevant processes for a workpiece and so does not give the minimum manufacturing time for a particular product. However, because of the diversity of jobs tackled with unique production it is extremely unlikely that any one arrangement of the processes would be capable of giving the minimum manufacturing time for every product.

With *layout by product*, the plant is laid out in the sequence of processes required by a particular product. The production sequence is thus specific to a particular product and so the layout is only suitable where large batches are involved, i.e. mass production. The layout is thus not flexible and adaptation to another product can be expensive. Failure of a particular machine in the production line can lead to a stoppage of the entire line. The workpiece travels from one particular machine to another particular machine and thus if there is a hold-up and the workpiece does not arrive at a machine then that machine is liable to remain idle. The system is thus inflexible, however it does reduce the time wasted by the workpiece in travel between machines and so gives the maximum through-put.

With a variety of products and perhaps small or medium batches, both the process layout and the product layout may be inappropriate, the process layout because too much time would be wasted and the product layout because it could not cope with changes in process sequence. Group layout may be more appropriate in such situations. *Group layout* involves the recognition that many of the products have similarities in their make-up. The range of products is considered and the similarities identified. This leads to an identification of a number of sets of process sequences. Within each set a product layout can be adopted. The result of this is that the layout consists of a number of short production lines, no one line generally being adequate for the manufacture of an entire component. The workpiece is then routed between these different short production lines according to the sequence of processes it requires. This group layout is more flexible than the product layout where a single layout is used and less flexible than the process layout. It does, however, have the advantage of requiring less travel time for the workpiece than the process layout.

Planning a layout

In planning a layout the main objective is usually to minimise the amount of movement of the workpiece, materials and people, hence a minimisation of cost. In the planning stage this invariably means a minimisation of distance. Visual aids tend to play a significant part in this planning stage, often with models being used and the various paths of the workpieces being indicated. Some computer programs are also available to assist in layout preparation.

A useful chart that can be constructed for a particular layout in order to show the amount of movement is the travel or cross chart. Figure 11.2 shows such a chart. The vertical axis of the chart indicates the starting points for a movement and the horizontal axis the finishing points. The entry made in a particular square refers to the number of items moved. Thus, for example, the chart indicates that 22 items were moved from the store to milling. Numbers appearing above the diagonal indicate the motion of items when following the sequence of operations listed on the chart; numbers below the diagonal occur when the movement is in the opposite direction to the listed sequence.

The comparative size of the numbers appearing in any square indicates the need for those operations to be kept close together if total transit times are to be

To / From	Stores	Milling	Turning	Assembly	Test	Despatch
Stores		22				
Milling			19	3		
Turning				16		
Assembly					11	
Test			1	2		8
Despatch						

Figure 11.2 Travel chart

kept low. Where the number in a square is low, or zero, there is less or no need for those operations to be kept close together.

11.3 Maintenance planning

There are essentially two types of maintenance policy that an organisation might adopt:

1 Breakdown maintenance, which involves waiting until the plant fails before repairing it.
2 Preventive maintenance, which involves anticipating breakdown and replacing or adjusting the plant before breakdown occurs.

The decision as to which form of maintenance policy to adopt will probably depend on the relative costs of the two. With breakdown maintenance a factor to be considered is the lost output or the cost of supplementary plant to be used while repairs are being made. A significant factor determining the cost of preventive maintenance is its level. Preventive maintenance involves inspection and servicing and the greater the frequency of the inspection the less likely there is to be failure but the greater the cost. Thus generally the level of preventive maintenance is often adjusted so that though there are some failures the cost is not too high.

Preventive maintenance is likely to be more cost effective than breakdown maintenance when:

1 The cost of preventive maintenance is less than the cost of the lost output and repair.
2 Plant failure can cause considerable lost output costs.
3 Plant failure can lead to safety problems.
4 The failure rate of plant starts to increase rapidly after a period during which it has been relatively low. Anticipating this point can be very cost effective.

When breakdown occurs there is both a lost production cost and a repair cost. Such costs can be minimised by:

1 Designing the plant so that repair is facilitated.
2 Designing the layout of the plant so that repair is facilitated.
3 Establishing an efficient repair facility so that repairs are effected rapidly.
4 Providing excess production capacity, i.e. back-up facilities, so that production can continue.
5 Providing buffer stocks so that a production line can still continue.

11.4 Capacity planning

The effective management of capacity is one of the most important responsibilities of operations management. The term *capacity* is used to describe the resources available for production purposes, including both facilities and people. The aim of capacity management is to match the level of production operations to the level of demand. The problem, however, is that the level of demand for which capacity has to be planned is often very uncertain. This uncertainty can result from uncertainty about the number of customer orders likely to be received and the type of resources necessary to meet them; or where production is for stock as opposed to direct customer order, there may be uncertainty about the level of stock necessary in order to meet orders. Demand forecasting is necessary.

Demand forecasting

Decisions regarding the capacity required by an organisation can be made utilising forecasts over a number of different periods of time and of differing degrees of precision. For planning the installation of major capital equipment, perhaps the setting up of a new production line for mass production of cars, a fairly long-term forecast for several years into the future will be required. For capacity decisions involving the allocation of existing machines and labour, short-term forecasts of demand may be required. Short-range forecasts can be fairly accurate and detailed but this is not likely to be true of long-term forecasts, such forecasts being easily upset by changes in perhaps the economic or political climate which are difficult to foretell.

A number of different methods can be used for demand forecasting, namely:

1 *Qualitative forecasting* based essentially on gathering the views of people.
2 The analysis of past data and then an extrapolation of past performance into the future (known as *time-series forecasting*).
3 Relating demand to some other variable and then using predictions of that variable to give, by comparison, demand forecasts (known as *causal forecasting*).

Qualitative forecasting can involve market surveys to gather opinions of customers, forecasts developed by panels of experts giving their opinions, sales staff and sales managers giving their opinions, etc. Forecasts made by individuals on the basis of their experience are, however, found to give, in general, only poor to fair accuracy, whether short term or long term. Market surveys can give much better predictions, but the cost of making this type of prediction is much higher than the cost of asking individuals, such as the salespeople, to state their views.

Time-series methods are used to analyse past demand patterns and then project these patterns into the future. The basic assumptions of this method is that the demand pattern can be broken down into a number of components. Thus, for instance, the demand might vary with time in such a way that it can be considered to have essentially just two components, a level demand with a seasonal pattern superimposed on it, as in Figure 11.3, with some random variations. Once the components have been identified they, excepting the random component, are projected forward into the future.

A simple method of carrying out a time-series analysis is to use the *moving-*

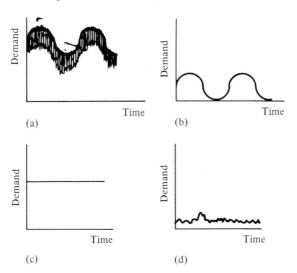

Figure 11.3 Time-series analysis. (a) The demand pattern. (b) The seasonal component. (c) The level component. (d) A random component

average method. This assumes that the time series consists of just a level component and a random component. The demand data are available for a number of periods of past time and the average is calculated over a number N of past periods.

$$\text{Average} = \frac{D_t + D_{t-1} + D_{t-2} + \cdots}{N}$$

D_t is the demand for the time t, D_{t-1} is the demand for the preceding period of time t minus one time unit, D_{t-2} is the demand for the period of time t minus two time units. On this basis the forecast for the period of time $(t + 1)$, i.e. the next period in the future, is the same as the average obtained for the periods up to time t. As time progresses, new demand data are generated and the procedure adopted is to maintain the same N periods of demand in the calculation and let the average move along.

The following is an example to illustrate the moving-average method. A three-period moving average is considered, i.e. $N = 3$.

Past data at time t

Period	March	April	May
Demand	15	17	16

The average is $(15 + 17 + 16) \div 3 = 16$. Thus the forecast for the next period of time, i.e. June, is 16. When the June data are obtained the average is revised and the three-period average calculated using the April, May and June data.

Period	April	May	June
Demand	17	16	21

The average is now $(17 + 16 + 21) \div 3 = 18$. The forecast for July is thus 18.

The greater the number of periods of time N used in calculating the moving average, the less responsive the average is to changes in demand and the more stable is the value obtained.

With the moving-average method, equal emphasis is placed on the demand data for each period, regardless of whether it is a more recent or an earlier period. An alternative method, *exponential smoothing*, places more weight on recent data. How this is done can be illustrated by considering the data used above for the moving-average example. For March, April and May the average is 16. In June the demand is found to be 21. To obtain the forecast for the next period using exponential smoothing, we do not consider this demand to be of the same significance as the demand data for March, April or May but, perhaps, we decide it is more significant and weight it. So in the calculation of the forecast for the next month we give it more weighting. The equation used for this is:

$$\text{New average} = \alpha D_t + (1 - \alpha)A_{t-1}$$

where D_t is the demand in period t, A_{t-1} is the average up to and including period $(t-1)$, i.e. the old average, and α is a weighting constant, sometimes referred to as smoothing constant, that we choose. α has a value between 0 and 1. Thus, if for our example we choose α to be 0.6, then the new average is

$$0.6 \times 21 + (1 - 0.6)16 = 19$$

This is then forecast for the month of July. More weight has been given to the June data than to the data for the previous months.

An alternative way of weighting data is to use the moving-average method but to weight each data period differently. Thus

$$\text{average} = \frac{W_1 D_t + W_2 D_{t-1} + W_3 D_{t-2} + \cdots}{N}$$

Where W_1, W_2, W_3, etc. are weighting factors such that $W_1 + W_2 + W_3 + \cdots$ $W_N = 1$. This method has, however, a disadvantage when compared with the exponential smoothing method in that the entire demand history for N periods must be calculated each time and it requires the adjustment of the weighting factors each time a new period is added. Exponential smoothing is a simpler method.

For long-term forecasting both moving averages, either simple or weighted, and exponential smoothing give poor forecasts. However, for short-term forecasting exponential smothing can give good results, better than moving averages. Other, more mathematically complex methods can be used with time series to give better results for short and medium term forecasts.

For many products there can be a close relationship between the demand for them and the sales of other products, general economic indicators or some other factor. For example, the demand for car accessories could be related to past sales of cars, the demand for computer games related to the past sales of computers. To use this method of forecasting for a particular product, a relationship has to be found with one or more other variables whose fluctuations are afterwards reflected in changes in demand for the product concerned. The method employed is to collect data for the demand for the product and values of the other variables and then look for relationship between them. A relationship between the demand and some variable can be expressed by what is known as a *simple regression* equation; where the demand is related to more than one variable then *multiple regression* is involved.

To illustrate the methods employed, consider the relationship between the demand for a car accessory and past sales of cars. The data might be:

Month	Car accessory sales	Car sales
1	800	1200
2	850	1250
3	870	1400
4	930	2100
5	1210	4000

A relationship exists between car accessory sales in any one month A_t and the car sales in the previous month C_{t-1}. The following is the relationship

$$A_t = 0.4C_{t-1} + 370$$

This is the regression equation (it is the standard straight line equation, between two variables y and x, of $y = mx + c$). Now if in any one month the car sales are, say, 5200 then the equation can be used to make a prediction of the number of car accessories that are likely to be demanded the next month, in this case

$$A_t = 0.4 \times 5200 + 370$$
$$= 2450$$

The above is an artificial case invented for the purpose of illustrating the method; in practice the data points would not precisely fit the relationship and there would be some scatter.

11.5 Aggregate planning

The term *aggregate* is used to indicate that the capacity planning under this heading is not concerned with details of individual products nor with details of scheduling plant and personnel but with the overall output required, i.e. all the product demands aggregated or collected together. Aggregate planning is usually concerned with medium-term planning, i.e. plans for up to one or two years in the future. Such planning is a comparision between the capacity available and what the demands will require. It is assumed, since only medium-term planning is involved, that there can be no major new capacity within such a time and that the demand has to be met by using or adjusting existing capacity.

Aggregate planning can be considered to involve the following sequence of operations.

1 Forecast demand for the products over the time concerned.
2 Express the demand for the different products in some common capacity-related units. This is necessary if all the demands are to be aggregated together. Thus a brewery might use litres of beer as its aggregated capacity measure, despite producing a variety of beers in a variety of different volume containers. Volume is often used in process production where chemical processes are

involved. In a highly mechanised production department machine hours might be used.

3 Produce an aggregated demand.

4 Consider the demand data and how best the capacity can be utilised to achieve those demands. There are essentially three ways this can be done: chase demand, level capacity or a mixed plan. With *chase demand* the output is directly linked to the demand and as demand changes so there will have to be capacity changes. *Level capacity* involves smoothing out the demand fluctuations and maintaining a level use of capacity over the time concerned (Figure 11.4). This can be done by, perhaps, producing for stock during low demand periods and making use of that stock during high demand periods. Another possibility is to allow a backlog of orders to develop at high demand periods. A *mixed plan* involves a mixture of chase demand, at perhaps the main peak and troughs of demand, and a level capacity approach to most of the demand.

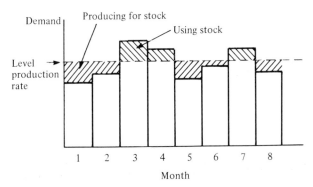

Figure 11.4 A level-capacity production plan

5 Consider what changes in capacity will be needed and how they can be obtained. This might involve hiring and firing employees, using overtime or lay-offs, producing for stock or utilising stocks, subcontracting, buying-in, adjusting the demand by changing the price or advertising, allowing backlogs to develop etc.

6 Finally put it all together as an aggregate plan.

11.6 Scheduling

Capacity and aggregate planning are concerned with establishing what resources are likely to be required; *scheduling* is concerned with the planning of activities to make the most effective and efficient use of available resources. Thus capacity and aggregate planning can be considered as determining the resources that should be required for scheduling. Scheduling can therefore be considered to be the last aspect in the sequence of decision-making concerned with production planning.

Where an assembly line is being used and there is just one product made by that line then there is no scheduling problem, the output being determined completely by the design of the line. However, where more than one product is made by the line, e.g. it might be that the line makes a variety of models of the same basic product, then there is a scheduling problem. Each product is generally made in batches, with a changeover to the next product involving costs associated with perhaps setting up or even retooling. The problem then becomes the determination of the most economic batch size. The production of small batches means fewer stocks to be held but more frequent resetting and associated costs; large batches, however, mean more stocks to be held but less frequent resetting and so, over a period of time, lower resetting costs. The problem is thus one of inventory management.

11.7 Inventory management

An *inventory* can be defined as the stock of material used to facilitate production or meet customer demand. Thus inventories can include raw materials, work in process and finished goods. A stock of finished goods can provide a way of meeting fluctuations in demand without having to change production capacity, i.e. accommodate short-term differences between demand and production output rate. Work-in-process stocks can be used to accommodate fluctuations in input and output rates of the various production stages. A stock of raw materials can be used to protect against uncertainties in the supply of the materials and also to permit fluctuations in the demand for the materials from the production stages that follow. A simple way of describing the above in terms of a model is shown in Figure 11.5. The size of the inventory is represented by the level of water in a tank, the tank having an input and an output. If the input and output rates are the same then the inventory level remains constant. If, however, there is a change in one then the inventory level can either rise or fall to accommodate the fluctuation and still give the required

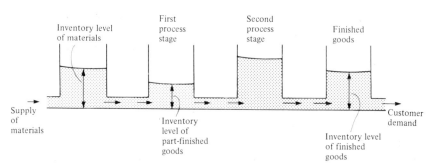

Figure 11.5 Inventory management

output. The existence of the inventories thus enables the production manager to maintain an even rate of production flow.

Principal advantages of holding inventories

1 A raw materials inventory allows the organisation to:
Accommodate fluctuations in the supply of the materials.
Provide strategic stocks for materials that might for some reason become in short supply.
Take advantage of quantity discounts.
Guard against anticipated price increases.
Accommodate fluctuations in the demand for materials for the production processes.
2 Work-in-progress inventories allow the organisation to:
Decouple the different stages of production and so maintain a more even rate of production flow.
Improve flexibility in production scheduling.
Improve the utilisation of plant and labour.
Cope better with plant breakdown and maintain a production flow.
3 A finished goods inventory allows the organisation to:
Achieve a steady supply to customers though production may be intermittent.
Cope with fluctuations in demand without making changes in the rate of production.
Achieve a supply to customers if plant breaks down or some other factor stops production.
Supply customers off-the-shelf without delay.

Inventories incur costs. These can be considered under the following headings:

1 *Item cost.*
2 *Ordering costs.* In the case of raw materials there will be a cost associated with the replenishment of stocks, the cost including such items as the paperwork, transportation, etc.
3 *Holding costs.* This cost is associated with the keeping of items in inventory for a period of time and includes the costs associated with storage, e.g. the facilities and insurance, costs associated with obsolescence, deterioration and loss of items, and the costs of the capital tied up in the inventories, such capital not therefore being available for other purposes.

There is also another cost factor that has to be considered in relation to inventories and that is the cost of running out of stock, showing itself as a hold-up in production or sales. This is referred to as a stockout cost.

Organisations tend to need inventories and since inventories cost money there is the obvious question – what is the right amount of inventory? An inventory holding in an item in excess of what is required is effectively money earning no useful return. An inventory holding inadequate to what is required can cost money in, for example, lost production, i.e. the stockout cost. The control of inventories is thus vital.

Control of inventories

A vital factor in the control of inventories is whether the demand for a particular inventory is independent or dependent. An *independent demand* is one that is independent of the production operations but depends only on external factors. Thus the demand for finished goods is independent. *Dependent demand* is where it is related to the demand for another item and is not independent of production operations. Thus the demand for part-finished items is determined by the demand for the finished goods and so is a dependent demand. The type of control system used will depend on whether the demand for the inventory is dependent or independent and, as an organisation is likely to have inventories which fall into both categories, more than one type of control system will have to be employed.

Inventory control for independent demand is based on forecasts, often evolving from past demand. The questions that have to be considered in relation to such inventory control are—when should an order be placed to replenish the inventory and how much should be ordered? One method could be to reorder whenever the inventory falls to a particular level. The term *two-bin system* is used for this method. This is because we can consider the inventory to consist of two bins, with one bin containing items which are issued while the second bin remains sealed. When the items in the first bin have all been issued an order is placed for fresh stock and the second bin should contain enough items to last out until a new delivery of items occurs. The term *single-bin* system could be used for a method whereby the inventory is allowed to run down to zero before an order is placed, i.e. there is nothing kept in a second bin to cover the time taken for the new stock to be delivered. An alternative to methods involving ordering when the inventory reaches a particular level, even

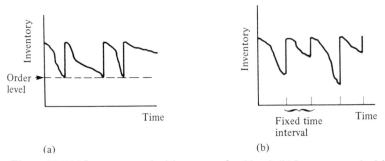

Figure 11.6 (a) Inventory replenishment at a fixed level. (b) Inventory replenishment at fixed time intervals

if the level is zero, is to order at fixed time intervals. Such orders are generally placed so as to bring the inventory up to a particular level.

Figure 11.6 illustrates the differences in these two approaches. For convenience it has been assumed for the figures that restocking is instantaneous following ordering.

Inventory control with independent demand is generally concerned with batch ordering and the minimising of the total variable costs, i.e. the minimisation of the total of the holding and ordering costs. Figure 11.7 shows how these costs might vary with the size of batch ordered. The batch order size at the minimum total cost is called the *economic order quantity* (EOQ). Various equations have been developed for calculating the economic order quantity and tables and charts to enable these equations to be easily used are available.

The following shows how an equation is developed and the conditions pertaining to its use. With the two-bin method, a constant demand, the items ordered or produced in batches are all put into the inventory at one time, there is a constant and known time from order to delivery, and no stockouts, and the inventory level will vary with time in the way shown in Figure 11.8. The quantity ordered each time and put into the inventory is Q. This then results in an average inventory level of $\frac{1}{2}Q$. If c is the unit cost of the item then the average cost of the inventory is $\frac{1}{2}Qc$. If C is the carrying cost rate per unit time, i.e. effectively the rate of interest lost on the capital, then the holding cost for that time is $\frac{1}{2}QcC$. This is the equation of the holding cost line in Figure 11.7. If the consumption rate per unit time is r then the number of items consumed in that unit time will be r/Q. If the cost of placing an order or setting up production is C_s, then the ordering cost in that time is rC_s/Q. This is the equation of the ordering cost line in Figure 11.7. Hence, as the total cost is the sum of the holding cost and the ordering cost,

Total cost per unit time $C_t = \frac{1}{2}QcC + rC_s/Q$

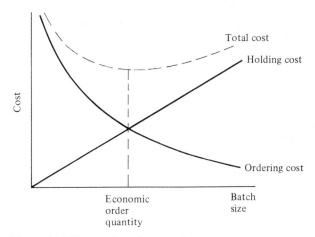

Figure 11.7 Economic order quantity

The minimum value of this total cost can be obtained by using calculus or by realising that the minimum occurs when the holding cost equals the ordering cost, i.e.

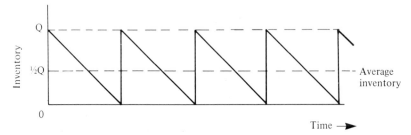

Figure 11.8 Inventory levels vary with time

$$\tfrac{1}{2}QcC = rC_s/Q$$

Hence

$$Q^2 = \frac{2rC_s}{cC}$$

This is the economic order quantity, hence:

$$EOQ = \sqrt{\frac{2rC_s}{cC}}$$

To illustrate the use of the above equation, consider a company which makes an item for which the yearly sales are forecast at 200, each item costing £150. The carrying-cost rate per year is 25% and the cost of setting up for production is £1000. If we assume that there is a constant demand and we can use the above equation then the economic order quantity is:

$$EOQ = \sqrt{\left(\frac{2 \times 200 \times 1000}{150 \times 0.25}\right)}$$
$$= 103$$

Thus the company should produce the items in batches of 103 if costs are to be kept to a minimum. The cost, over the above the production cost of the item, for such batch sizes is, using the earlier equation for C_t:

Total cost per year $C_t = \tfrac{1}{2} \times 103 \times 150 \times 0.25 + 200 \times 1000/103$
$$= £3873$$

In this example the order has effectively been considered to be from the sales department for its inventory of finished goods to the production department. A similar approach can be used for a production department ordering raw materials.

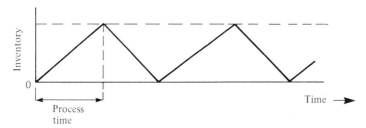

Figure 11.9 Inventory levels with continual stock delivery

In the above equations it was assumed that all the items would be delivered into inventory at the same time. The equations are, however, modified if the items are delivered into stock continually throughout the process period, as illustrated by Figure 11.9. The average inventory is then $\frac{1}{2}Q(1 - r/q)$, where r is consumption rate per unit of time and q the process rate per unit of time. Thus:

$$C_t = \tfrac{1}{2}QcC(1 - r/q) + rC_s/Q$$

and, as before

$$EOQ = \sqrt{\left| \frac{2rC_s}{cC(1 - r/q)} \right|}$$

Another factor that can be taken into account is the holding of buffer stocks. In Figures 11.8 and 11.9 the inventory is shown as falling to zero. If there is a buffer stock level B then, as shown in Figure 11.10, the effect is to increase the average inventory to $(B + \tfrac{1}{2}Q)$ and so the total cost per unit time to:

$$C_t = (B + \tfrac{1}{2}Q)cC + rC_s/Q$$

In calculating the minimum cost, the term involving B has no effect on the value and thus the economic order quantity is unchanged, as:

$$EOQ = \sqrt{\frac{2rC_s}{cC}}$$

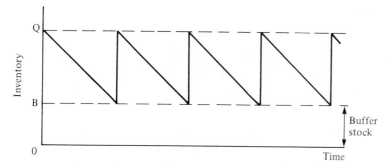

Figure 11.10 Effect of buffer stocks on the inventory

Scheduling for batch processing

We have considered above the optimum batch size or *economic order quantity*. If an organisation is producing more than one product on the same machines there can be a clash of requirements, unless the different batches occur in sequence. This is often not the case: the situation may well be as described in Figure 11.11, where the required process times for two products overlap. A technique that can be used to deal with scheduling in such situations is the line of balance method. This is considered later in this chapter.

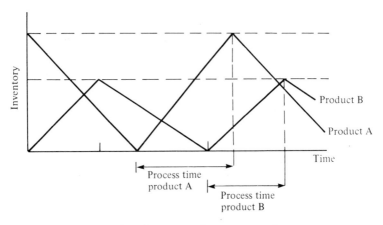

Figure 11.11 Process time for two products may overlap

11.8 Materials requirements planning

Where items are assembled in batches, certain components will be required at particular times to suit the batch assembly schedule. Thus, to maintain an even production flow, inventories of these items are likely to be held. The demand for these items is dependent, i.e. not subject to market forces but dependent on the demand for other items in the production operations. For example, the drilling operation for a particular product cannot take place until the machining operation is complete. Thus, to avoid the drill operators running out of work if there is a delay in the machining, inventories of machined parts can be held. The level of such an inventory depends on the demand generated by the drill operators, not on any market forces. *Materials requirements planning* (MRP) is concerned with ensuring that items are available when required and also that unnecessarily high inventory levels are not held.

Figure 11.12 shows the essential features of materials requirements planning system. The purpose of the MRP system is to coordinate the entire planning operation and produce purchase orders and production orders at the appropriate times and in the appropriate quantities. Inputs to the system are details of the materials (resulting from the design data), a master schedule giving details of the quantities and types of products required in each time period, and inventory records giving the current inventory levels and details of ordering procedures. There is also feedback of information to the system on how it is functioning and hence the possibility of changes being made in plans. The system is almost invariably run on a computer, the MRP system being essentially the computer program.

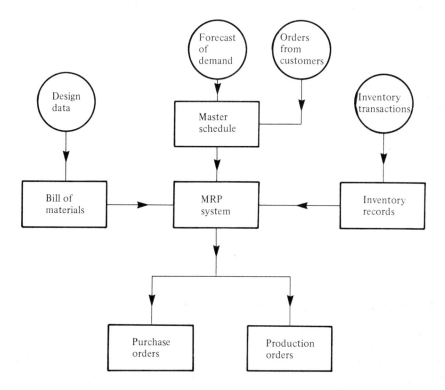

Figure 11.12 Essential features of a materials requirements planning (MRP) system

11.9 Schedules and loads

The term *job schedule* is used for a schedule that has been prepared for a single job. It will show the start and finish times for each activity involved in that job. A bar chart, sometimes referred to as a *Gannt chart*, can be used to show the schedule. The vertical axis of such a chart, see Figure 11.13, shows the activities and the horizontal axis time. The planned operations and their times are shown in the form of bars, the beginning of the bar indicating the start time and the end of the bar the finish time. Thus, for example, activity *B* is planned to start at the beginning of week 2 and finish half way through week 4. Activity

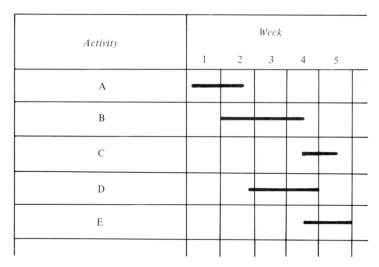

Figure 11.13 A schedule

A might be the designing, activity *B* the process planning, activity *C* the setting up, etc.

The term *factory* or *production schedule* is used where the schedule simultaneously considers a number of jobs. A bar chart can be used, as before, and in that case activity *A* might refer to product *A*, activity *B* to product *B*, etc. Production schedules and job schedules obviously interact.

Another form of schedule is the one that indicates the work assigned to each work centre. The term *load* is used for the work assigned to a particular work centre, perhaps a particular machine, and the term *loading* to describe the working out of the anticipated load profile for a work centre. Loading indicates the utilisation that is being planned for a work centre. There is an obvious interaction between such load schedules and both production and job schedules.

11.10 Network analysis

In both planning and scheduling activities a technique that is widely used is *network analysis*. Such a method can show the interdependence of and relationships between activities. The following represents the various stages in carrying out such an analysis in its simplest form:

1 Identify all the activities involved.
2 Construct a network diagram where each of these activities is represented by an arrow and the arrangement of the arrows represents the order in which the activities will occur and their interdependence. At the start and finish of each

arrow a small circle is drawn to indicate an 'event', i.e. the start or completion of the activity. Any number of activities can go into or out of an event. Activities occurring in the same linear path are sequential and thus directly dependent on each other; activities on different paths are termed parallel activities and can therefore take place at the same time. Figure 11.14 illustrates these principles.

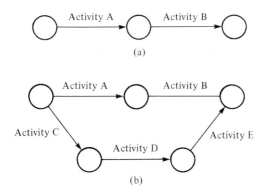

Figure 11.14 (a) Sequential activities: B cannot occur until A is complete. (b) Parallel activities: C, D and E can occur without relationships to A and B

3 Estimate the time taken for each activity and insert the times alongside the appropriate arrows.
4 Examine the various paths through the network and compute for each activity the earliest start time and the latest finish time.
5 Further analysis can then lead to the identification of the critical path, this being defined as the longest time path through the network. It is this path which determines the completion time and the activities along that path are critical since any delays in them will increase the overall time and delay the completion.

The above outlines the basic steps in carrying out network analysis. One type of activity that was not mentioned above is the *dummy activity*. Dummy activities consume no time and may be used to indicate a dependence or to

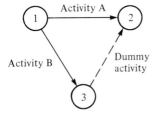

Figure 11.15 Dummy activities: no time spent going from event 3 to event 2

avoid having more than one activity with the same beginning and end events. Dotted lines are used for the arrows representing dummy activities, as in Figure 11.15.

To illustrate this network analysis approach consider the following example, making a cup of coffee.

	Activity	Activities on which dependent
A	Fill kettle with water	—
B	Plug kettle into mains	A
C	Water heats up	B
D	Pour hot water onto coffee in cup	C, G
E	Get cup and saucer	—
F	Get coffee	—
G	Put coffee in cup	E, F
H	Take cup with coffee to kettle	G
I	Get milk	—
J	Add milk to hot coffee	D, I
K	Get sugar	—
L	Put sugar in hot, milky coffee.	J

Figure 11.16 shows the network for the coffee making. Event 1 is where it all starts, event 12 where it finishes. An event marks the completion or start of an activity or activities. Thus, for example, event 4 is the completion of the activities of the water heating up and taking the cup with coffee to the kettle and the start of the activity of pouring the hot water onto the coffee in the cup. It is sometimes easier to draw a network by starting from the end and working backwards.

The numbers written under each arrow are the times taken for the activities, in this case in seconds (s). Thus, for example, activity A, which is the filling of the kettle with water, takes 25 s.

The earliest start time for each activity is calculated from the beginning of the network by adding together all the preceding activity durations. Where two or more activities lead into one event then the last of the activities to finish determines the start time for the subsequent activity. This is because this activity cannot start until all the activities leading to this event have been completed. Thus for event 3 the earliest start time is $25 + 3 = 28$ s. There are two paths to event 4, one having a total duration of $25 + 3 + 200 = 228$ s and the other $10 + 4 + 2 = 16$ s. Hence the earliest start time is 228 s. The earliest start times are shown on the network as the first of the two times in the rectangular boxes.

The latest finishing time is calculated from the end of the network by subtracting activity durations from the project finish time. Where two or more activities stem from one event, the earliest of the times determines the latest finishing time for the previous activities. Thus for event 11 the latest finishing

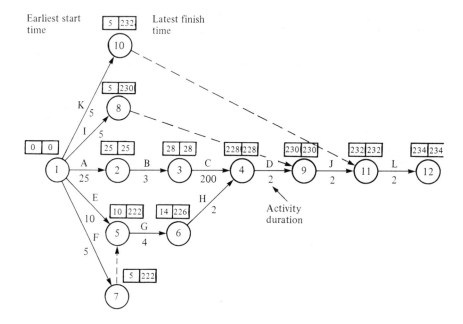

Figure 11.16 Making a cup of coffee

time is 234 − 2 = 232 s. For event 10 the latest finishing time is 232 − 0 = 232 s.

From the above information a number of other times can be calculated. Thus the earlier finish time for any activity is the sum of the earliest start time of the activity and its duration. Hence for activity *H* the earliest finish time is 14 + 2 = 16 s. The latest start time for any activity is the latest finish time for that activity minus its duration. Thus for activity *H* the latest start time is 228 − 2 = 226 s.

The earliest finish time for any activity, or for the entire project, is determined by the longest time path through the network to that event. Thus the earliest finish time for activity *D* is determined by the path through activities *A*, *B*, *C* and not the path through *E*, *G*, *H*. Therefore the shorter time path will have more time available to it than is required. The time available for an activity is the latest finish time minus the earliest start time. Thus for activity *H* the time available is 228 − 14 = 214 s. The actual time required is 2 s, thus there is 214 − 2 = 212 s spare, this spare time being referred to as the *total float*.

The total float is the amount of time by which an activity can slip without causing any change in the total time required for the project. However it should be noted that if there is slippage in the time for activity *E* in the path, *E*, *G*, *H*, then the time available for slippage in activity *H* will be reduced.

The critical path through the network is the longest time path. For the coffee making it is *A*, *B*, *C*, *D*, *J*, *L*. It is the path with minimum total float. Any delay in the activities along this path will delay the completion of the project. A delay along any other path will just use up some of the total float associated with that path.

11.11 Progressing

The term *progressing* is used to describe the checking that production is according to plan and the taking of remedial action where it is not. The term *progress chaser* or *expeditor* is used for the individual whose job it is to carry out this work.

Information on the progress of production is obtained in a number of ways. For example there might be counting or recording devices attached to machines. There are the operator's work record, a sheet on which an operator lists all the jobs he or she has done, and job cards, which are the instructions to an operator as to what jobs to do and which are filled in when jobs are completed. In addition there is much information to be obtained by progress chasers just walking round the production facilities and observing.

Production delays can occur for a variety of reasons, including those under the following headings.

1 *Personnel*. There may be absenteeism, disputes, high labour turnover, inadequately skilled operators, etc.
2 *Machines*. There may be breakdowns, tool problems, loading errors, etc.
3 *Materials*. Materials may be delivered late or of the wrong quality, there may be handling problems in the stores, etc.
4 *Design*. There may be design errors or perhaps an accuracy specified which presents production problems.
5 *Planning*. The plans may be too optimistic or contain errors.
6 *The customer*. The customer may make changes to the specification or delivery dates.

Progress information can be presented on the Gannt chart (Figure 11.17). As well as showing the planned times (as earlier shown in Figure 11.13) it can also show the progress towards meeting those plans. Thus if we wanted to consider progress at the end of the fourth day for the example given? Activity A is complete, Activity B is also complete—ahead of schedule, Activity C is still occurring but progress is ahead of schedule in that more is completed than was expected, activity D is behind schedule in that not as much has been completed as was planned.

11.12 Production control

Where mass production is involved and there is a small range of products, *production control* is often mainly concerned with ensuring that inventories are maintained at the appropriate levels. This is a simpler situation than in jobbing

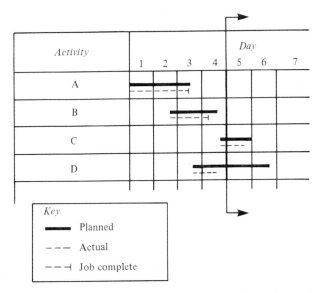

Figure 11.17 Presenting progress information using a Gannt chart

or small batch production, which involve a great number of variables, generally a wide range of products and often flexible plant. Progress chasing plays a big role in such situations.

Production control can be fairly straightforward if for each operation a batch is kept intact, the entire batch finishing one operation before moving on to the next operation. However this can result in a longer throughput time and inefficient use of plant. For this reason *batch splitting* is often used. In this, part of the batch is undergoing one operation while another part is undergoing a different operation. This, while decreasing throughput time and increasing the efficiency of plant usage it does make it more difficult to ascertain whether the entire batch is proceeding to plan. A technique that can be used is the *line of balance*.

To illustrate the use of this technique, consider a simple product on which just four operations A, B, C and D, are performed sequentially. Operation A takes 2 days; operation B, 3 days; operation C, 1 day and the final operation D, 2 days. This is a total time of 8 days. Suppose that the delivery requirements are 20 units per day, starting after the 8th day, i.e. the batch has been split. To meet this we must have 20 units through operation D by the end of the 8th day. Because operation C has a lead time over operation D of 2 days, sufficient units should have been completed through this operation at day 8 to meet the delivery schedules for both the 8th and 9th days, i.e. a total of 40 units. These are the cumulative completions by that 8th day.

For operation B the lead time over operation D is 3 days. Hence at day 8 we must have a cumulative completion for process B of 60 units, to meet the deliveries for days 8, 9 and 10. For operation A the lead time over operation D is 6 days. Hence the cumulative completion for process A must be 120 units, to meet deliveries for days 8, 9, 10, 11, 12 and 13. We can present this as a histogram, Figure 11.18.

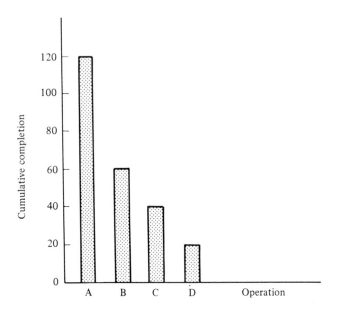

Figure 11.18 The planned completions

Control over actual production at the various operations is obtained by comparing the histogram of actual progress with the line representing the planned progress histogram, the line being referred to as the line of balance. Figure 11.19 shows such a comparison. Thus, for this example, cumulative completions from operation A are ahead of schedule, from operation B behind schedule.

Problems

1 Explain the differences between *job*, *small-batch*, *large-batch* and *process production*.
2 What type of product is likely to be produced by (a) job and (b) mass production?
3 What criteria are likely to be involved in the design of a plant layout?
4 Explain the differences between plant layouts by (a) process, (b) product and (c) group?
5 Which type of layout is likely to be the most efficient for (a) job production, (b) mass production?

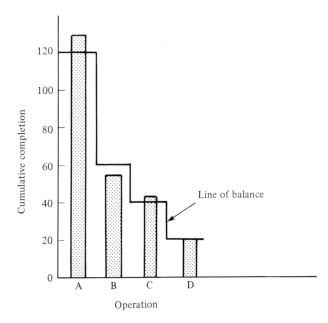

Figure 11.19 A progress chart

6 Explain the significance of a travel chart.
7 Explain the difference between *breakdown maintenance* and *preventive maintenance*.
8 Under what circumstances is preventive maintenance likely to be more cost effective than breakdown maintenance?
9 What is meant by capacity planning?
10 Explain what is meant by demand forecasting by (a) qualitative means, (b) time-series, and (c) causal methods.
11 Explain the principle of the *moving-average method* of demand forecasting.
12 How does the *exponential smoothing method* of forecasting differ in principle from the moving-average method?
13 Explain how causal methods of forecasting can be done using a *regression equation*.
14 What is meant by *aggregate planning*?
15 What is meant by *inventory management*?
16 What are the principal advantages to an organisation of holding inventories in (a) raw materials, (b) part finished goods and (c) finished goods?
17 What are the main types of costs that inventories can incur?
18 Inventory control can be considered in terms of *dependent* or *independent demand*. Explain these terms.
19 What is meant by the *two-bin system* of inventory control?

20 Derive the equation for the economic order quantity, $EOQ = \sqrt{(2rC_s/cC)}$, explaining the terms and the assumptions made.

21 Explain what is meant by the *economic order quantity*.

22 What is meant by the term *materials requirements planning?*

23 Explain the purpose of a Gannt chart and describe its form.

24 What is meant by *loading?*

25 Explain the significance of the following in relation to a network diagram, (a) an arrow, (b) a circle, (c) a dotted line arrow.

26 Define the following terms, (a) earliest start time, (b) latest finishing time, (c) earliest finish time, (d) latest start time, (e) total float, (f) critical path.

27 What is meant by *progressing?*

28 What is the function of a *progress chaser?*

29 What is meant by *batch splitting* and what are the reasons for doing it?

30 Explain how the *line of balance method* can be used for production control where there is batch splitting.

31 Investigate a plant layout in some company and comment on it in relation to the types of product that are produced.

32 The following is a travel chart that has been produced over a period of one month. Comment on the layout of the production facilities and suggest on improvement, giving the new travel chart for your proposal.

From \ To	Stores	Turning	Milling	Grinding	Assembly	Testing
Stores		20	25	2	2	1
Turning			18			
Milling				16	24	
Grinding					15	
Assembly						38
Testing				5	3	

33 Discuss the relationship that should exist between marketing and production strategies.

34 For each of the following present arguments for a type of production operation and a form of layout.
(a) Domestic vacuum cleaners.
(b) Furniture.
(c) Specialist machine tools made to order.
(d) A popular car model.

35 Compare line with small batch production processes, considering such factors as efficiency, flexibility, capital investment, labour skills, type of product, control and planning.

36 Present a reasoned argument for your proposal for a maintenance plan for (a) electric light bulbs throughout the company, (b) machines in a job production situation, (c) machines in a line production situation.

37 Discuss the relative merits, and problems, associated with the following methods of demand forecasting, (a) qualitative forecasting using the opinions of salesmen, (b) qualitative forecasting using market surveys, (c) the moving average method, (d) the weighted moving average method, (e) exponential smoothing, (f) causal methods and the development of a regression equation.

38 The following data show the monthly demand for a product. Use (a) five-period moving average method, (b) exponential smoothing with a weighting factor of 0.7 for the last month, to obtain forecasts of the demand in the next month.

Month	Jan	Feb	Mar	Apr	May	June	July
Demand	550	650	600	750	900	800	600

39 For the product data given in the previous question the demand in the month of August was found to be 500. Using the same methods as in that question, forecast the demand for September.

40 The following data show the number of breakdowns to which the service engineers have been called during the last few months. Prepare (a) a three-period moving average forecast for the next month, (b) a three-period weighted moving average forecast for the next month if $w_1 = 0.5$, $w_2 = 0.3$ and $w_3 = 0.2$.

Month	Jan	Feb	Mar	Apr	May	June
Breakdowns	52	65	60	70	65	80

41 Obtain some sales data for a product over a period of time and devise a forecasting method for use with it. Present reasoned arguments for your choice of method.

42 The sales manager believes that the demand per month for photographic film is related to the number of fine Sundays in the month. He therefore collects data for a number of months. Is his belief justified and if so what is the regression equation?

Number of fine Sundays	0	1	2	3	4	5
Sales in thousands	3	5	7	9	11	13

43 A hotel manager wants an aggregate plan. Based on the following data produce one. No changes are expected year by year.

Month	Jan	Feb	Mar	Apr	May	Jun	Jly	Aug	Sept	Oct	Nov	Dec
Room demand	15	15	25	30	35	50	50	50	45	35	15	30

The manager requires one employee for each 10 rooms used, each being paid £500 per month. There is the possibility of overtime, up to 20% at time-and-a-half. Temporary staff can be hired at a cost of £300 per month. To fire an employee would cost £1000, to fire a temporary worker has zero cost.

44 The following is a forecast of the number of units production department will have to produce.

Week	1	2	3	4	5	6	7	8
Units	3500	2500	2500	3500	5000	6500	5000	5000

Produce a level capacity aggregate plan.

45 Determine the economic order quanitity to be ordered in each of the following situations, assuming that the demand for the inventory is independent and constant, delivery being in complete batches.

(a) Demand per year 4000, price per unit £20, carrying cost rate per year 30%, ordering cost per unit £10.

(b) Demand per year 20 000, price per unit £12, carrying cost per unit per year £0.50, ordering cost per unit £10.

46 A company makes a furniture item for which the yearly sales are fairly constant at 800, each item costing £250. How many of these items should be made in a batch if the carrying cost rate per year is 20% and the cost of setting up production £500?

47 Draw a Gannt chart for the operations giving the following data.

Activity		Activities on which dependent	Duration in days
A	Obtaining materials	—	0.5
B	Machining	A	1.5
C	Drilling	B	1.0
D	Grinding	C	1.0
E	Testing	D	0.5

48 Draw a Gannt chart for the project giving the following data

Activity	Activities on which dependent	Duration in days
A	—	2
B	A	1
C	B	3
D	A	2
E	D	3
F	C, E	2
G	F	1

49 Draw a network for the project described in Problem 48. (a) What is the critical path? (b) What is the minimum project completion time?

50 Draw a network for the following project, identifying earliest start and latest finish times for each event and determining the critical path.

Activity	Activities on which dependent	Duration in days
A	—	4
B	—	2
C	A	1
D	C	5
E	B	1
F	D, E	3
G	F	1
H	F	3
I	G, H	2

51 Draw a network for the following project, identifying the earliest start and latest finish times for each event. Determine the total floats and the critical path. Comment on the significance of this path and the floats.

Activity	Activities on which dependent	Duration in days
A	—	1
B	—	2
C	—	3
D	A	2
E	C	2
F	B, D, E	3
G	C	2
H	F, G	1

12

Quality control

12.1 Quality

The quality of a product can be defined as its fitness for the purpose for which it is required. The term 'quality' should not be considered to imply high price, high value, high precision, etc. Quality is a measure of whether a product is up to standard, what the customer wants, what was designed, etc. Any product, however lowly, can be of acceptable quality if it meets the prescribed requirements.

Quality can be considered as a measure related to various aspects of production.

1 *Design quality*. This is the degree to which the design specification of a product satisfies the customer's design requirements.
2 *Manufacture quality*. This is the degree to which the manufactured product conforms to the design specifications.
3 *Product quality*. This embraces both the design and manufacture quality and is the degree to which the product satisfies the customer's requirements.

Linked with the concept of quality is *reliability* in that generally a high quality product is more likely to be reliable than a low quality product. Reliability can be defined as the ability of a product to function as required, where required, when required and for as long as is required.

The cost of quality

The cost of quality can be considered to be made of two components, *control costs* and *failure costs*.

Control costs
These are the costs concerned with ensuring that products are of the required quality. This can be considered as being made up of two elements, *prevention costs* and *appraisal costs*. Prevention costs are those associated with those activities designed to prevent poor quality products being produced, e.g.

training, and appraisal costs are those associated with inspection of the products in order to remove those that are not of the quality required by the customer.

Failure costs

These are the costs associated with the product failing to meet the required quality, either during the production process or after the product has been received by the customer. The cost incurred as a result of failure to meet the required quality during the production process includes such costs as those of the labour and materials for a product which cannot be sold, of reworking the product to bring it up to quality, etc. The cost incurred as a result of failure after the product has been received by the customer includes the cost of refunds, repairing or replacing the product, the cost of dealing with complaints, etc.

The more money that is spent on prevention the fewer will be the number of failures, either during production or after the customer has received the product. The more money that is spent on appraisal the fewer will be the number of failures, either after the customer has received the product or during manufacture if appraisal occurs at a number of points in the manufacturing process. Thus, the higher the cost of control the fewer the failures.

The greater the number of failures, either during production or after the customer has received the product, the greater the failure costs. Thus the greater the money devoted to control, the smaller the cost of failures.

Figure 12.1 shows, in a simplified way, how both the control and the failure costs might vary with the number of failures. Adding together these two costs gives the total cost of quality, this showing a minimum value at some particular

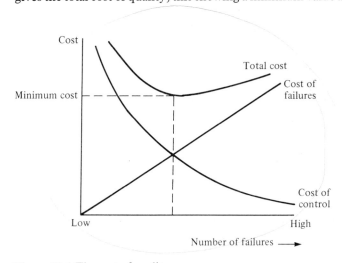

Figure 12.1 The cost of quality

number of failures. Thus the total cost of quality can be minimised if a certain number of failures are accepted as occurring.

12.2 Quality control

Inspection is a necessary feature of any quality control system. The critical points in a manufacturing process where inspection can most effectively occur are:

1 *Incoming materials.* Are the materials in accord with the specifications? Are bought-in parts to the required specifications? Checking at this stage can eliminate materials of the wrong specification and avoid an entire batch of products later in the manufacturing process failing to be of the right quality.
2 *In-process.* This is inspection of work in process. The identification of failures part way through the manufacturing process can be less costly than identifying them only at the end.
3 *Finished product.* This is the final inspection prior to sale to a customer.

The type of measurement to be made at each inspection determines the type of quality control to be exercised. There are essentially just two types of measurement, namely:

(a) *A measurement of variables.* A product may have some property that varies, e.g. chemical composition, length, hardness, etc. The inspection can then be of this variable. *Control by variables* is the term used.
(b) *A measurement of attributes.* A product may have some attribute which is either right or wrong, e.g. an item fits into a hole or it does not, the colour is red or it is white. The inspection can then be based on acceptance or rejection techniques. *Control by attributes* is the term used.

When to inspect and what to measure at the inspection are two key features of any quality control system. The third feature is concerned with how much inspection. The choice is between an inspection involving every item being inspected or one involving an inspection of just a sample. The greater the number of items inspected the greater the cost but the less the chance that a customer will receive a defective item. There is also the point that inspection, for some items, might cause damage or even complete destruction, e.g. fuses, and so inspection of just a sample is all that is possible. Statistical methods are thus widely used to determine not only how many items to inspect but also what deductions can be made from the results with regard to the entire batch of items from which the sample was taken.

Two types of statistical methods are used:

1 *Acceptance sampling.* A random sample is taken from a batch and on the basis of an inspection of that sample the entire batch is either accepted or rejected.
2 *Process control.* This involves inspection of items during the production

process with a view to determining how the process is proceeding, e.g. whether a machine is developing a fault.

With both statistical methods the inspection method used could be either by variables or attributes.

12.3 Acceptance sampling

Acceptance sampling involves taking a random sample from a batch of products or materials and on the basis of inspecting that sample, very often by attributes, either rejecting or accepting the entire batch. There are a number of different ways we can take samples but for the moment consider just a single sample to be taken. Thus if there is a batch size of, say, 100 then we might consider taking a sample of 10 and inspecting them. Suppose we find 1 defective item in that 10, i.e. 10% of the sample is defective. Does this mean that 10% of the entire batch is defective? Or could it be less than 10%, or perhaps more? If we had taken 10 samples of 10 items, would we have obtained 10% defective in each sample?

The answers to these questions depend on the laws of probability. In fact, if there were 10 defective items in 100, i.e. 10% defective, then there is, in taking a sample of 10 from the 100, a 35% chance that we will find a sample with no defective items, a 39% chance of a sample with 1 defective item and a 26% chance we will find a sample with more than 1 defective item. If we had been rejecting, on the basis of the sample, any batch where the inspection revealed more than 1 defective item in 10 then there would have been a 26% chance that the sample taken would have been pessimistic and the entire batch would have been rejected. There would, conversely, have been a 74% chance that we would have obtained a sample with no or just one defective item and so rightly accepted the batch.

If we take a sample size of 10 and an acceptance plan that we will accept the batch if the sample contains one or less defective items, then a graph can be drawn relating the chance of accepting a batch with the actual percentage of defective items in the batch. Such a graph is known as the *operating characteristic*, Figure 12.2 showing the one for this data. Hence, using the graph, we can determine that if the actual percentage of defective items in the batch was 10% then there would be a 74% chance we would have accepted the batch on the basis of our acceptance plan. However, the operating characteristic does tell us that the acceptance plan could result in our accepting batches with percentages of defective items greater or less than 10%. For example there is a chance of about 15% that the actual percentage of defective items in the batch would be 30%.

The operating characteristic curve depends on the size of the sample and the acceptance plan adopted. Tables and charts are available to enable the operating characteristic to be drawn for different sample sizes and different

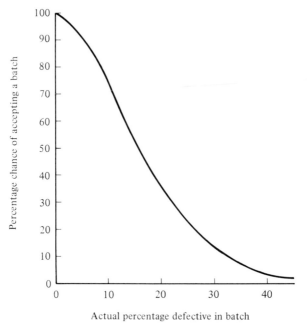

Figure 12.2 An operating characteristic when sample size in 10 and number of defectives in sample is 1 or less

acceptance plans. Figure 12.3 shows the operating characteristic when the sample size is 10 but the acceptance plan is that a batch will be accepted if the sample contains two or less defective items. Compare this graph with that in Figure 12.2. Now, if the actual percentage of defective items is 10% there is a 93% chance we would have accepted the batch. The chance of the batch having 30% defective items is, however, 38%. Thus this acceptance plan where the number of defective items can be as high as 2 leads us to accept batches which can have considerably higher percentage of defective items than the plan where we only accepted if the sample contained one or less defective items.

The lower the number of defective items in a sample which results in acceptance of a batch, the tighter the inspection. Similarly, the bigger the size of the sample the tighter the inspection (Figure 12.4). The tighter the inspection, the smaller is the chance that the sampling will result in a batch that should have been rejected being accepted.

The term *acceptable quality level* (AQL) defines the highest percentage of rejects in a batch which the receiver of the goods will accept. Thus we might, for instance, have an AQL at 10%. This would mean that the receiver of the goods would want there to be a sampling plan which ends up rejecting all batches with more than 10% defective items. However this can only be such a certainty if all the items in the batch are inspected, i.e. a 100% sample. Any

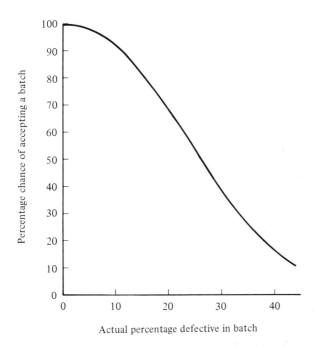

Figure 12.3 An operating characteristic when sample size is 10 and number of defectives in sample is 2 or less

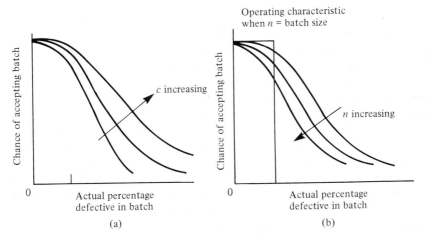

Figure 12.4 (a) A constant sample size (n). (b) A constant number of defectives in sample (c)

sample which is less than 100% will mean there is some chance that a batch with more than 10% defective items will get through.

The term *lot tolerance percentage defective* (LTPD) defines the percentage of defective items which is unacceptable to the receiver of the goods. This is the percentage value above which we want to be almost certain that no batch will get through the inspection stage and be accepted.

An operating characteristic and hence a sampling plan can be derived from a specification of the acceptable quality level and the lot tolerance percentage defective. Figure 12.5 shows an operating characteristic and these two points. Typically the AQL is around 1% or 2% and the LTPD about 10%. Such values give an operating characteristic where the chance of accepting a batch with the AQL is about 99.5% and the chance of accepting a batch with the LTPD is about 5%.

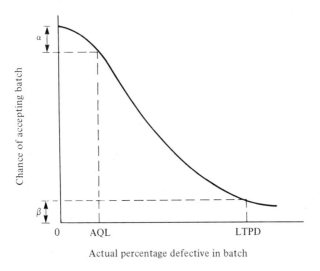

Figure 12.5 Operating characteristic showing AQL and LTPD

The customer, i.e. consumer, wants to have a low chance of accepting batches including too many defective items and thus wants a low value of β, the term indicated in Figure 12.5. The term β is called the consumer's risk. The producer wants to have as low a chance of rejecting a batch containing an acceptable number of defective items, i.e. α as low as possible. The term α is called the producer's risk.

Thus by specifying the AQL and LTPD, and hence also α and β, an operating characteristic can be derived and a sampling plan determined.

In the discussion so far concerning acceptance sampling a decision whether to accept or reject a batch has been made on the basis of a single random sample taken from a batch. Other sampling methods, however, can be used. *Double sampling* involves the following sequence:

1 A random sample is taken from the batch and inspected.

2 If the number of defective items is less than or equal to some acceptance number c_1 then the entire batch is accepted with no further sample being taken.

3 If the number of defective items is greater than some acceptance number c_2 then the entire batch is rejected with no further sample being taken.

4 If the number of defective items is between c_1 and c_2 then a second sample is taken and inspected. If the number of defective items in this sample plus the number found in the first sample is greater than c_2 then the batch is rejected, below c_2 it is accepted.

Double sampling involves two samples possibly being taken, *multiple sampling* is a similar process with more than two samples possibly being taken. Double and multiple sampling allows smaller-sized samples to be taken than would be the case with single sampling.

12.4 Process control

No matter how carefully controlled a manufacturing process is or however skilled the operators are, the items produced will show variations. In any process there is invariably some inherent variability. Thus, for example, with a machine filling containers, exactly the same weight of material will not be put in each container; some variation in the weight will occur. Generally this variation will occur around some average value. *Process control* is concerned with establishing the degree of variation that is inevitable with a process, any variation which is then greater than this natural variation can be attributed to other causes and remedies sought to bring the production back within the natural variation.

Variations can thus be considered as falling into two categories:

1 Due to the inherent variability of a process, such variations being in a random manner about some average value.

2 Due to some specific reason, such as tool wear, operator error, faulty material, etc.

The variations resulting from the inherent variability of the process occur randomly and can be described by a *frequency distribution* curve, the curve often describing what is termed a *normal distribution*. To illustrate what is meant by a frequency distribution, consider a rod being machined to some particular length. Inspection of the machined lengths will indicate some variability and, if this is due purely to the inherent variability of the process, these variations will be randomly scattered about some mean. We might thus have, when a sample of 40 rods is measured:

1 rod with length in the range   99.86 to   99.90 mm
7 rods with lengths in the range 99.91 to   99.95 mm
12 rods with lengths in the range 99.96 to 100.00 mm

11 rods with lengths in the range 100.01 to 100.05 mm
8 rods with lengths in the range 100.06 to 100.10 mm
1 rod with length in the range   100.11 to 100.15 mm

The number of rods within a particular length interval is called the frequency. Figure 12.6 shows the histogram produced by plotting this data. With larger amounts of data the histogram would be smoothed out to give a graph like that shown in Figure 12.7, this graph being the frequency distribution. The bell shape of this graph is a very common form of distribution that occurs when the variations are due to the inherent variability of a process, this type of distribution being referred to as the normal distribution.

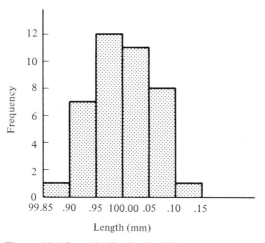

Figure 12.6 Length distribution histogram

With a normal distribution the curve is symmetrical about the average value of the variable, the length in Figure 12.7. With a normal distribution, regardless of the variable or the frequency values, the shape is always the same and thus equations developed for the curve can be applied to all processes that give this type of distribution. On this basis *control charts* have been developed. They aim to determine whether inspection data fits the distribution pattern that could be expected of the random variations of a process or whether it does not fit and so there is some fault in the process.

There are two types of control chart: an average or mean chart and a range chart. The data for both charts are obtained by taking a sample and inspecting it. In the case of a control chart for variables, the variable is measured for each item in the sample. Thus there might be, for a length variable, lengths of:

10.1, 10.0, 10.2, 10.2, 10.0 mm

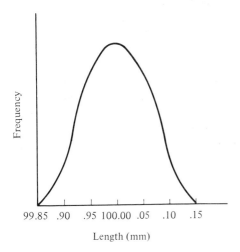

99.85   .90   .95 100.00 .05   .10   .15

Length (mm)

Figure 12.7 Frequency distribution of length

The average, or mean, length is given by summing the lengths and then dividing by the number of items, i.e. 50.5 ÷ 5 = 10.1 mm. The range for that sample is the difference between the greatest and the smallest values, i.e. 10.2 − 10.0 = 0.2 mm. These calculations are repeated for each sample and the average and range values plotted on the appropriate control charts, as in Figure 12.8.

The average control chart shows lines representing the *design limits* and the *control limits*. The design limits are those imposed by the design of the item concerned. The control limits are drawn so that they would only be exceeded once in every 40 times if random variations are occurring, their positions being determined by the properties of the normal distribution. The range control chart shows just a control limit, this line is drawn so that it is only expected to be exceeded once in every thousand times if random variations are occurring.

If the control charts indicate that samples are going beyond the control limits more than can be expected by random chance, then the deduction made is that there is some other effect causing it; for example:

1 The average chart goes out of limit for hardness measurements – the reason being perhaps that the components were in the furnace for the wrong period of time or the furnace temperature was incorrect.

2 The average chart goes suddenly out of limit for length of a machined component – the reason being perhaps that the machine jammed and so altered its setting or a new operator started or there was a change in properties of the materials being machined.

3 The average chart drifts out of limit for length of a machined component – the reason being perhaps tool wear or the tool moving out of alignment.

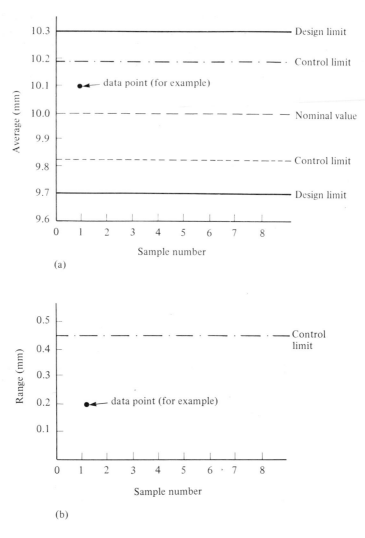

Figure 12.8 (a) Average control chart for controlling to a design limit. (b) Range control chart

4 The range chart goes out of limit for hardness measurements – the reason being perhaps that the batch was of variable quality before the heat treatment or there was an uneven temperature distribution in the furnace.
5 The range chart goes out of limit for length for a machined component – the reason being perhaps that play is occurring in a bearing or the material being machined is of variable quality.

Some control charts are used not in relation to some design limit but in order to control to the best that a process is capable of. For this the average control chart is drawn with two pairs of control limits, called action and warning limits. The *action limits* are drawn either side of the average of all, or a large number of, samples so that there is only a 1 in 1000 chance they will be exceeded for a process conforming to the normal distribution. The *warning limits* are drawn either side of the average so that there is only a 1 in 40 chance they will be exceeded. Figure 12.9 shows such a chart. The range chart is drawn as before (Figure 12.8(b)).

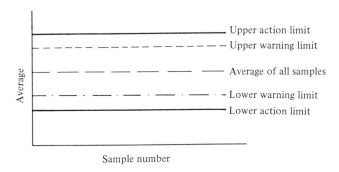

Figure 12.9 Average control chart for process capability

Control charts so far considered refer to control involving inspection of variables. Where the inspection is of attributes the charts are similar to those already described but have a vertical axis of either number of defects or proportion or percentage of defective items (see Figures 12.12 and 12.13 in the later discussion on control limits).

Control limits

The following explains how the control limits used with control charts can be determined. The methods used depend on whether the control is by variables or attributes, and also whether the control is to design specifications or the capability of the process.

Control limits to design specifications
(a) Inspect ten or more samples.
(b) Calculate the range for each sample.
(c) Calculate the average range.
(d) Multiply the average range by the appropriate value of the factor $A''_{0.025}$, the value depending on the size of sample used. The factor is for a chance of 25 in 1000, or 1 in 40.

Sample size	4	5	6	7	8
$A''_{0.025}$	1.02	1.95	0.90	0.87	0.84

(e) Draw control limits on the average control chart a distance equal to the product obtained in (d) inwards from each design limit (Figure 12.10).

(f) The control limit for the range control chart is the product of the average range and the appropriate value of the factor $D'_{0.001}$, the value depending on the size of sample used. The factor is for a chance of 1 in 1000.

Sample size	4	5	6	7	8
$D'_{0.001}$	2.57	2.34	2.21	2.11	2.04

(g) Draw on the range control chart the limit a distance equal to the product obtained in (f) above the zero range value.

Control limits for process capability
(a) Inspect ten or more samples.
(b) Calculate the range for each sample.
(c) Calculate the average range.
(d) Multiply the average range by the appropriate value of the factor $A'_{0.025}$,

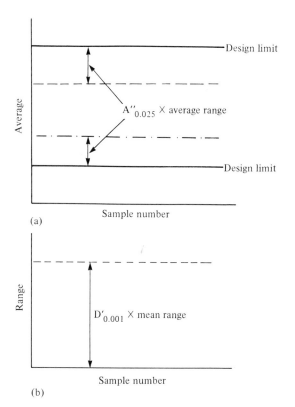

(a)

(b)

Figure 12.10 Charts to design limits. (a) Average control chart control limits. (b) Range control chart control limit

the value depending on the size of sample used. The factor is for a chance of 25 in 1000, or 1 in 40.

Sample size	4	5	6	7	8
$A'_{0.025}$	0.48	0.38	0.32	0.27	0.24

(e) Calculate the average for each sample.

(f) Calculate the average of the averages.

(g) One warning control limit on the average control chart is obtained by adding to this average of the averages the value of the product obtained in (d), the other warning limit by subtracting (Figure 12.11).

(h) Multiply the average range by the appropriate value of the factor $A'_{0.001}$, the value depending on the size of the sample used. The factor is for a chance of 1 in 1000.

Sample size	4	5	6	7	8
$A'_{0.001}$	0.75	0.59	0.50	0.43	0.38

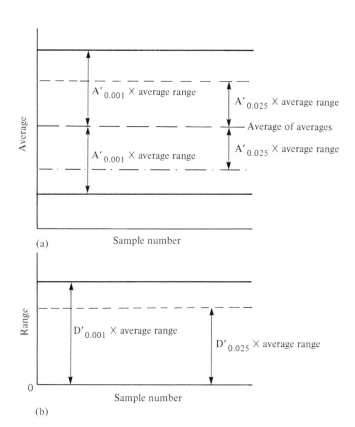

Figure 12.11 Charts to process capability. (a) Average control chart limits. (b) Range control chart control limit

(i) One action control limit on the average control chart is obtained by adding to the average of the averages (f) the value of the product obtained in (h), the other by subtracting.

(j) For the range control chart the action control limit is obtained by multiplying the average range (c) by the appropriate value of the factor $D'_{0.001}$, the value depending on the size of the sample used. The factor is for a chance of 1 in 1000 and values are given in the sequence for the calculation of control limits to design specifications.

(k) For the range control chart the warning control limit is obtained by multiplying the average range (c) by the appropriate value of the factor $D'_{0.025}$, the value depending on the size of the sample used. The factor is for a chance of 25 in 1000, in 1 in 40.

Sample size	4	5	6	7	8
$D'_{0.025}$	1.93	1.81	1.72	1.66	1.62

Control limits for attributes with number defective

(a) Inspect ten or more samples and ascertain the number defective in each sample.

(b) Calculate the average number of defective items per sample ($\bar{c}$).

(c) Calculate the standard deviation L, where

$$L = \sqrt{\{\bar{c}\,[1 - (\bar{c}/n)]\}}$$
n is the sample size.

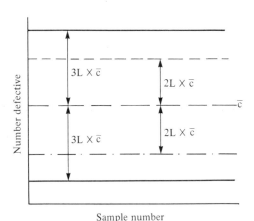

Sample number

Figure 12.12 Control limits for attributes with number defective

(d) One action limit is obtained by adding $3L$ to $\bar{c}$, the other by subtracting (Figure 12.12).

(e) One warning limit is obtained by adding $2L$ to $\bar{c}$, the other by subtracting.

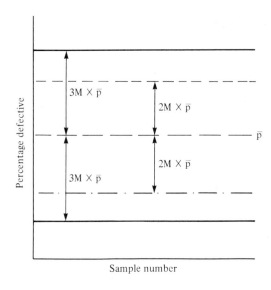

Figure 12.13 Control limits for attributes with percentage defective

Control limits for attributes with percentage defective
(a) Inspect 10 or more samples and ascertain the number defective in each sample.
(b) Calculate the percentage defective in each sample.
(c) Calculate the average percentage defective ($\bar{p}$).
(d) Calculate the average sample size ($\bar{n}$).
(e) Calculate the standard deviation M, where

$$M = \sqrt{[\bar{p}(100 - \bar{p})/\bar{n}]}$$

(f) One action limit is obtained by adding $3M$ to $\bar{p}$, the other by subtracting (Figure 12.13).
(g) One warning limit is obtained by adding $2M$ to $\bar{p}$, the other by subtracting.

Problems

1 What is meant by the term *quality* in the manufacturing industry?
2 What is meant by *reliability*?
3 Describe the components making up the cost of quality.
4 At what points in the manufacturing process can inspection most effectively occur?
5 Explain the difference between the measurement of variables and of attributes.
6 Explain the difference between *acceptance sampling* and *process control*.
7 Sketch an operating characteristic curve and explain its significance.

8 Define the terms *acceptable quality level*, *lot tolerance percentage defective*, *consumer's risk*, *producer's risk*.

9 Explain how double sampling differs from single sampling and its possible advantage.

10 What are the two types of control chart used with control by variables?

11 What is the average value and the range for the following inspection data:

Length (mm)	80.2	79.9	80.0	80.1	80.1	79.9

12 How do action limits differ from warning limits on control charts?

13 For some product or products produced by a company, establish the various points at which inspections occur and critically comment on the justification for inspections at those points.

14 Write a report arguing the case for some particular level of inspection and hence quality control. In your report consider both control and failure costs.

15 Present a case for acceptance sampling and indicate the factors that will determine the size of sample taken.

16 The following data relate to quality control with variables to process capability. Construct the quality control charts with the appropriate control limits. Each sample consists of four items.

Sample number	1	2	3	4	5	6	7	8	9	10
Lengths (mm)	10.1	10.1	9.9	10.0	9.9	10.2	10.0	9.9	9.9	10.0
	9.9	10.1	10.0	9.8	9.9	10.0	9.9	9.9	10.0	10.1
	10.0	10.0	10.1	10.0	10.0	10.1	10.0	10.0	9.9	9.9
	10.0	9.9	10.0	10.0	10.1	9.9	10.1	10.0	9.8	10.1

17 For the quality control of the items giving the data quoted in the previous problem, the following results were obtained. Comment on their significance.

Sample number	11	12	13
Lengths (mm)	10.1	10.2	10.1
	10.0	10.1	10.2
	10.1	10.2	10.3
	10.2	10.1	10.1

18 The following data relate to quality control by attributes. Construct the quality control chart with the appropriate control limits. Each sample consists of ten items.

Sample number	1	2	3	4	5	6	7	8	9	10
No. defective	2	2	1	3	2	2	1	0	2	1

19 For the quality control of the items giving the data quoted in the previous problem, the following results were obtained. Comment on their significance.

Sample number	11	12	13
No. defective	2	3	3

20 Devise a quality control system for letters typed in a typing pool.

13

Work design

13.1 Work study

The term *work study* is used to describe the scientific study of work, the term *time and motion study* being used in America. The aims of work study can be considered to include:

1 Making the most effective use of economic resources, i.e. people, machines, space, capital, that are available.
2 Establishing the most economical way of doing a job, making this then the standard method.
3 Establishing the time required for a suitably trained worker to do this standard job when working at a defined level of performance, this then becoming the standard time for the job.
4 Improving planning and control as a result of using standard methods and times.

Work study embraces method study and work measurement. *Method study* is the systematic recording and critical examination of existing and proposed ways of doing work, as a means of developing and applying easier and more effective methods and reducing costs. *Work measurement* is the application of techniques designed to establish the time for a qualified worker to carry out a specified job at a defined level of performance.

13.2 Method study

Method study is concerned with finding the best way to do jobs. Seven steps have been identified.

1 *Select*. Select the job to be investigated. The most appropriate jobs are those where significant savings might be produced as the result of the investigation, e.g. where bottlenecks exist in the production process. Thus not all jobs merit being investigated, the cost saving would not be worthwhile in view of the expense of the investigation.

2 *Record*. Observe the job and record all aspects of it. Many techniques exist for recording existing work methods. One method is a flow diagram. This shows the locations of specific activities carried out by workers, their sequence, and the routes followed by the workers, materials and equipment. Figure 13.1 shows an example of such a chart (a travel chart, as in Figure 11.2 might also be used).

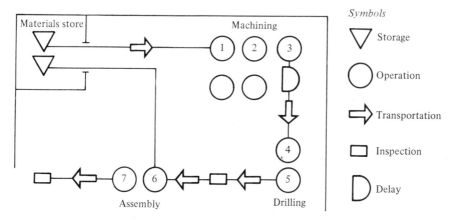

Figure 13.1 Method study flow diagram

Another method is a process chart. A variety of types of process charts are used. For example, there might be an outline process chart giving an overall view of the sequence of processes used for a particular product; another process chart might give the sequence of events associated with a particular worker; yet another might give the record of all the events associated with material. Figure 13.2 shows an example. Another method of recording a job involves the use of photography, either still or cine, and video.

3 *Analyse*. Study the record with a view to determining the best method.

4 *Develop*. Develop from the analysis the most efficient or optimum method.

5 *Define*. Describe the new method that has been developed.

6 *Install*. Set up this method as the standard practice, retraining workers, providing appropriate equipment, materials, etc.

7 *Maintain*. Ensure that the new method continues to be used, reviewing its operation periodically.

The above sequence has been written to fit an existing job where improvement is desired. If a new job is involved, the sequence is the same but rather than existing practice being recorded a record is made of a number of alternative methods tried for the job in order that the optimum one can be selected.

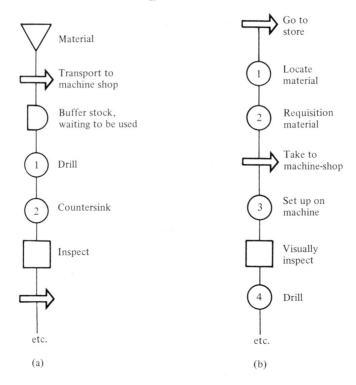

Figure 13.2 (a) Outline process chart. (b) Worker process chart

Work measurement

The main objective of work measurement can be considered to be the establishment of standard times for jobs. The *standard time* is defined as the time needed for a qualified and trained worker to carry out the job at a defined level of performance. Standard times are used in the costing of jobs (see Chapter 14), production planning, determining the manning required for a job, incentive payments for workers, etc.

The methods used for work measurement can be classified into two categories: *direct* and *indirect work measurement*. Within each category a number of different methods can be used.

Direct work measurement
(a) *Time study*. This method requires an investigator to record directly the times, and rates of working, for each element of a job. The instrument generally used is the stopwatch. The rating is an estimation by the investigator of the general tempo of the work and involves a consideration of such factors as speed of movement and effectiveness of the worker. The basic times of each element are calculated from

$$\text{basic time} = \text{observed time} \times \frac{\text{observed rating}}{\text{standard rating}}$$

The standard rating is the performance expected from an average qualified and motivated worker working at a natural rate. The basic times for all the elements are added up and when combined with an allowance, for such things as relaxation, a standard time is obtained.

(b) *Activity sampling*. Time study, while appropriate for defined short-cycle repetitive work, is not so appropriate for long irregular-cycle work, for instance, the amount of time supervisors spend on paperwork. For such situations *activity sampling* can be used. This involves determining the percentage of time spent on a particular activity and is determined from a large number of instantaneous observations made over a period of time, each observation recording what is happening at that instant. The percentage of observations recorded for a particular activity is a measure of the percentage of time during which that activity occurs.

Indirect work measurement

(a) *Synthetic timing*. Many jobs are likely to contain a number of elements common to other jobs. Thus, by breaking a job into its constituent elements, a time can be obtained for the job on the basis of a knowledge of the times needed for these elements in other situations.

(b) *Predetermined motion time systems (PMTS)*. This method involves the breaking down of a job into its basic human motions, the times for these motions being obtained from tables. Basic human motions are such things as reaching, moving, turning, grasping, releasing, eye travel, etc. With each such motion there is further subdivision to quantify, for instance, the type of reaching motion involved, e.g. reaching for an object in a fixed location, reaching for an object jumbled up with other objects, etc. In order to determine the time of a job a considerable amount of detailed information is required.

(c) *Analytical timing*. Standard times are determined using the data available for some of the elements in a job and the other elements are estimated on the basis of experience. This method is often used for non-repetitive work, e.g. maintenance.

13.3 Ergonomics

Ergonomics is the study of humans in the working situation. The design of a machine or a work station that has to be operated by a human being must take account of the characteristics of the human being. The man–machine interface has to be considered. It has been said that the controls of a lathe are situated in such a way that the ideal operator to interface with them would be 1.37 m ($4\frac{1}{2}$ ft) tall, 0.61 m (2 ft) across the shoulders and have an arm span of 2.35 m (8 ft).

Since this is not the description of the average man, most men will not find the controls convenient to use and this can cause fatigue as well as generally increasing the time required to carry out a job with the lathe.

Using controls, reading displays, and environmental factors, such as noise, are all factors that have to be considered in the man–machine interface. Thus, for example, in considering the type of control to be used consideration should be given to the speed, accuracy and range of use that the operator will be involved in. A lever is capable of fast operation but gives poor accuracy and a poor range. A handwheel is not capable of high speed but can give good accuracy and a fair range. A push button gives high speed but is completely unsuitable for anything other than an on–off control, thus providing no accuracy or range.

Displays can take a wide variety of forms, e.g. lights which are either on or off, various positions of a switch, pointers on dials, dials moving past pointers, and counters. The characteristics of these various forms of display need to be considered. If an operator needs to see at a glance whether the power is on or off then a warning light would be better than a pointer moving across a scale, since time is needed to read this form of display. Counters enable quantitative readings to be taken without ambiguity, this not being true of pointers on dials, but such a display is not very good for checking purposes. With a pointer on a dial, marks can be put on the dial to indicate crucial values and so checking values is better with a pointer on a dial than with a counter.

The three main environmental factors that can affect workers are illumination, noise and the climate. If these are not at acceptable levels then fatigue, discomfort and possibly physical damage can occur. All have an effect on worker performance. An adequate lighting system should provide:

1 Sufficient brightness for the job concerned, fine work requiring higher levels than coarse work.
2 Uniform illumination.
3 Some contrast between the brightness of the job and the background. This can aid the eye to detect detail.
4 No glare, e.g. unwanted reflection of light from polished or glass surfaces. This can, for instance, make it very difficult to read meters.

Noise not only can have a nuisance value but can cause interference with communications, and give rise to stress and hearing damage. Noise levels can be controlled by putting barriers between workers and the source, providing protective devices for workers, e.g. ear plugs or muffs, controlling the source of the noise or rearranging the locations of the workers and the noise sources.

The climate experienced by a worker depends on factors such as

air temperature
air movement
relative humidity
air pollution.

While workers are highly adaptable to such environmental factors, fatigue, discomfort and poor performance are obvious consequences of a poor working climate.

13.4 Principles of motion economy

In order both to reduce fatigue and to keep motion to a minimum, a number of principles of *motion economy* have been derived from observations of workers. The following is one version of these principles, the original list produced by Gilbreth having been modified by later researchers.

Use of the human body

1 The two hands should preferably begin as well as complete their motions at the same time.
2 The two hands should not be idle at the same time except during rest periods.
3 The movements of the arms should be in opposite directions and symmetrical directions, rather than in the same direction, and should be made simultaneously.
4 Continuous curved motions are preferable to straight-line motions involving sudden and sharp changes in direction.
5 Hand, arm and body movements should be confined to the lowest possible classification with which it is possible to perform the task. The following list gives the classification with the lowest, i.e. least tiring and most economical, movements first.
(a) Finger motions.
(b) Finger and wrist motions.
(c) Finger, wrist and lower arm motions.
(d) Finger, wrist, lower and upper arm motions.
(e) Finger, wrist, lower and upper arm and shoulder motions.
6 Momentum should be employed to assist the worker wherever possible, and it should be reduced to a minimum if it must be overcome by muscular effort.
7 Ballistic movements are faster, easier, and more accurate than restricted (fixation) or 'controlled' movements.
8 Work should be arranged to permit natural and habitual movements.
9 The need to fix and focus the eyes on an object should be minimised and when necessary the occasions should occur as close together as possible.

Arrangement of the workplace

10 There should be a definite and fixed place for all tools and materials.
11 All tools, equipment and materials should be located as close to the point of use as possible.

12 Gravity-fed bins and containers should be used to deliver the material as close to the point of assembly in use as possible.

13 Drop deliveries, whereby the worker may deliver the finished article without moving to dispose it, should be used wherever possible.

14 Tools, equipment and materials should be located to permit the best sequence of motions.

15 All materials and tools should be located within the normal group area.

16 Illumination levels and brightness contrasts between objects and the surroundings should be arranged to avoid or alleviate visual fatigue.

17 Noise and vibration should be minimised.

18 The height of the workplace and the seating should enable comfortable sitting or standing during work.

19 Seating should permit good posture and adequate coverage of the work area.

20 All work activities should permit the worker to adopt several different but equally healthy and safe, postures without reducing his or her capability to do the work.

Design of tools and equipment

21 Clamps, jigs or fixtures rather than hands should be used to hold work.

22 Two or more tools should be combined wherever possible.

23 Tools and materials should be pre-positioned wherever possible.

24 The loads should be distributed among the limbs according to their capacities.

25 Wheels, levers, switches, etc. should be located in such positions that the operator can manipulate them with the least change in body position.

13.5 Job design

Job design and redesign involve the consideration of individual tasks or activities and bringing them together into a job which is assigned to an individual worker or group of workers. This problem can be approached in a number of different ways. One approach is that of scientific management, as developed by F. W. Taylor. In relation to job design, this approach leads to:

1 A separation of those engaged in mental and in manual work.

2 The maximum decomposition of work tasks so that any one job consists of a very small number of tasks or activities.

3 The minimisation of the skill requirements of any task or activity.

4 No consideration of psychological or social factors, i.e. consideration of the worker not as a human but rather as a machine.

5 The assumption that workers work purely for financial reward.

A consequence of adopting such an approach can be considered to be the system established by Henry Ford for the mass production of cars, the famous Model T. This system had four basic elements:

1 The product was standardised. This meant not only the car itself but the components used in the car. Thus a car was not made by making separately all the pieces for the particular car concerned so that they ended up fitting only that one car. Components were standardised so that any one of a batch of components could be used in any one of the cars.
2 Flow-line production methods were used, i.e. the assembly line, with each worker tied to a particular job in a particular position along the line.
3 An extensive use of machine tools. This made many of the jobs semi-automatic and de-skilled many jobs.
4 The use of the principles of scientific management, e.g. the maximum decomposition of work tasks.

Job design was thus not influenced by any real consideration of the individual worker's abilities or indeed any consideration of the worker as being an individual. More recent considerations of job design have, however, involved a consideration of the worker as an individual and how to make best use of the abilities of individual workers. An example of such a development is the work of Herzberg in 1959 when he, and his co-workers, considered the factors which lead to job satisfaction and dissatisfaction, *motivator* and *hygiene* factors. Motivator factors include achievement, recognition, the work itself, responsibility and advancement. Hygiene factors, those that can lead directly to dissatisfaction, include company policy and administration, supervision, salary, interpersonal relations and working conditions. Improving hygiene factors can get rid of dissatisfaction but will not lead to motivation: for that, the motivator factors must be right. Increasing the motivator factors is known as 'job enrichment'.

Job enrichment can involve:

1 Combining tasks to give greater skill variety.
2 Forming natural work units so that the unit can develop an identity.
3 Increasing job responsibilities and autonomy.

The aim is (a) to give to the workers higher internally generated work motivation and higher satisfaction from their work, and (b) to give to the organisation work performance of higher quality together with lower absenteeism and higher turnover.

Other developments in job design are job rotation and job enlargement. *Job rotation* describes the procedure whereby workers within a group can exchange jobs at prescribed times or when freely chosen. The prime aim of this is to reduce strain and monotony that can occur when a worker carries out the same job every day. *Job enlargement* describes the procedure whereby several similar work steps or tasks are joined together to increase the scope of the work.

13.6 Health and safety of workers

In the design of any work situation and task the health and safety of the workers has to be considered, along with that of any other person who might be affected by the work activity or the product of the work. To this end health and safety legislation exists, the predominant Act in Great Britain being the Health and Safety at Work Act, 1974.

The Act consists of four parts, Part I being concerned with health, safety and welfare in connection with work and the control of dangerous substances and certain emissions into the atmosphere; Part II is concerned with the Employment Medical Advisory Service; Part III with the Building Regulations and Amendment of Building (Scotland) Act, 1959 and Part IV with miscellaneous and general matters. The Act did not immediately replace earlier legislation and regulations but allowed them to remain in force until revoked and replaced by new regulations or Codes of Practice issued under the Act.

Part I of the Act has four basic objectives:

1 To secure the health, safety and welfare of people at work.
2 To protect persons other than persons at work against risks to health or safety arising out of or in connection with the activities of those at work.
3 To control the keeping and use of explosive or highly flammable or otherwise dangerous substances and generally prevent the unlawful acquisition, possession and use of such substances.
4 To control the emission into the atmosphere of noxious or offfensive substances.

Part I of the Act establishes the general duties of employers.

2 – (1) It shall be the duty of every employer to ensure, so far as is reasonably practicable, the health, safety and welfare at work of all his employees.
(2) Without prejudice to the generality of an employer's duty under the preceding subsection, the matters to which that duty extends include in particular –
(a) the provision and maintenance of plant and systems of work that are, so far as is reasonably practicable, safe and without risk to health;
(b) arrangements for ensuring, so far as is reasonably practicable, safety and absence of risks to health in connection with the use, handling, storage and transport of articles and substances;
(c) the provision of such information, instruction, training and supervision as is necessary to ensure, so far as is reasonably practicable, the health and safety at work of his employees;
(d) so far as is reasonably practicable as regards any place of work under the employer's control, the maintenance of it in a condition that is safe and without risks to health and the provision and maintenance of means of access and egress from it that are safe and without such risks;

(e) the provision and maintenance of a working environment for his employees that is, so far as it reasonably practicable, safe, without risks to health, and adequate as regards facilities and arrangements of their welfare at work.

In addition the employer must also undertake other duties:

1 He must provide a written statement of the organisation's general policy with respect to the health and safety of the employees and how it will be carried out.
2 Safety representatives elected from and by the employees have to be consulted by the employer on safety matters.
3 If the safety representatives so request, a safety committee must be established.

Part I also specifies that employers have general duties to conduct their business in such a way that persons not in their employment are not thereby exposed to risks to their health and safety. In addition, those responsible for places of work have duties with respect to those using the premises who are not their employees, e.g. visiting workers carrying out maintenance or repair.

Also a general duty is imposed on those who design, manufacture, import or supply any article for use at work:

(a) to ensure, so far as is reasonably practicable, that the article is so designed and constructed as to be safe and without risks to health when properly used;
(b) to carry out or arrange for the carrying out of such testing and examination as may be necessary for the performance of the duty imposed on him by the preceding paragraph;
(c) to take such steps as are necessary to secure that there will be available in connection with the use of the article at work adequate information about the use for which it is designed and has been tested, and about any conditions necessary to ensure that, when put to that use, it will be safe and without risks to health.

Employees also have duties with respect to health and safety at work.

7 It shall be the duty of every employee while at work –
(a) to take reasonable care for the health and safety of himself and of other persons who may be affected by his acts or omissions at work; and
(b) as regards any duty or requirement imposed on his employer or any other person by or under any of the relevant statutory provisions, to cooperate with him so far as is necessary to enable that duty or requirement to be performed or complied with.

The term 'employees' covers everybody who is employed by an organisation, i.e. managers, supervisors and workers. This section of the Act could thus be invoked in the case of a supervisor who failed to implement some safe system of working or an operative who endangered others by the reckless use of some machine.

Two other general duties are imposed:

8 No person shall intentionally or recklessly interfere with or misuse anything provided in the interests of health, safety or welfare in pursuance of any of the relevant statutory provisions.

9 No employer shall levy or permit to be levied on any employee of his any charge in respect of anything done or provided in pursuance of any specific requirement of the relevant statutory provisions.

Thus, for instance, an employee removing the safety guards from a machine or playing about with a fire extinguisher could be prosecuted under this part of the Act.

Two bodies were set up by the Act: the Health and Safety Commission and the Health and Safety Executive. The Commission consists of a chairman appointed by the Secretary of State, three members resulting from consultations with organisations representing employers, three from consultations with organisations representing employees and up to three other members drawn from other activities. The Executive consists of three people, one appointed by the Commission to be the director of the Executive and the others appointed by the Commission after consulting the director.

The Commission is charged with the task of making arrangements for the carrying out of research, the provision of training and information, advisory services and the development of regulations. The Executive exercises on behalf of the Commission such of its functions as the Commission directs it to exercise, e.g. carrying out investigations, and acting as a source of information.

An important aspect of the work of the Commission is the issuing of regulations and codes of practice. Regulations are legally binding; codes of practice are guidance. Failure to observe a code of practice does not render a person liable to criminal or civil proceedings but where that person has broken a general duty in the Act or regulations, the fact that he or she has failed to observe a code of practice is liable to be taken as conclusive evidence that he or she did not do all that was reasonably practicable to ensure the health and safety of those at work. The task of enforcing the statutory provisions is given to the Executive in cooperation with local authorities and other enforcement bodies.

For more details with regard to the Act and its interpretation the reader is referred to more specialist texts, e.g. Howells, R. and Barrett, B. The Health and Safety at Work Act (Institute of Personnel Management).

The Factories Act, 1961

The Health and Safety at Work Act, 1974 is, as new regulations and codes of practice come into effect, steadily replacing previous legislation. Until all has been replaced, the older Acts are still in force. An important Act in this context is the Factories Act, 1961. The following are some of the provisions included within that Act:

1 An abstract of the Act and certain prescribed information must be kept posted at the principal entrance of a factory.

2 Every factory must be kept clean and free from drain smells; dirt and refuse to be removed daily and workroom floors cleaned weekly.

3 Each worker should have a minimum space of 400 cubic feet, no space higher than 14 feet above the floor being counted. This rule is to avoid overcrowding.

4 After the first hour a temperature of 60°F (15.5°C) is to be maintained where work people are sitting down most of the time. This has later been amended to include an upper limit of 66.2°F or 19°C.

5 Adequate ventilation and suitable lighting must be provided.

6 There must be sufficient and suitable conveniences provided.

7 Adequate drinking water must be provided with facilities for washing.

8 Accommodation for clothing not worn during working hours must be made available.

9 First aid equipment must be provided and maintained.

10 All stairs, passages and gangways are to be kept free from obstruction.

11 Safe means of access must be provided to every workplace.

12 Also included are provisions regarding the safe use of machinery, e.g. all dangerous parts of machinery should be securely fenced.

Problems

1 What is meant by the terms work study, method study and work measurement?

2 Describe the seven steps necessary in method study.

3 Distinguish between a flow diagram and a process chart in method study.

4 Define basic time and standard rating.

5 Explain the principles of (a) time study, (b) activity sampling, (c) synthetic timing, (d) predetermined motion time systems, (e) analytical timing, methods of work measurement.

6 Explain the term ergonomics.

7 List some of the factors that have to be considered in the design of a control to be operated by a human being.

8 List five forms of display.

9 What are the three main environmental factors that can affect workers?

10 What is meant by the principles of motion economy?

11 A small company making one-off or small batches of items has never bothered with any work measurement, basing its charges to customers and the rates it pays its workers on the manager's experience. Write a report to the manager giving the advantages, limitations, possible problems that might be encountered, a possible method or methods that might be used, etc. if work measurement were to be used. Also consider the cost factor.

12 Consider the ergonomics of some human-operated machine, e.g. a lathe, a

domestic electric cooker, a car or a motorbike. Consider the controls and the reading displays. List the types that are used and consider whether they are the most appropriate for the activity concerned. Also consider the location of the controls and displays. Hence write a report on the ergonomics of the machine concerned.

13 Write a critical analysis of the different types of reading displays and the various situations under which they are most appropriate.

14 Analyse some task carried out by a worker, e.g. somebody engaged in an assembly job or perhaps a typist, and comment on the motions involved in terms of the principles of motion economy.

15 Write a report advocating job enrichment for a particular group of workers. In your report discuss the advantages, and possible disadvantages, and how it might be organised.

16 What were the basic objectives of Part I of the Health and Safety at Work Act, 1974?

17 What were the general duties imposed by the Health and Safety at Work Act, 1974 on (a) the employers and (b) the employees?

18 What are the functions of (a) the Health and Safety Commission and (b) the Health and Safety Executive?

14

Costing

14.1 Cost accounting

The term *cost accounting* can be defined, according to the Institute of Cost and Management Accountants, as the application of accounting and costing principles, methods and techniques in the ascertainment of costs and the analysis of savings and/or excesses as compared with previous experience or standards.

The term *historical costing* is used where the costs are determined after the production of goods has occurred, the term *standard costing* being used where costs are estimated before production on the basis of standards that have been predetermined for materials and labour. With historical costing the costs are determined from records taken during the production of the materials, labour and machines used. With standard costing the standards to be adopted for the costs of the various parts of the process are obtained from past experience of manufacturing the same or similar work. Thus standard costing is not the actual cost of the production but an estimate of what the cost will be; historical costing gives the actual cost. The advantage to management of using standard costing is that it enables cost control to be exercised in that there is a cost forecast before production and this can be compared with the actual cost and the factors responsible for any discrepancy in costs identified.

Standard costing is widely used where there is repetitive manufacture, i.e. where the processes and materials used are close enough to past experience to enable standards to be determined. With one-off production, or where a new process is being used or a new product being produced, it may not be possible to use standard costing as there may be insufficient experience to make standards feasible.

The price of a product

The selling price of a product can be considered to be made up of a number of elements:

1 *Prime costs*. These are the costs of the material and labour directly involved

in the production of the product. They are referred to as *direct costs* when they can be separately identified as relating directly to the product concerned. Hence the direct material cost is the cost of the materials consumed or incorporated in the production of the product, the direct labour cost is the cost of the labour directly involved in the process.

2 *Production overheads.* These are the costs involved in the production process which cannot be directly attributed to the product concerned, e.g. the cost of the machinery, maintenance, storekeeping, etc. This type of cost is called *indirect cost.*

Elements 1 and 2 above represent the total production costs for a product. This, however, is not the selling price in that further costs have to be included as well as a profit element.

3 *General administration and selling overheads.* These include the costs of distribution and marketing and part of the general administration costs of the organisation.

4 *Profit.* Elements 1, 2 and 3 above constitute the total product cost. Added to that to give the selling price is the profit element.

14.2 Direct labour costs

With historical costing the direct labour costs can be found from the records of each person's involvement in the production process and the labour rates. Thus there might be two hours by an operator on machine X at a labour rate of £5 per hour and then one hour by an operator on machine Y at a labour rate of £6 per hour, and finally half-an-hour on machine Z at a labour rate of £4 per hour. The total direct labour cost is, in this example, £10 + £6 + £2 = £18.

With standard costing, the direct labour cost is found from a consideration of the standard times specified for each part of the process, adjusted for the target levels of operator performance, and the labour rates. The standard time for a job is defined as the time needed for a qualified worker to carry out the job at the defined level of performance. Such times can be determined by work measurement in which investigators observe workers carrying out the process concerned and time the various parts of the job. Effectively this defines a 'standard worker' in that the work measurement investigators adjust the times they determine to take account of the general tempo of the work to give time values corresponding to the standard worker. *Operator performance* is a measure of how a particular worker compares with the performance expected of this standard worker, an operator rated at 100 being equivalent to the standard worker while one rated at less than 100 performs at a higher rate and one at more than 100 at a slower rate.

Hence the standard cost for a job is given by

Standard cost = standard time × operator rating × labour rate

Thus for a job for which the standard time is 2 hours, the operator rating 110 and the labour rate £6 per hour, the standard cost is

Standard cost $= 2 \times (110 \div 100) \times 6$
$= £13.20$

14.3 Direct materials costs

Where materials are bought specifically for a particular job, an actual materials cost can be put against that job. However, in many instances the materials used on a particular job are drawn from stocks held by the company. Materials are thus bought for the store and then issued from the store as required. Since the materials drawn from the store for a particular job may have been bought at different times and at different prices, it is not always feasible to use the actual cost of the materials in arriving at the direct materials cost. A number of methods are used to arrive at a cost figure for materials in such cases.

1 *Standard cost.* The material issued from the store for a job is charged at a standard cost. This is a cost that has been chosen for that particular material and is not necessarily the real price paid for it.
2 *First in, first out (FIFO).* The cost of the oldest of that type of material in the store is used, it being assumed that the material that was first in is first used.
3 *Last in, first out (LIFO).* The cost of the most recent purchase of the material is used, it being assumed that the material last into the store is the first used.
4 *Highest in, first out.* The cost is taken as being that of the most highly priced purchase, regardless of the date at which the material was bought.
5 *Average cost.* The average price of the stock held of a particular material is used as the cost.
6 *Market price.* The cost is taken as the market price of the material on the date the material is taken from the store, regardless of the price actually paid for the material.
7 *Replacement pricing.* The cost is taken as the price that it is anticipated will have to be paid to replace the material drawn from the store.

In general, the method most used where standard costing is used for a job is standard cost for the materials; however, in other situations the average cost is widely used.

14.4 Overhead costs

The term *overheads* is used to describe the costs which cannot be specifically allocated to any particular job or product, i.e. the indirect costs of materials, labour and other expenses. Indirect materials are those materials which are used to further the manufacturing process but which cannot be directly identified in the end product, e.g. cutting oil. Indirect labour consists of all wages and salaries paid to those people not directly concerned with the production of a product, e.g. supervisors, managers, clerks, typists, salespeople, etc. Indirect expenses include all the expenses incurred by the company in carrying out their business activities which are not capable of being directly identified with a specific job or product, e.g. rent, electric power, insurance, depreciation, etc.

The procedure generally used, under what is termed *absorption costings*, for recovering the overhead costs is:

1 The company is divided into cost centres.
2 The overhead costs are apportioned between these cost centres.
3 Overhead costs within service costs centres are transferred to producing cost centres and then the collective overhead cost is apportioned in some way to the units of product or jobs and so absorbed and passed on to the customer.

The term *cost centre* is used for any section of an organisation for which costs are separately identified. Thus the production department might be a cost centre, or there might be a number of production cost centres according to the different products produced. The justification for a particular cost centre is that financial control is aided by separately identifying such costs.

Two types of cost centres can be identified within a manufacturing company, *service cost centres* and *producing cost centres*. service cost centres are those which do not actually make products, e.g. personnel, stores, etc. Producing cost centres are those concerned with actually making the products.

Apportioning overhead costs to cost centres

The overhead costs have to be apportioned in some equitable way between the cost centres. Thus, for example, the rent cost may be allocated between the cost centres according to the floor areas they occupy. Lighting costs may be allocated according to the number of electric light fittings in each cost centre. Some overhead costs may be able to be directly attributed to one or more cost centres, e.g. materials used in maintenance can be put as an overhead cost to the maintenance cost centre, depreciation of machines used in production can be put as an overhead cost to the production cost centres using those machines. Table 14.1 indicates how some of the overhead costs might be apportioned between cost centres.

By taking into account all the overhead cost elements, the total overhead costs can be apportioned between the cost centres. Service departments, such as stores, maintenance, personnel, security, etc. are not directly involved with the productive process and, as such costs need to be passed on to the buyers of the company's products, their overhead costs have to be allocated to the producing cost centres. This allocation can be made in a variety of ways. Thus the overheads of the stores cost centre can be divided between the producing cost centres according to the fraction of the cost of direct materials each centre uses. Maintenance can be divided between the producing cost centres according to the number of labour hours in each centre. Security might be

Table 14.1 *Typical apportionment of overhead costs*

Overhead item	Total cost	Service cost centres			Producing cost centres		
		Stores	Main-tenance	Per-sonnel	A	B	C
	£	£	£	£	£	£	£
Rent (allocated according to area)	50 000	10 000	5 000	5 000	10 000	15 000	5 000
Light (according to number of light fittings)	10 000	500	500	800	3 000	4 000	1 200
Heat (according to cubic capacity, or perhaps number of radiators)	12 000	500	800	1 000	3 500	4 500	1 700
Insurance (according to valuation)	2 000	400	100	300	300	500	400
Materials for maintenance	2 000	—	2 000	—	—	—	—

divided between the producing cost centres according to the floor area
occupied by them.

The end result of this allocation of the service cost centre overheads is that
there is an overhead cost for each producing cost centre and that the total of all
these overhead costs is the total overhead cost of the company.

Absorbing the overhead costs

The overhead costs for a particular production cost centre have to be, in some
way, allocated to the products produced at that cost centre so that the
overheads can be passed on to the purchaser of the product and hence
recouped. A variety of methods are used to apportion the overheads among the
products, or jobs. Probably the four main methods used are:

1 *According to the direct labour costs.* The greater the direct labour costs of a
product, the greater its share of the overheads.
2 *According to the direct labour hours.* The greater the number of labour hours
needed for a product, the greater its share of the overheads.
3 *According to the machine hours involved.* The greater the number of machine
hours needed to produce a product, the greater its share of the overheads.
4 *According to the number of units of product produced.* The more units of a
particular product produced, the greater its share of the overheads.

Where the overheads are apportioned according to the direct labour costs,
the fraction of the total direct labour costs over some period of time that can be
attributed to a particular product is calculated. Thus if the total overhead cost
is £50 000 and the total direct labour costs for a particular cost centre are
£10 000, for every £1 of direct labour an overhead cost of £5 is to be levied. This

amount is known as the *direct labour cost rate*. Thus if a particular product has direct labour costs of £400 against this cost centre then the overhead cost is £2000. This method of apportioning overheads is often used where the production process is labour intensive.

Where the overheads are apportioned according to the direct labour hours the fraction of the total direct labour hours over some period of time that can be attributed to a particular product is calculated. Thus if the total overhead cost is £50 000 and the total direct labour hours for a particular cost centre are 5000 hours then for every hour of direct labour the share of the overhead cost is £10. This is known as the *direct labour hour rate*. Thus if a product has direct labour hours of 50 then the overhead cost is £500.

Where the overheads are apportioned according to the machine hours, the overheads that are to be charged against a particular machine or group of machines are calculated and the machine hour rate is determined. This is overheads per hour of machine time. The overheads for a particular machine can be calculated on the basis of a fraction of the rents, rates, heating, lighting, etc. according to the fraction of the total area of floor space occupied by the machine, due allowance being made for gangways. In addition, specific machine overheads are included, e.g. machine depreciation, maintenance, tooling costs, etc. The total number of machine hours used in the calculation of the machine hour rate is either the number of hours for which the machine is expected to operate or the number of hours for which it could be operated if used at normal capacity. Thus if the overheads are, say, £10 000 and the total number of machine hours 5000 then the machine hour rate is £2. Hence if a particular product requires three hours on this particular machine then the overhead cost is £6. This method of apportioning overheads is often used where the production processes are machine intensive. A high percentage of the overheads in such a case is likely to be due to the depreciation of the machinery and tooling costs.

Where the overheads are apportioned according to the number of units of product produced, the fraction of the total number of units produced over some period of time that can be attributed to a particular product is calculated. Thus if the total overhead cost is £50 000 and the total number of units of product produced is 25 000, then the overhead for unit of product is £2.

When standard costing is used a *standard overhead cost rate* is calculated, using whichever of the above methods is most appropriate. The standard overhead cost rate is defined as the rate determined by dividing the expected overhead cost attributable to a particular cost centre by the predetermined quantity of the base, e.g. machine hours, to which the rate is applied. The difference between the standard and historical overhead cost rates is that the standard is calculated in terms of the expected overheads while the historical one is in terms of what has occurred.

14.5 Depreciation

Depreciation can be defined as the diminution in the intrinsic value of an asset due to use and the passage of time. Thus a particular machine might have an initial cost C, this including not only the cost of the machine but its installation and other related costs, and a final scrap value S. Then the depreciation during the life of that machine is $(C - S)$. If the life of the machine is reckoned as being n years then the average depreciation per year is $(C - S)/n$.

A machine costs, when fully installed, £12 000 and is expected to have a life of five years, at the end of that time having a scrap value of £500. For this particular example the average depreciation is thus

$$\frac{£(12\,000 - 500)}{5} = £2300 \text{ per year.}$$

Table 14.2   *The straight line method of calculating depreciation*

Year	Value at start of year £	Depreciation £	Value at end of year £
1	12000	2300	9700
2	9700	2300	7400
3	7400	2300	5100
4	5100	2300	2800
5	2800	2300	500

The method of calculating depreciation per year shown in Table 14.2, known as the *straight line method*, assumes that an item depreciates by equal amounts in each year. An alternative way of calculating depreciation is to assume that a greater amount of depreciation occurs in earlier than later years. Think, for example, of the depreciation in value of a new car. One way of calculating depreciation by this approach is called the *reducing balance method* and provides for depreciation by means of charges calculated as a constant proportion of the balance of the value of the asset after deducting the previously calculated depreciation. Thus, for the machine referred to in the example above, where the life is reckoned as being five years, we can take 47 per cent of its value as the amount of depreciation each year. This is demonstrated in Table 14.3.

The value at the end of the fifth year is, allowing for rounding errors in the calculations, the scrap value of £500. The value of the percentage to be used in calculating the above depreciations depends on the ratio of the scrap-to-cost value, i.e. $S \div C$, and the number of years n of useful life. The following equation indicates how the percentage r is calculated:

$$r = \left[1 - \left(\frac{S}{C} \right)^{1/n} \right] \times 100\%$$

Table 14.3 *The reducing balance method of calculating depreciation*

Year	Value at start of year £	Depreciation £	Value at end of year £
1	12,000	47% of 12,000 = 5640	6360
2	6360	47% of 6360 = 2989	3371
3	3371	47% of 3371 = 1584	1787
4	1787	47% of 1787 = 840	947
5	947	47% of 947 = 445	502

This gives a value for r which means that at the end of the life the value will be the scrap value.

The above are just two ways in which depreciation can be calculated and so allowed for in the determination of the value of assets. Other methods are also used. One method involves valuing the assets at the end of each period of time; the depreciation is then the difference between the initial and the end values for that period of time.

14.6 Job costing

Job costing involves the allocation of the various cost elements against a job. Each job is considered separately and the material, labour and overhead costs are determined. This generally means that a significant clerical effort is required to collect together all the forms used to log the various elements of the job undertaken by different workers and all the materials requisition forms used to extract materials for the job from the store. It is generally a historical system of costing with all the details being gathered after the event. This method is widely used for one-off orders.

The following outlines the types of information gathered and the way the job is costed.

1 A works order is issued. This gives the details of the order and a job number. All future work and materials for the job quote this number.
2 To obtain materials, a materials requisition sheet is sent to the materials store. After executing the order, the storekeeper passes the sheet to the costing department. On the basis of this information the direct materials cost can be determined.
3 Each worker, on completing work on the job, fills in the relevant part of the job card. When the job is complete, the card passes to the costing department. On the basis of this information the direct labour cost can be determined.
4 The costing department then computes the overhead cost and, by adding together the direct materials cost, the direct labour cost and the overheads, obtains the cost of the job.

14.7 Batch costing

Batch costing is just another form of job costing; however, instead of the unit against which the costs are levied being a single job, a group or batch of components is considered. Thus the job costing is concerned with the specific order of a customer whereas a batch costing is generally likely to be concerned with a quantity of components being made for the store. The difference is essentially one of making a component against a specific order or making components in anticipation of orders.

The type of documentation used to obtain information for batch costing can be the same as that used for job costing. There may be a major cost item involved in the initial setting of the necessary jigs and tools, this cost then being defrayed over the total number of components produced in the batch. Thus there can be a valid argument for big batches, however against this has to be considered the capital tied up in the stock so produced (see Chapter 11 on inventories).

14.8 Process costing

Process costing is often used where continuous runs of identical products are produced, e.g. mass production, or where the process is continuous, as in the production of chemical or food products. The aim is to produce an average cost of a product. This is done by accumulating the direct materials and labour costs, and the overhead costs, and then dividing that sum by the number of units produced during the period of time concerned.

The procedure adopted for process costing is to divide the company into a number of cost centres, and possibly in connection with a production department cost centre into a number of sub-cost centres. The reason for cost centres, and sub-cost centres, is that management consider it necessary to identify the costs, and output, by such centres in order to exercise control effectively. The reason for sub-cost centres in a production department is generally that it is useful to identify costs, and output, in relation to the types of process. Thus there might, for instance, be a cost centre for machining and another one for heat treatment.

For each cost centre the direct materials and direct labour costs for a process are ascertained from records, e.g. material requisition forms and time sheets. Overhead charges are also allocated and so a total cost can be ascertained for a particular period of time for a process. Separate costings are carried out for service and production cost centres and at the end of the accounting period concerned the costs of the service cost centres are transferred, in some equitable way, to production cost centres so that the costs can be levied against products and hence passed on to the purchaser of those products.

For the accounting period concerned there need to be records kept of opening stocks, uncompleted units and completed units. Where there are uncompleted units, i.e. work in progress, an estimate needs to be made of the state of these units and their value. Hence the number of equivalent, completed units can be estimated. Uncompleted units in one accounting period will form the opening stocks in the next accounting period. These opening stocks are considered in the same way as the uncompleted units and an estimate is made of the equivalent, completed units that are produced in the accounting period concerned. The sum of the completed units and the equivalent completed units in the time gives the effective output from the cost centre concerned.

The above represents a simplified account of process costing, the aim being to indicate the principle rather than the detail of how to carry out such a costing. For more details the reader is referred to specialised texts on cost accounting, e.g. *Introduction to Cost and Management Accounting* by M. Houlton (Heinemann).

14.9 Marginal costing

The conventional method of dealing with overheads apportions all the overhead costs to the product, as described earlier in this chapter, and is known as absorption costing. Overhead costs, however, can be considered to be made up of fixed and variable overhead costs. Fixed overhead costs do not vary with output and are incurred in respect to a definite period of time, e.g. security or rent, while variable overhead costs are related to the output, e.g. indirect materials such as cutting oil and the electrical power used by the machines (but probably not the electrical power used for the lighting and heating which is likely to be a fixed overhead). There is thus an argument that only the variable overhead costs should be directly allocated to a product and that the difference between the total cost arrived at with this level of overheads and the selling price should give a return which covers both the fixed overheads and profit.

This method of costing is known as *marginal costing*, the term marginal cost being used for the total cost arrived at by adding together the prime costs, i.e. direct materials and direct labour, and the variable overhead costs. The difference between the value of the sales and the marginal costs is known as the *contribution*. This contribution is used to cover the fixed overhead costs and profit.

Contribution = Sales value − (prime costs + variable overheads)
      Profit = Contribution − fixed overheads

Marginal costing can simplify pricing of products in that the price is more directly related to the costs incurred in the production of the product. This is because the calculation of the marginal cost only involves quantities related to the output of the product, and there is no difference in costs due to differences in the quantities produced in different accounting periods. With absorption costing this is not the case, since the fixed overheads are spread over the quantity of product produced and so a variable output leads to a variable total cost.

There is also the advantage with marginal costing that it identifies the contribution a product makes and thus decisions can be taken more easily as to whether to continue that line of product. Management can decide, for example, to set the price of a product so that it perhaps makes only a small contribution, or none at all, and balance this by some other product making a bigger contribution.

The following example illustrates, in very simple terms, the differences between absorption and marginal costing.

Absorption costing:

Direct materials costs	£200
Direct labour costs	£120

The total overheads for that cost centre are £500. The overheads are apportioned among the various products according to the number of machine hours and this results in an allocation of one tenth of the overheads to this particular product. Hence the overheads are £50. Thus the total cost of the product is £370. A profit element of £30 is added to the cost to give a selling price of £400.

Marginal costing:

Direct materials costs	£200
Direct labour costs	£120

The total variable overheads for that cost centre are £120 and the fixed overheads £380. The variable overheads are apportioned among the various products according to the number of machine hours and this results in one tenth of the overheads being allocated to this product, i.e. a variable overhead of £12. Thus the total marginal cost is £342. If the sales price is fixed as £400 then the product makes a contribution of £58 which can be used to cover fixed overheads and profit.

14.10 Break-even analysis

Break-even analysis can be used to show how revenue and costs relate to provide a profit, or loss, at different volumes of sales and the point, termed the *break-even point*, at which there is neither profit or loss.

Figure 14.1 shows how a break-even chart is constructed. The vertical axis of the chart shows costs and revenue, the horizontal axis shows volume of sales or volume of production, either being in cash value or number of units. The costs consist of essentially just two elements, fixed costs which are independent of the volume of sales or production and variable costs which depend on the volume. The total cost is the sum of these fixed and variable costs. Hence, by combining these, a graph can be drawn of costs against volume. As long as there are fixed costs this graph will not pass through the origin, i.e. zero cost–zero volume point, in that there will always be some cost even when the volume sold or produced is zero. On the same chart the graph is drawn of revenue against volume. This graph will pass through the origin in that there is no revenue with no sales or production. The point where these two graphs, cost–volume and revenue–volume, cross is the break-even point. Where the cost–

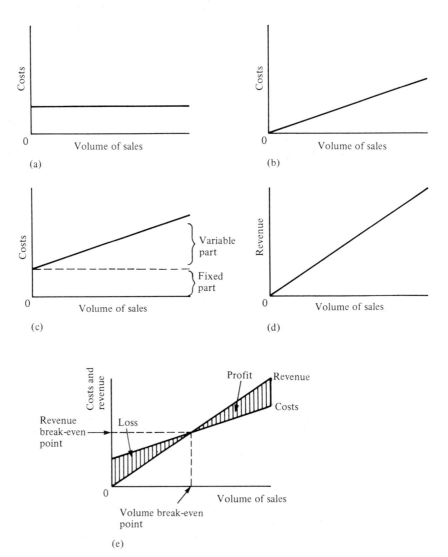

Figure 14.1 The steps in producing a break-even chart. (a) Fixed costs. (b) Variable costs. (c) Total costs. (d) Sales revenue. (e) The composite chart

volume graph, at a particular volume, is greater than the revenue–volume graph, there is a loss; where the revenue–volume graph is greater than the cost–volume graph, there is a profit.

The break-even point can be obtained from the chart or by the use of the following equations:

$$\text{Volume break-even point} = \frac{\text{total fixed cost}}{\text{unit selling price} - \text{unit variable cost}}$$

$$\text{Revenue break-even point} = \frac{\text{total fixed cost} \times \text{total revenue}}{\text{total revenue} - \text{total variable cost}}$$

The above equations are just alternative ways of expressing the break-even point.

To illustrate the use of the above equations, consider the situation where the variable cost for a unit is £1 and the unit selling price is £4. If the total fixed costs for the product are £6000 then the break-even point is 6000 ÷ (4 − 1) = 2000. This is the number of units that have to be sold before a profit can be made; for sales below 2000 there will be a loss.

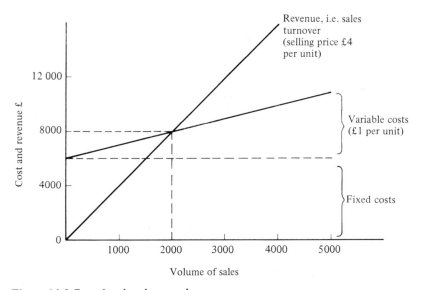

Figure 14.2 Complete break-even chart

If for a sales turnover, i.e. total revenue, of £4000, the total variable costs are £1000 and the total fixed costs £6000 then the break-even point is 6000 × 4000 ÷ (4000 − 1000) = £8000. This is what the sales turnover will have to exceed if a profit is to be made.

Figure 14.2 shows the break-even chart for the above data. The profit or loss at any particular sales volume is the difference between the revenue and the total cost graphs. Thus, for the data given in Figure 14.2, at a sales volume of 1000 units there is a loss of (£7000 − £4000) = £3000. At a sales volume of 3000 units there is a profit of (£12 000 − £9000) = £3000. The break-even point is 2000 units with sales turnover (revenue) of £6000. Such profit and loss data, as a function of sales volume, can be plotted as another graph (Figure 14.3). This graph shows how the profits and losses vary with the volume of sales.

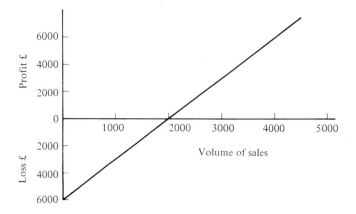

Figure 14.3 Profit graph relating to Figure 14.2

14.11 Selection of process

A number of vital questions need to be posed before a decision is made as regards to the manufacturing process to be used for a product.

1 *What is the material?* The type of material to be used affects the choice of processing method. Thus, for example, if casting is to be used and the material has a high melting point then the process must be either sand casting or investment casting.

2 *What is the shape?* The shape of the product is generally a vital factor in determining which type of process can be used. Thus, for example, a product in the form of a tube could be produced by centrifugal casting, drawing or extrusion but not generally by other methods.

3 *What type of detail is involved?* Has the product to have holes, threads, inserts, hollow sections, fine detail etc? Thus, for example, forging could not be used if there was a requirement for hollow sections.

4 *What dimensional accuracy and tolerances are required?* High accuracy would rule out sand casting, though investment casting might well be suitable.

5 *Are any finishing processes to be used?* Has the process to be used to give the

product in its final finished state or will there have to be an extra finishing process. Thus, for example, planing will not produce as smooth a surface as grinding.

6 *What quantities are involved?* Is the product a one-off, a small batch, a large batch, continuous production? While some processes are economic for small quantities, others do not become economic until large quantities are involved. Thus open die forging could be economic for small numbers but closed die forging would not be economic unless large numbers were produced.

The cost aspects of process selection

With sand casting a new mould has to be made for each product manufactured. With die casting the same mould can be used for a large number of components but the initial die cost is high. Which process would be the cheapest if say 10 products were required, or perhaps 1000 products?

The manufacturing cost, for any process, can be considered to be made up of two elements—fixed costs and variable costs. Figure 14.4 shows the graphs of costs against quantity for the two processes and how their total costs compare. For the die casting there is a higher fixed cost than for the sand casting, due to the cost of the die. The sand casting has, however, a greater cost per item produced. The comparison graph shows that, up to a quantity N, sand casting

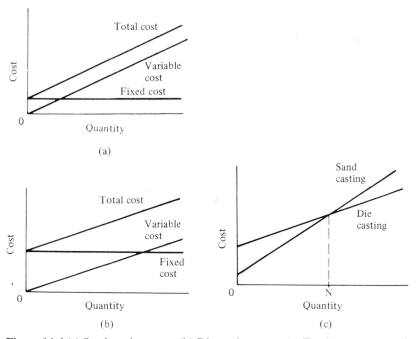

Figure 14.4 (a) Sand casting costs. (b) Die casting costs. (c) Total costs compared

is the cheaper but that for quantities greater than N *die* casting is the cheaper process.

Capital costs for installations, e.g. the cost of a machine or even a foundry, are usually defrayed over an expected lifetime of the installation. However, in any one year there will be depreciation of the asset and this is the capital cost that is defrayed against the output of the product in that year. Thus if the capital expenditure needed to purchase a machine and instal it was, say, £50 000, then in one year a depreciation of 10% might be used and thus £5000 defrayed as capital cost against the quantity of product produced in that year.

Consider, as an example, a product to be made from a thermoplastic. It is to be hollow and of fairly simple form. It can be made by injection moulding or rotational moulding (other methods are also possible but for the purpose of this example only these two alternatives are considered). The cost of the installation for injection moulding is £90 000, the cost for the rotational moulding, using a fairly simple form of the equipment, is £10 000. Thus if these installations depreciate at 10% per year then the costs per year are—injection moulding £9000 and rotational moulding £1000. If these installations are to be used in the year to make only the product we are concerned with, then these charges in their entirety constitute an element of the fixed costs.

Another element of the fixed costs is the cost of the dies or tools needed specifically for the product concerned. In the case of the injection moulding the cost is £2000 and for the rotational moulding £500.

Other factors we could include in the fixed costs are plant maintenance and tool or die overhaul. However, considering just the above two charges, for the installation when we assume that it only produces the above product per year and for the specific tooling if only used for the one production run, we have:

	Injection moulding	Rotational moulding
Installation cost	£9000	£1000
Die costs	£2000	£500
Total fixed cost	£11 000	£1500

The variable costs are the material costs, the labour costs, the power costs, and any finishing costs required. These might appear, per unit produced, as follows:

	Injection moulding	Rotational moulding
Direct labour cost	£0.30	£1.40
Other labour costs; e.g. supervision	£0.15	£0.30
Power costs	£0.06	£0.15
Finishing costs	£1.00	zero
Material costs	£0.40	£0.40
Total variable cost	£1.91	£2.25

If 1000 units are required then the costs will be:

	Injection moulding	Rotational moulding
Fixed costs	£11 000	£1500
Variable costs	£1900	£2250
Total cost	£12 910	£3750

On the basis of the above costings, the rotational moulding is considerably cheaper than the injection moulding. The injection moulding installation can, however, produce 50 items per hour and the use in one year of this to produce just 1000 items is a considerable under-use of the asset. The rotational moulding installation can only produce 2 items per hour.

Thus if the rotational moulding installation were perhaps only used for this product but the injection moulding installation were used for a number of products, then the injection moulding installation cost would be defrayed over a greater number of products and thus that element of the fixed cost defrayed against the product would be less.

The question to be asked when costing a product are:

1 *Is the installation to be used solely for the product concerned?*
The purpose of this question is to determine whether the entire capital cost has to be written off against the product or whether it can be spread over a number of products.

2 *Is the tooling to be used solely for the product concerned?*
If specific tooling has to be developed then the entire cost will have to be put against the product.

3 *What are the direct labour costs per item?*

4 *What are the other labour costs involved?*
In this category consider supervision costs, inspection costs, labouring costs, setting costs, etc.

5 *What is the power cost?*

6 *Are there any finishing processes required—if so, what are their costs?*

7 *What are the materials costs?*

8 *Are any overhead costs to be included?*

Make or buy decisions

In considering the production of a product, a decision that may be needed is whether components should be made within the company or bought in from outside suppliers. It may be that such suppliers will have to make the components specially to the specified design, or that the components are standard and can be purchased from stock. The deciding factor is likely to be the cost. If the component can be bought more cheaply than it can be made, then it is likely that it will be bought rather than made.

Figure 14.5 illustrates on a cost–quantity graph the effects on cost of making or buying in a particular example. The cost of making involves a high fixed cost (for the plant required) and a variable cost for each item produced. Buying, if a standard item, involves no fixed cost but only a variable cost. This variable cost per item is greater than the variable cost per item if the component was made in-company. At a particular quantity N, the costs for making and buying are the same, below that point the buying cost is less than the making cost, above that point the making cost is less than the buying cost. The vital questions to be posed are:

1 *Is it cheaper to make or buy the component for the quantity required?*
2 *If it is cheaper to buy, are there any other reasons why you should not buy?*

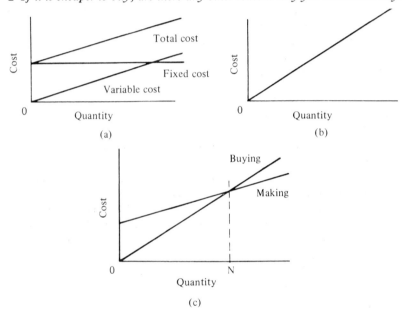

Figure 14.5 Making or buying? (a) Making costs. (b) Buying costs. (c) Costs compared

Quality

In general, the higher the quality the greater the costs. With any particular process there is a technological limit to the accuracy that can be achieved and the nearer that process is used to that limit the more it will cost. A different process might be more cost effective at such accuracies because it is not near its technological limit. With any process there will be a variation in the quality of the output, some processes tending to have a greater variability than others. Thus the requirement for high quality might only be met by having a large number of rejects, as not reaching the required quality. This will have cost implications.

Quality control also costs money. A high quality-requirement is likely to mean a high level of inspection, to ensure that the quality is maintained. The obvious questions that have to be posed of any product design are:

1 *Is the specified quality really necessary?*
2 *Do all aspects of the design have to be to the specified quality?*—e.g. accurate independent locations of three holes may not be necessary, but just accuracy in positioning the holes in relation to each other.

Problems

1 Distinguish between *standard* and *historical costing*.
2 What are the various elements that constitute the selling price?
3 What is meant by the terms *direct labour costs* and *direct materials costs* and how do they differ from *indirect costs*?
4 How can direct labour and direct materials costs be found for a particular product if historical costing is used?
5 Explain how the direct labour cost is found for a product if standard costing is used.
6 What are *overhead costs*?
7 What is meant by a *cost centre*?
8 With *absorption costing* explain how the overhead costs can be recovered.
9 With absorption costing how are the overheads of service departments recovered?
10 With absorption costing, how can the overheads be allocated to the various products produced at a production cost centre?
11 What is *depreciation* and how can it be determined if the depreciation is (a) assumed to be constant for each year of the life of a machine and (b) assumed to change by a constant proportion of the balance of the value of the machine each year?
12 Explain how (a) a job, (b) a batch and (c) a process can be costed.
13 What is meant by *marginal costing* and how does it differ from absorption costing?
14 Explain what is meant by *break-even analysis*.
15 Explain how a break-even chart can be plotted.
16 A production department produces three products, A, B and C, during an accounting period. Calculate the cost for each during this period if the overheads are absorbed according to (a) direct labour costs, (b) machine hours involved.

Costs	Product A	Product B	Product C
Total labour hours	200	800	300
Total machine hours	100	200	600
Direct materials	800	500	600

The labour rate for all the operations is £5 per hour and the total overheads £1200.

17 Calculate the total direct cost for the following product using standard costing.

4 hours process A, operator rating 110, labour rate £4.00 per hour
2 hours process B, operator rating 100, labour rate £5.00 per hour
3 hours process C, operator rating 95, labour rate £4.00 per hour
2.0 kg material A, standard material cost £0.40 per kg
1.0 kg material B, standard material cost £1.50 per kg
3.5 kg material C, standard material cost £2.00 per kg

18 Calculate the total direct cost for the following product.

Process	Cost centre	Labour hours	Machine hours
Machining	A	2.0	2.0
Drilling	A	0.5	0.5
Finishing	B	0.5	4.0
Packing	C	0.5	0

Direct materials cost £25.00
Cost centre A, labour rate £5.00 per hour, machine hour rate £10.00
Cost centre B, labour rate £4.00 per hour, machine hour rate £3.00
Cost centre C, labour rate £3.00 per hour

19 Investigate the methods used in a materials store in some company and report on the documentation used and the way materials are costed.

20 A machine costs, when fully installed, £10 000 and is expected to have a life of six years. At the end of this time it will have a scrap value of £1500. Calculate the value of the machine at the end of each year if the depreciation is to be reckoned according to (a) equal amount per year, (b) the reducing balance method by which a constant proportion of the balance of the value of the asset is deducted each year.

21 Obtain information on the accounting method used to depreciate the value of some asset, quoting not only the method but actual values.

22 A production department in mass producing a product incurs costs of £12 000 for direct materials, £20 000 for direct labour and £15 000 for overheads. If 40 000 units are produced, what is the average cost per unit?

23 How would the answer to Problem 22 have to be modified if there had been 500 units at the beginning of the accounting period half-finished and at the end of the period there were 800 units half-finished, the 40 000 representing the units completed? Take it that a half-finished product uses all the materials, as they are required at the beginning of the process, but only half the labour.

24 Using *marginal costing* calculate the contribution the following product makes:

Direct materials cost	£1000
Direct labour cost	£ 800
Fixed overheads	£ 500
Variable overheads	£ 200

25 Write a report arguing the merits of marginal costing in preference to absorption costing.

26 Construct a break-even chart of the following product data and determine the break-even sales volume.

Fixed costs £2000

Variable costs £2 per unit

Revenue £7 per unit

27 Calculate the break-even sales turnover for the following product:

Total fixed costs £8000

Total variable costs £2000

Total sales turnover £13 000

28 A product has fixed costs of £5000 and variable costs of £5 per unit. At present the product is sold for £12 per unit. However it is proposed to reduce the selling price to £11 per unit. What effect will this have on (a) the break-even point and (b) the profit when 1000 units are sold?

29 Plot a profit–sales volume graph for the product given the data (a) in Problem 26 (b) in Problem 27.

30 A company is introducing a new product. The annual capacity of the production department for this product is 35 000 units with fixed costs of £20 000. The variable costs are £3 per unit. Sales are anticipated as being 30 000 units at a selling price of £7 per unit.

(a) What will be the profit at that volume of sales?

(b) Draw profit–sales volume graphs showing how the profit at the above selling price depends on the volume. What is the minimum number of units that must be sold if a profit is to be realised?

(c) What would be the effect on the answer to (b) if the selling price were reduced to £5?

(d) With this reduced selling price, what number of units would need to be sold to realise the same profit as with the £7 selling price? What comment can you make on this result?

31 The following are the costs that would be incurred for two alternative processes for the production of a hollow thermoplastic container:

Costs	Injection moulding £	Rotational moulding £
Fixed: installation	9000	1000
: die	2000	500
Variable (per unit)		
: indirect labour	0.15	0.30
: power	0.06	0.15
Direct labour (per unit)	0.30	1.40
Direct materials (per unit)	0.40	0.40
Finishing	1.00	0

(a) Which process would be the more cost-effective if (i) 1000 (ii) 100 000 units were required?

(b) The injection moulding process can produce fifty units per hour, the rotational moulding only two per hour. What implication might this have for using the equipment to produce a wider range of goods and so affect the choice of process?

32 The following data refer to three alternative processes by which a gearwheel can be produced (based on data given in *Engineering Materials: An Introduction* (The Open University T252 Unit 2). On the basis of the data, which process would you propose as the most cost effective if the output required was (a) 100, (b) 10 000 units?

Gravity die casting method, costs per unit
Materials: 10 g aluminium alloy plus 5 g scrap, at £1.00 per kg
Special tooling: a die at £500 which has a life of 100 000 gears
Labour: casting 0.010 hours at £4.00 per hour
finishing 0.001 hours at £4.00 per hour
Overheads to be charged at 200% of labour cost

Pressure die casting method, costs per unit
Materials: 23 g zinc alloy plus 2 g scrap, at £0.45 per kg
Special tooling: a die at £4000 which has a life of 100 000 gears
Labour: casting 0.0015 hours at £4.00 per hour
finishing 0.001 hours at £4.00 per hour
Overheads to be charged at 200% of labour cost

Injection moulding method, costs per unit
Materials: 4 g nylon plus 0.4 g scrap, at £2.00 per kg
Special tooling: a die at £4000 which has a life of 100 000 gears
Labour: casting 0.006 hours at £4.00 per hour
Overheads to be charged at 200% of labour cost

33 Figure 14.6 shows the basic design form for a container for concentrated still drinks. It is required in very large quantities, the proposed material

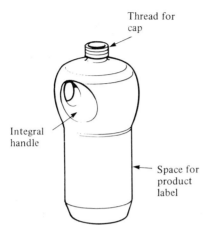

Figure 14.6 Container for liquids

being high density polyethylene. Blow moulding and rotational moulding have been proposed.

(a) Compare the two proposed processes. In view of the large numbers required, which of them would be likely to be the more economic?

(b) What restrictions are imposed on the design by these processes?

34 Figure 14.7 shows a simple cap designed to lock on to the end of small shafts, the barb being to prevent the cap coming off. The cap is to be made from metal.

(a) What processes can be used to manufacture the push-cap?

(b) If the cap is wanted in large quantities, which would be the most economic method?

(c) Would your answer to (b) be changed if only small quantities were required?

Barb

Figure 14.7 Push-cap

35 A proposed design for an electrical screwdriver involves a steel blade mounted in a plastic handle, cellulose acetate—a transparent thermoplastic being proposed for the handle. An important consideration is that when the handle of the screwdriver is rotated the blade rotates with the handle and not within the handle. For this reason the blade has a flattened cross-section for that part within the handle (Figure 14.8). The screwdrivers are

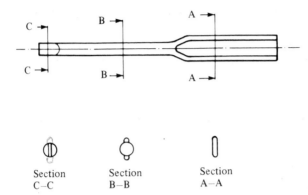

Figure 14.8 Screwdriver blade

to be produced in very large quantities and must be at a low cost in order to compete in the market-place.

(a) What are the restrictions imposed by the design on the manufacturing processes for the blade and the handle?

(b) Suggest possible processes for manufacturing the blade and the handle.

(c) Which of the processes would you consider on cost and quality considerations to be the optimum ones?

(d) How will the blade be fixed in the handle?

36 The frame of a bicycle could be made of tubular mild steel welded together or produced in one piece, perhaps by casting.

(a) Compare these two methods of producing bicycle frames. Consider all the processes that would be involved in going from raw material to finished goods.

(b) Which method is likely to be the more economic?

(c) If you were a manufacturer of bicycle frames and opted for the welding together of tubular mild steel, what factors would you need to take into account in deciding whether to make your own tube or buy in the tube?

37 Figure 14.9 shows a simple reinforcement plate made out of sheet mild steel. Shearing and machining are two possible methods of producing the item from sheet metal.

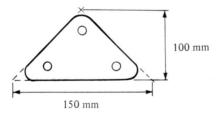

Figure 14.9 Reinforcement plate

(a) Compare these methods, bearing in mind the cost factor, if large quantities are to be produced.

(b) In the case of shearing, how can the amount of scrap be minimised?

(c) If a few non-standard versions of the plate were required, which method would be the more economic?

38 Forging, casting or machining from the solid have been proposed as methods that could be used to produce spanners.

(a) With forging, what would be the optimum method if the spanners were required in large quantities?

(b) With casting, what would be the optimum method if the spanners were required in large quantities?

(c) Compare forging, casting and machining for the production of spanners where they are required in large quantities. Which method would be likely to give the cheapest product?

(d) Which of the processes would lead to the most scrap?

39 Your company makes a large number of items by machining, some of them being relatively simple jobs and some complex, involving compound curves and intricate surface features. It has been proposed that powder techniques should be used for much of the work that is currently machined. It is suggested that the relationship between cost and quantity of any product manufactured is, for the various processes, of the form shown in Figure 14.10.

(a) Present arguments justifying, or rejecting, the graphs shown in Figure 14.10.

(b) What limitations on the designs possible are posed by the use of powder techniques?

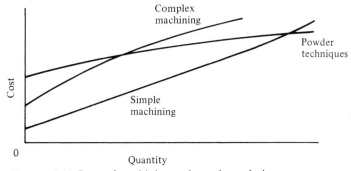

Figure 14.10 Costs of machining and powder techniques

40 Figure 14.11 shows the proposed design for a retaining clip, of hardened and tempered carbon steel. The surfaces must be smooth with no sharp corners on the inner circumferential edges.

(a) What processes could be used to manufacture the item?

(b) What would be the most economic method if 4000 items were required?

41 A car dashboard has to be formed to a variety of contours, have holes for instruments, various control knobs and a glove compartment. Both metal and plastic can be used.
(a) What processes can be used to form the dashboard? State for each process whether the dashboard can be made in one piece, in one operation.
(b) Discuss the economics of the various processes.
(c) For the components that are attached to a dashboard, what fixing methods can be used?

600 mm

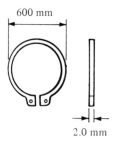

2.0 mm

Figure 14.11 A retainer ring. Only the nominal dimensions, to give an order of size, have been given

42 A base for a special machine tool has to be made, only one having to be produced. There are two proposals: one to make a grey iron casting, and the other to make it of steel and weld together the various parts after machining. The casting will be twice the weight of the fabricated base.
(a) Compare the two methods proposed for the base.
(b) For the casting, the cost of making a sand pattern is £400, the labour costs, with overheads, for the making of the casting is £840, and the materials cost £250. For the fabricated base, the cost of the machining, including overheads, is £300, the cost of the welding, including overheads, is £500 with an extra £50 being required for the cost of the electrode material consumed during the welding, and the cost of the material is £120. Which method is the more economical?
(c) Before the base can be made, it is realised that there could be a market for a small number of these bases. Plot graphs showing how the cost varies with quantity and comment on the significance of the graphs in the determination of the optimum manufacturing method.
43 It is proposed to make long lengths of a suitable section which can then be cut to appropriate lengths and joined to make metal window frames. The proposal is to use steel.
(a) What processes could be used to produce such a section?
(b) What processes could be used to join the sections at the corners to form the window frames? Do you envisage that special corner pieces will be needed? If so, how will they be made?
(c) Compare the processes you suggest in (a) and (b) for quality, fixed and variable costs, and facilities required.

Index